'If you want to understand the precepts and improvement potential of Total Quality Management then this fourth edition is for you! Written with students and practitioners in mind John Oakland brings quality concepts alive in his own inimitable style. Unparalleled case studies – a must read!'

Edel O'Neill, *Reader in Management, Ulster Business School, University of Ulster, Northern Ireland*

Total Quality Management and Operational Excellence

Text with cases

FOURTH EDITION

John S. Oakland

LONDON AND NEW YORK

First edition 1995
Third edition 2003
Fourth edition 2014
by Routledge
2 Park Square, Milton Park, Abingdon, Oxon OX14 4RN

and by Routledge
711 Third Avenue, New York, NY 10017

Routledge is an imprint of the Taylor & Francis Group, an informa business

British Library Cataloguing in Publication Data
A catalogue record for this book is available from the British Library

Library of Congress Cataloging in Publication Data
    Oakland, John S.
    Total quality management and operational excellence: text with cases/
    John Oakland.—Fourth Edition.
    pages cm
    Includes bibliographical references and index.
    1. Total quality management.   2. Total quality management—Case studies.
    I. Title.
    HD62.15.O173 2013
    658.4′013—dc23
    2013036192

ISBN: 978-0-415-63549-3 (hbk)
ISBN: 978-0-415-63550-9 (pbk)
ISBN: 978-1-315-81572-5 (ebk)

Typeset in Palatino and Univers
by Florence Production Ltd, Stoodleigh, Devon, UK
Printed and bound by CPI Group (UK) Ltd, Croydon, CR0 4YY

For Susan

CONTENTS

FIGURES

Tables

PREFACE

When I wrote the first edition of *Total Quality Management* in 1988, there were very few books on the subject. Since its publication the interest in TQM and business performance improvement has exploded. There are now many texts on TQM and its various aspects, including business/operational excellence, business process management, Six Sigma and Lean manufacturing based approaches.

So much has been learned during the last 30 years of TQM implementation that it has been necessary to rewrite the book and revise it again. In essence this is the sixth edition since TQM was first published in 1989 (two editions of *Total Quality Management* and four editions of *TQM Text & Cases*). The content and case studies in this edition have been changed substantially to reflect the developments, current understanding, and experience gained of TQM and a completely new chapter on lean systems has been included. The result is a new book: *Total Quality Management and Operational Excellence* (TQM & OpEx).

Increasing the satisfaction of customers and other stakeholders through effective goal deployment, cost reduction, process improvement, people involvement and supply chain development has proved essential for organizations to stay in existence in the twenty-first century. We cannot avoid seeing how quality has developed into a most important competitive weapon, and many organizations have realized that TQM and its relatives is *the* way of managing for the future. Neglect of product and service quality can have disastrous consequences, as we have seen repeatedly in recent years around the globe. Consequential reputational damage is deeper and quicker now than ever before because information, opinion, and ultimately consumer choice, is affected at scale due to the nature of modern communication technologies. Of course, TQM is far wider in its application than assuring product or service quality – it is a way of managing organizations to improve every aspect of performance, both internally and externally.

This book is about how to manage in a total quality way through excellent operation. It is structured in the main around four parts of what has become known as the 'Oakland model for TQM' – improving *Performance* through better *Planning* and management of *People* and *Processes* in which they work. The core of the model will always be performance in the eyes of the customer, but this must be extended to include performance measures for all stakeholders. This core still needs to be surrounded by *Commitment* to quality and meeting customer requirements, *Communication* of the quality message and recognition of the need in many cases to change the *Culture* of most organizations to create total quality. These three Cs are the 'soft foundations' which must encase the 'hard management necessities' of the four Ps.

Under these headings the essential steps for the successful implementation of TQM & OpEx are set out in what I hope is still a meaningful and practical way. The book should guide the reader through the language of TQM and all the recent developments, and set down a clear way to manage change – new material has been included on these aspects. The Oakland DRIVER improvement methodology is a major contribution to this.

Many of the new approaches related to quality and improving performance appear to present different theories. In reality they are talking the same 'language' but may use different 'dialects'; the basic principles of defining quality and taking it into account throughout all activities of the 'business' are common. Quality has to be managed, it does not just happen. Understanding and commitment by senior management, effective leadership, teamwork, good process management and sound improvement methods and tools are fundamental parts of the recipe for success. I have tried to use my extensive research and consultancy experience to take what, to many, is a jigsaw puzzle and assemble a comprehensive, pragmatic, working model for total quality. Moreover, I have tried to show how holistic TQM now is, embracing the most recent models of 'excellence', Six Sigma, Lean and a host of other management methods and teachings.

To support the 19 chapters of text are ten case studies, eight of which are brand new. I have again presented these together at the end of the book as many overlap different topics in the chapters, but I have offered guidance on which parts are illustrated by the particular cases.

The book should meet the requirements of the increasing number of 'students' who need to understand the part TQM & OpEx may play in their studies of science, engineering or business and management. I hope that those engaged in the pursuit of professional qualifications in the management of quality, such as the Chartered Quality Institute, the American Society for Quality and similar ones around the world, will make this book an essential part of their library. With its companion book, *Statistical Process Control* (now in its sixth edition), *Total Quality Management and Operational Excellence: Text & Cases* documents a comprehensive approach, one that has been used successfully in many organizations across the globe. In the interest of brevity, the terms Total Quality Management (TQM) and Total Management & Operational Excellence (TQM & OpEx) will be interchangeable throughout the book; generally the term TQM will be used.

I would like to thank my colleagues in Oakland Consulting for the sharing of ideas and help in their development. The book is the result of many man-years collaboration in assisting organizations to introduce good methods of management and embrace the concepts of total quality. I am also most grateful to the busy senior managers in the case study organizations for their contributions in pulling together the cases and for obtaining permission for their publication.

John S. Oakland

Part I

The foundations of TQM

Good order is the foundation of all good things.

Edmund Burke, 1729–1797,
from 'Reflections on the Revolution in France'

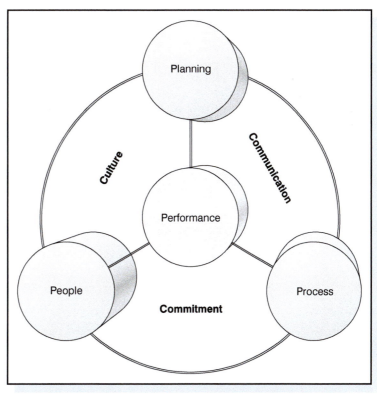

Chapter 1

Understanding quality

QUALITY, COMPETITIVENESS AND CUSTOMERS

In recent times, organizations have experienced a period of great change in their markets and operations. International as well as domestic competition has meant that many organizations have faced an increasingly turbulent and hostile environment. The pace of technological change has quickened to lightening speed, customers have become more demanding and competition has become more intense and sophisticated. Regulators and consumer groups have also added to these pressures.

Good quality performance has always been a key strategic factor for business success but it is now more than ever required to compete successfully in the global markets of the twenty-first century. Many organizations have adopted a range of improvement approaches in response to these forces. We have seen the growing adoption of a range of quality and management systems standards, the emergence of total quality management (TQM), business process re-engineering (BPR), business excellence, performance excellence, Lean thinking, Six Sigma, statistical process control, etc, etc. The battle weary could be excused from taking a rather jaundiced view of this ever-lengthening list of 'quality' offers but, by and large, they share many of the principles and elements that are found in TQM.

Whatever type of organization you work in these days – a bank, a hospital, a university, an airline, an insurance company, local government, a factory – competition is rife: competition for customers, for students, for patients, for resources, for funds. Any organization basically competes on its *reputation* – for quality, reliability, price and delivery – and most people now recognize that quality is the key to achieving sustained competitive advantage. If you doubt that, just look at the way some organizations, even whole industries in certain countries, have used quality strategically to win customers, obtain business resources or funding and be competitive. Moreover, this sort of attention to quality improves performance in reliability, delivery and price.

Reputations for poor quality last for a long time and good or bad reputations can become national or international. Yet the management of quality can be learned and used to improve reputation. For any organization, there are several aspects of reputation which are important:

1. It is built upon the competitive elements of being 'On-Quality; On-Time; On-Cost'.
2. Once an organization acquires a poor reputation for product or service quality or reliability, it takes a very long time to change it.
3. Reputations, good or bad, can quickly become national reputations.
4. The management of the competitive weapons, such as quality, can be learned like any other skill and used to turn round a poor reputation.

Before anyone will buy the idea that quality is an important consideration, they would have to know what was meant by it.

What is quality?

Quality starts with understanding customer needs and ends when those needs are satisfied. 'Is this a quality watch?' Pointing to my wrist, I ask this question of a class of students – undergraduates, postgraduates, experienced managers – it matters not who. The answers vary:

- 'No, it's made in Japan.'
- 'No, it's cheap.'
- 'No, the face is scratched.'
- 'How reliable is it?'
- 'I wouldn't wear it.'

My watch has been insulted all over the world – London, New York, Paris, Sydney, Dubai, Brussels, Amsterdam, Leeds! Clearly, the quality of a watch depends on what the wearer requires from a watch – perhaps a piece of jewellery to give an impression of wealth; a timepiece that gives the required data, including the date, in digital form; or one with the ability to perform at 50 metres under the sea? These requirements determine the quality.

Quality is often used to signify 'excellence' of a product or service – people talk about 'Rolls-Royce quality' and 'top quality'. In some manufacturing companies the word may be used to indicate that a piece of material or equipment conforms to certain physical dimensional characteristics often set down in the form of a particularly 'tight' specification. In a hospital it might be used to indicate some sort of 'professionalism'. If we are to define quality in a way that is useful in its *management*, then we must recognize the need to include in the assessment of quality the true requirements of the 'customer' – the needs and expectations.

Quality then is simply *meeting the customer requirements and* this has been expressed in many ways by other authors:

- 'Fitness for purpose or use' – Juran, an early doyen of quality management.
- 'The totality of features and characteristics of a product or service that bear on its ability to satisfy stated or implied needs' – BS 4778. 1987 (ISO 8402, 1986) *Quality Vocabulary; Part 1, International Terms.*

- 'Quality should be aimed at the needs of the consumer, present and future' – Deming, another early doyen of quality management.
- 'The total composite product and service characteristics of marketing, engineering, manufacture and maintenance through which the product and service in use will meet the expectation by the customer' – Feigenbaum, the first man to publish a book with 'Total Quality' in the title.
- 'Conformance to requirements' – Crosby, an American consultant famous in the 1980s.
- 'Degree to which a set of inherent characteristics fulfils requirements' – ISO (EN) 9000:2000 *Quality Management Systems – Fundamentals and Vocabulary*.

Another word that we should define properly is *reliability*. 'Why do you buy a Japanese car?' 'Quality and reliability' comes back the answer. The two are used synonymously, often in a totally confused way. Clearly, part of the acceptability of a product or service will depend on its ability to function satisfactorily *over a period of time* and it is this aspect of performance that is given the name *reliability*. It is the ability of the product or service to *continue* to meet the customer requirements. Reliability ranks with quality in importance, since it is a key factor in many purchasing decisions where alternatives are being considered. Many of the general management issues related to achieving product or service quality are also applicable to reliability.

It is important to realize that the 'meeting the customer requirements' definition of quality is not restrictive to the functional characteristics of products or services. Anyone with children knows that the quality of some of the products they purchase is more associated with *satisfaction in ownership* than some functional property. This is also true of many items, from antiques to certain items of clothing. The requirements for status symbols account for the sale of some executive cars, certain bank accounts and charge cards, and even hospital beds! The requirements are of paramount importance in the assessment of the quality of any product or service.

By *consistently* meeting customer requirements, we can move to a different plane of satisfaction – *delighting the customer*. There is no doubt that many organizations have so well ordered their capability to meet their customers' requirements, time and time again, that this has created a reputation for 'excellence'. A development of this thinking regarding customers and their satisfaction is *customer loyalty*, an important variable in an organization's success. Research shows that focus on customer loyalty can provide several commercial advantages:

- Customers cost less to retain than acquire.
- The longer the relationship with the customer, the higher the profitability.
- A loyal customer will commit more spend to its chosen supplier.
- About half of new customers come through referrals from existing clients (indirectly reducing acquisition costs).

Many companies use measures of customer loyalty to identify customers which are 'completely satisfied', would 'definitely recommend' and would 'definitely repurchase'.

The ability to meet the customer requirements is vital, not only between two separate organizations, but within the same organization.

When the air stewardess pulled back the curtain across the aisle and set off with a trolley full of breakfasts to feed the early morning travellers on the short domestic flight into an international airport, she was not thinking of quality problems. Having stopped at the row of seats marked 1ABC, she passed the first tray onto the lap of the man sitting by the window. By the time the second tray had reached the lady beside him, the first tray was on its way back to the hostess with a complaint that the bread roll and jam were missing. She calmly replaced it in her trolley and reached for another – which also had no roll and jam.

The calm exterior of the girl began to evaporate as she discovered two more trays without a complete breakfast. Then she found a good one and, thankfully, passed it over. This search for complete breakfast trays continued down the aeroplane, causing inevitable delays, so much so that several passengers did not receive their breakfasts until the plane had begun its descent. At the rear of the plane could be heard the mutterings of discontent. 'Aren't they slow with breakfast this morning?' 'What is she doing with those trays?' 'We will have indigestion by the time we've landed.'

The problem was perceived by many on the aeroplane to be one of delivery or service. They could smell food but they weren't getting any of it, and they were getting really wound up! The air hostess, who had suffered the embarrassment of being the purveyor of defective product and service, was quite wound up and flushed herself, as she returned to the curtain and almost ripped it from the hooks in her haste to hide. She was heard to say through clenched teeth, 'What a bloody mess!'

A problem of quality? Yes, of course, requirements not being met, but where? The passengers or customers suffered from it on the aircraft, but in part of another organization there was a man whose job it was to assemble the breakfast trays. On this day the system had broken down – perhaps he ran out of bread rolls, perhaps he was called away to refuel the aircraft (it was a small airport!), perhaps he didn't know or understand, perhaps he didn't care.

Three hundred miles away in a chemical factory . . . 'What the hell is Quality Control doing? We've just sent 15,000 litres of lawn weed killer to CIC and there it is back at our gate – they've returned it as out of spec.' This was followed by an avalanche of verbal abuse, which will not be repeated here, but poured all over the shrinking quality control manager as he backed through his office door, followed by a red faced technical director advancing menacingly from behind the bottles of sulphuric acid racked across the adjoining laboratory.

'Yes, what is QC doing?' thought the production manager, who was behind a door two offices along the corridor, but could hear the torrent of language now being used to beat the QC man into an admission of guilt. He knew the poor devil couldn't possibly do anything about the rubbish that had been produced except test it, but why should he volunteer for the unpleasant and embarrassing ritual now being experienced by his colleague – for the second time this month. No wonder the QC manager had been studying the middle pages of the *Telegraph* on Thursday – what a job!

Do you recognize these two situations? Do they not happen every day of the week – possibly every minute somewhere in manufacturing or the service industries? Is it any different in banking, insurance, health services? The inquisition of checkers and testers is the last bastion of desperate systems trying in vain to catch mistakes, stop defectives, hold lousy materials, before they reach the external customer – and woe betide the idiot who lets them pass through!

Two everyday incidents, but why are events like these so common? The answer is the acceptance of one thing – *failure*. Not doing it right the first time at every stage of the process.

Why do we accept failure in the production of artefacts, the provision of a service, or even the transfer of information? In many walks of life we do not accept it. We do not say, 'Well, the nurse is bound to drop the odd baby in a thousand – it's just going to happen'. We do not accept that!

In each department, each office, even each household, there are a series of suppliers and customers. The PA is a supplier to the boss. Are the requirements being met? Does the boss receive error-free information set out as it is wanted, when it is wanted? If so, then we have a quality PA service. Does the air steward receive from the supplier to the airline the correct food trays in the right quantity, at the right time?

Throughout and beyond all organizations, whether they be manufacturing concerns, banks, retail stores, universities, hospitals or hotels, there is a series of *quality chains* of customers and suppliers (Figure 1.1) that may be broken at any point by one person or one piece of equipment not meeting the requirements of the customer, internal or external. The interesting point is that this failure usually finds its way to the interface between the organization and its outside customers, and the people who

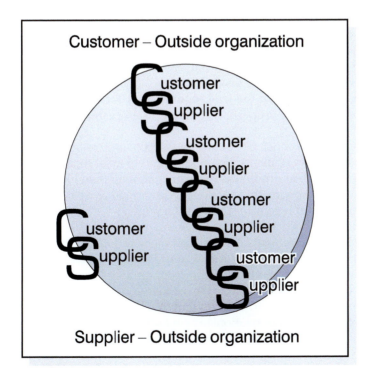

Figure 1.1
The quality chains

operate at that interface – like the air hostess – usually experience the ramifications. The concept of internal and external customers-suppliers forms the *core* of total quality management.

A great deal is written and spoken about employee motivations as a separate issue. In fact the key to motivation *and* quality is for everyone in the organization to have well-defined customers – an extension of the word beyond the outsider that actually purchases or uses the ultimate product or service to anyone to whom an individual gives a part, a service, information – in other words the results of his or her work.

Quality has to be managed – it will not just happen. Clearly it must involve everyone in the process and be applied throughout the organization. Many people in the support functions of organizations never see, experience or touch the products or services that their organizations buy or provide, but they do handle or produce things like purchase orders or invoices. If every fourth invoice carries at least one error, what image of quality is transmitted to customers?

Failure to meet the requirements in any part of a quality chain has a way of multiplying and a failure in one part of the system creates problems elsewhere, leading to yet more failure, more problems and so on. The price of quality is the continual examination of the requirements and our ability to meet them. This alone will lead to a 'continuing improvement' philosophy. The benefits of making sure the requirements are met at every stage, every time, are truly enormous in terms of increased competitiveness and market share, reduced costs, improved productivity and delivery performance, and the elimination of waste.

Meeting the requirements

If quality is meeting the customer requirements, then this has wide implications. The requirements may include availability, delivery, reliability, maintainability and cost-effectiveness, among many other features. The first item on the list of things to do is find out what the requirements are. If we are dealing with customer-supplier relationship crossing two organizations, then the supplier must establish a 'marketing' activity or process charged with this task.

Marketing processes establish the true requirements for the product or service. These must be communicated properly throughout the organization in the form of specifications.

The marketing process must of course understand not only the needs of the customer but also the ability of their own organization to meet them. If my customer places a requirement on me to run 1,500 metres in 4 minutes, then I know I am unable to meet this demand, unless something is done to improve my running performance. Of course I may never be able to achieve this requirement.

Within organizations, between internal customers and suppliers, the transfer of information regarding requirements is frequently poor to totally absent. How many executives really bother to find out what their customers' – their PA's or secretary's – requirements are? Can their handwriting be read, do they leave clear instructions, does the PA/secretary always know where the boss is? Equally, does the PA/secretary establish what the boss needs – error-free word processing, clear messages, a tidy office? Internal supplier-customer relationships are often the most difficult to manage in terms of establishing the requirements. To achieve quality throughout an organization, each person in the quality chain must interrogate every interface as follows:

Customers

- Who are my immediate customers?
- What are their true requirements?
- How do or can I find out what the requirements are?
- How can I measure my ability to meet the requirements?
- Do I have the necessary capability to meet the requirements?
  (If not, then what must change to improve the capability?)
- Do I continually meet the requirements?
  (If not, then what prevents this from happening, when the capability exists?)
- How do I monitor changes in the requirements?

Suppliers

- Who are my immediate suppliers?
- What are my true requirements?
- How do I communicate my requirements?
- How do I, or they, measure their ability to meet the requirements?
- Do my suppliers have the capability to meet the requirements?
- Do my suppliers continually meet the requirements?
- How do I inform them of changes in the requirements?

The measurement of capability is extremely important if the quality chains are to be formed within and without an organization. Each person in the organization must also realize that the supplier's needs and expectations must be respected if the requirements are to be fully satisfied.

To understand how quality may be built into a product or service, at any stage, it is necessary to examine the two distinct, but interrelated aspects of quality:

- Quality of design
- Quality of conformance to design.

Quality of design

We are all familiar with the old story of the tree swing (Figure 1.2), but in how many places in how many organizations is this chain of activities taking place? To discuss the quality of – say – a chair it is necessary to describe its purpose. What it is to be used for? If it is to be used for watching TV for 3 hours at a stretch, then the typical office chair will not meet this requirement. The difference between the quality of the TV chair and the office chair is not a function of how it was manufactured, but its *design*.

Quality of design is a measure of how well the product or service is designed to achieve the agreed requirements. The beautifully presented gourmet meal will not necessarily please the recipient if he or she is travelling on the motorway and has stopped for a quick bite to eat. The most important feature of the design, with regard to achieving quality, is the specification. Specifications must also exist at the internal supplier-customer interfaces if one is to achieve a total quality performance. For example, the company lawyer asked to draw up a contract by the sales manager requires a specification as to its content:

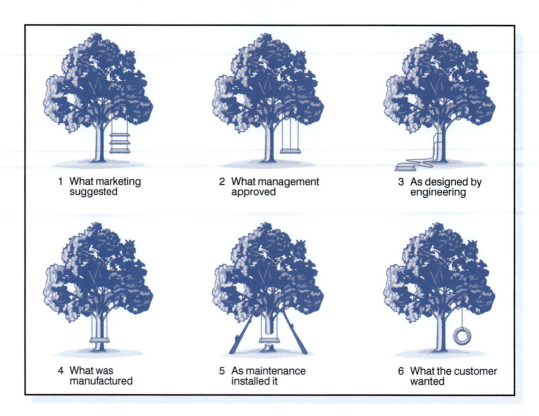

Figure 1.2
Quality of design

1. Is it a sales, processing or consulting type of contract?
2. Who are the contracting parties?
3. In which countries are the parties located?
4. What are the products involved (if any)?
5. What is the volume?
6. What are the financial aspects, e.g. price escalation?

The financial controller must issue a specification of the information he or she needs, and when, to ensure that foreign exchange fluctuations do not cripple the company's finances. The business of sitting down and agreeing a specification at every interface will clarify the true requirements and capabilities. It is the vital first stage for a successful total-quality effort.

There must be a corporate understanding of the organization's quality position in the market place. It is not sufficient that marketing specifies the product or service 'because that is what the customer wants'. There must be an agreement that the operating departments can achieve that requirement. Should they be incapable of doing so, then one of two things must happen: either the organization finds a different position in the market place or substantially changes the operational facilities.

Quality of conformance to design

This is the extent to which the product or service achieves the quality of design. What the customer actually receives should conform to the design, and operating costs are tied firmly to the level of conformance achieved. Quality cannot be inspected into products or services; the customer satisfaction must be designed into the whole system. The conformance check then makes sure that things go according to plan.

A high level of inspection or checking at the end is often indicative of attempts to inspect in quality. This may well result in spiralling costs and decreasing viability. The area of conformance to design is concerned largely with the quality performance of the actual operations. It may be salutary for organizations to use the simple matrix of Figure 1.3 to assess how much time they spend doing the right things right. A lot of people, often through no fault of their own, spend a good proportion of the available time doing the right things wrong. There are people (and organizations) who spend time doing the wrong things very well, and even those who occupy themselves doing the wrong things wrong, which can be very confusing!

MANAGING QUALITY

Every day two men who work in a certain factory scrutinize the results of the examination of the previous day's production, and begin the ritual battle over whether the material is suitable for despatch to the customer. One is called the production manager, the other the quality control manager. They argue and debate the evidence before them, the rights and wrongs of the specification, and each tries to convince the other of the validity of his argument. Sometimes they nearly start fighting.

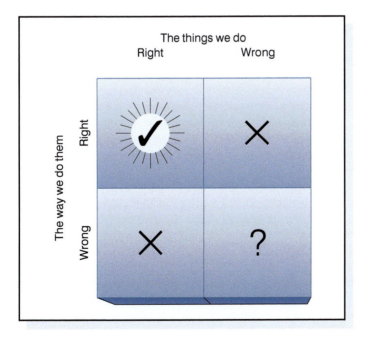

Figure 1.3
How much time is spent doing the right things right?

This ritual is associated with trying to answer the question, '*Have we done the job correctly?*', correctly being a flexible word, depending on the interpretation given to the specification on that particular day. This is not quality *control*, it is *detection* – wasteful detection of bad product before it hits the customer. There is still a belief in some quarters that to achieve quality we must check, test, inspect or measure – the ritual pouring on of quality at the end of the process. This is nonsense, but it is frequently practised. In the office one finds staff checking other people's work because they expect errors, validating computer data, checking invoices, word processing etc. There is also quite a lot of looking for things, chasing why things are late, apologizing to customers for lateness and so on. Waste, waste, waste!

To get away from the natural tendency to rush into the detection mode, it is necessary to ask different questions in the first place. We should not ask whether the job has been done correctly, we should ask first '*Are we capable of doing the job correctly?*' This question has wide implications, and this book is devoted largely to the various activities necessary to ensure that the answer is yes. However, we should realize straight away that such an answer will only be obtained by means of satisfactory methods, materials, equipment, skills and instruction, and a satisfactory 'process'.

Quality and processes

As we have seen, quality chains can be traced right through the business or service processes used by any organization. A process is the transformation of a set of inputs into outputs that satisfy customer needs and expectations, in the form of products, information or services. Everything we do is a process, so in each area or function of an organization there will be many processes taking place. For example, a finance department may be engaged in budgeting processes, accounting processes, salary and wage processes, costing processes, etc. Each process in each department or area can be analysed by an examination of the inputs and outputs. This will determine some of the actions necessary to improve quality. There are also cross-functional processes.

The output from a process is that which is transferred to somewhere or to someone – the *customer*. Clearly to produce an output that meets the requirements of the customer, it is necessary to define, monitor and control the inputs to the process, which in turn may be supplied as output from an earlier process. At every supplier-customer interface then there resides a transformation process (Figure 1.4), and every single task throughout an organization must be viewed as a process in this way. The so-called 'voice of the customer' is a fundamental requirement to good process management and the 'voice of the process' provides key feedback to the supply side of the process equation: right Suppliers + correct Inputs = correct Outputs + satisfied Customers (SIPOC).

Once we have established that our process is capable of meeting the requirements, we can address the next question, '*Do we continue to do the job correctly?*' which brings a requirement to monitor the process and the controls on it. If we now re-examine the first question, 'Have we done the job correctly?' we can see that, if we have been able to answer the other two questions with a yes, we *must* have done the job correctly. Any other outcome would be illogical. By asking the questions in the right order, we have moved the need to ask the 'inspection' question and replaced a strategy of *detection* with one of *prevention*. This concentrates attention on the front end of any process – the inputs – and changes the emphasis to making sure the inputs are capable

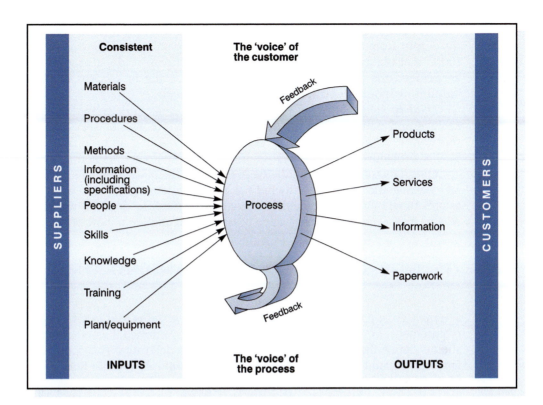

Figure 1.4
A process – SIPOC

of meeting the requirements of the process. This is a managerial responsibility and is discharged by efficiently organizing the inputs and its resources and controlling the processes.

These ideas apply to every transformation process; they all must be subject to the same scrutiny of the methods, the people, skills, equipment and so on to make sure they are correct for the job. A person giving a conference presentation whose audio/visual equipment will not focus correctly, or whose material is not appropriate, will soon discover how difficult it is to make a presentation that meets the requirements of the audience.

In every organization there are some very large processes – groups of smaller processes often called *core business processes*. These are activities the organization must carry out especially well if its mission and objectives are to be achieved. The area will be dealt with in some detail later on in the book. It is crucial if the management of quality is to be integrated into the strategy for the organization.

The *control* of quality can only take place at the point of operation or production – where the letter is word-processed, the sales call made, the patient admitted or the chemical manufactured. The act of *inspection is not quality control.* When the answer to 'Have we done the job correctly?' is given indirectly by answering the questions of capability and control, then we have *assured* quality, and the activity of checking becomes one of *quality assurance* – making sure that the product or service represents the output from an effective *system* to ensure capability and control. It is frequently found that organizational barriers between functional or departmental empires has encouraged the development of testing and checking of services or products in a vacuum, without interaction with other departments.

Quality control then is essentially the activities and techniques employed to achieve and maintain the quality of a product, process, or service. It includes a monitoring activity, but is also concerned with finding and eliminating causes of quality problems so that the requirements of the customer are continually met.

Quality assurance is broadly the prevention of quality problems through planned and systematic activities (including documentation). These will include the establishment of a good quality management system and the assessment of its adequacy, the audit of the operation of the system and the review of the system itself.

QUALITY STARTS WITH UNDERSTANDING THE NEEDS

The marketing processes of an organization must take the lead in establishing the true requirements for the product or service. Having determined the need, the organization should define the market sector and demand, to determine such product or service features as grade, price, quality, timing, etc. For example, a major hotel chain thinking of opening a new hotel or refurbishing an old one will need to consider its location and accessibility before deciding whether it will be predominantly a budget, first-class, business or family hotel.

The organization will also need to establish customer requirements by reviewing the market needs, particularly in terms of unclear or unstated expectations or preconceived ideas held by customers. It is central to identify the key characteristics that determine the suitability of the product or service in the eyes of the customer. This may, of course, call for the use of market research techniques, data gathering and analysis of customer complaints. If necessary, quasi-quantitative methods may be employed, giving proxy variables that can be used to grade the characteristics in importance, and decide in which areas superiority over competitors exists. It is often useful to compare these findings with internal perceptions.

Excellent communication between customers and suppliers is the key to a total quality performance; it will eradicate the 'demanding nuisance/idiot' view of customers, which even now still pervades some organizations. Poor communications often occur in the supply chains between organizations, when neither party realizes how poor they are. Feedback from both customers and suppliers needs to be improved where dissatisfied customers and suppliers do not communicate their problems. In such cases, non-conformance of purchased products or services is often due to customers' inability to communicate their requirements clearly. If these ideas are also used within an organization, then the internal supplier/customer interfaces will operate much more smoothly.

All the efforts devoted to finding the nature and timing of the demand will be pointless if there are failures in communicating the requirements throughout the organization promptly, clearly and accurately. The marketing processes should be capable of producing a formal statement or outline of the requirements for each product or service. This constitutes a preliminary set of *specifications*, which can be used as the basis for service or product design. The information requirements include:

1. Characteristics of performance and reliability – these must make reference to the conditions of use and any environmental factors that may be important.
2. Aesthetic characteristics, such as style, colour, smell, task, feel, etc.
3. Any obligatory regulations or standards governing the nature of the product or service.

The organization must also establish systems for feedback of customer information and reaction, and these systems should be designed on a continuous monitoring basis. Any information pertinent to the product or service should be collected and collated, interpreted, analysed and communicated, to improve the response to customer experience and expectations. These same principles must also be applied inside the organization if continuous improvement at every transformation process interface is to be achieved. If one function or department in a company has problems recruiting the correct sort of staff, for example, and HR has not established mechanisms for gathering, analysing and responding to information on new employees, then frustration and conflict will replace communication and co-operation.

One aspect of the analysis of market demand that extends back into the organization is the review of market readiness of a new product or service. Items that require some attention include assessment of:

1. The suitability of the distribution and customer-service processes.
2. Training of personnel in the 'field'.
3. Availability of 'spare parts' or support staff.
4. Evidence that the organization is capable of meeting customer requirements.

All organizations receive a wide range of information from customers through invoices, payments, requests for information, complaints, responses to advertisements and promotion, etc. An essential component of a 'customer relationship management' system for the analysis of demand is that this data is channelled quickly into the appropriate areas for action and, if necessary, response.

There are various techniques of research, which are outside the scope of this book, but have been well documented elsewhere. It is worth listing some of the most common and useful general methods that should be considered, both externally and internally:

- Surveys – questionnaires, etc.
- Panel or focus group techniques
- In-depth interviews
- Brainstorming and discussions
- Role rehearsal and reversal
- Interrogation of trade associations.

The number of methods and techniques for researching the market is limited only by imagination and funds. The important point to stress is that the supplier, whether the internal individual or the external organization, keeps very close to the customer. Good research, coupled with analysis of CRM data, is an essential part of finding out what the requirements are, and breaking out from the obsession with inward scrutiny that bedevils quality.

QUALITY IN ALL FUNCTIONS

For an organization to be truly effective, each component of it must work properly together. Each part, each activity, each person in the organization affects and is in turn affected by others. Errors have a way of multiplying, and failure to meet the requirements in one part or area creates problems elsewhere, leading to yet more errors, yet more problems and so on. The benefits of getting it right first time everywhere are enormous.

Everyone experiences – almost accepts – problems in working life. This causes people to spend a large part of their time on useless activities – correcting errors, looking for things, finding out why things are late, checking suspect information, rectifying and reworking, apologizing to customers for mistakes, poor quality and lateness. The list is endless, and it is estimated that about one-third of our efforts are still wasted in this way. In the service sector it can be much higher.

Quality, the way we have defined it as meeting the customer requirements, gives people in different functions of an organization a common language for improvement. It enables all the people, with different abilities and priorities, to communicate readily with one another, in pursuit of a common goal. When business and industry were local, the craftsman could manage more or less on his own. Business is now so complex and employs so many different specialist skills that everyone has to rely on the activities of others in doing their jobs.

Some of the most exciting applications of TQM have materialized from groups of people that could see little relevance when first introduced to its concepts. Following training, many different parts of organizations can show the usefulness of the techniques. Sales staff can monitor and increase successful sales calls, office staff have used TQM methods to prevent errors in word-processing and improve inputting to computers, customer-service people have monitored and reduced complaints, distribution has controlled lateness and disruption in deliveries.

It is worthy of mention that the first points of contact for some outside customers are the telephone operator, the security people at the gate or the person in reception. Equally the e-business, paperwork and support services associated with the product, such as websites, invoices and sales literature and their handlers, must match the needs of the customer. Clearly TQM cannot be restricted to the 'production' or 'operations' areas without losing great opportunities to gain maximum benefit.

Managements that rely heavily on exhortation of the workforce to 'do the right job right the first time', or 'accept that quality is your responsibility', will not only fail to achieve quality but may create division and conflict. These calls for improvement infer that faults are caused only by the workforce and that problems are departmental or functional when, in fact, the opposite is true – most problems are inter-departmental. The commitment of all members of an organization is a requirement of 'organization-wide quality improvement'. Everyone must work together at every interface to achieve improved performance and that can only happen if the top management is really committed.

BIBLIOGRAPHY

Beckford, J., *Quality, A Critical Introduction* (3rd edn), Routledge, London, 2007.

Crosby, P.B., *Quality is Free*, McGraw-Hill, New York, 1979.

Crosby, P.B., *Quality Without Tears*, McGraw-Hill, New York, 1984.

Dale, B.G. (ed.), *Managing Quality* (5th edn), Philip Alan, Hemel Hempstead, 2007.

Dahlgaard, J.J., Khanji, G.K. and Kristensen, K., *Fundamentals of Total Quality Management*, Taylor & Francis, Oxford, 2007.

Deming, W.E., *Out of the Crisis*, MIT, Cambridge, Mass., 1982.

Deming, W.E., *The New Economies*, MIT, Cambridge, Mass., 1993.

Feigenbaum, A.V., *Total Quality Control* (4th edn), McGraw-Hill, New York, 2004.

Garvin, D.A., *Managing Quality: The Strategic Competitive Edge*, The Free Press (Macmillan), New York, 1988.

Ishikawa, K. (translated by D.J. Lu), *What is Total Quality Control? The Japanese Way*, Prentice-Hall, Englewood Cliffs, NJ, 1985.

Juran, J.M. and DeFeo, J.A., *Juran's Quality Handbook*, McGraw-Hill, New York, 2010.

Kehoe, D.F., *The Fundamentals of Quality Management*, Springer (Chapman & Hall), London, 1996.

Murphy, J.A., *Quality in Practice* (3rd edn), Gill and MacMillan, Dublin, 2000.

Price, F., *Right Every Time*, Gower, Aldershot, 1990.

Stahl, M.J. (ed), *Perspectives in Total Quality*, Quality Press, Milwaukee, 1999.

CHAPTER HIGHLIGHTS

Quality, competitiveness and customers

- The reputation enjoyed by an organization is built by quality, reliability, delivery and price. Quality is perhaps the most important of these competitive weapons.
- Reputations for poor quality last for a long time, and good or bad reputations can become national or international. The management of quality can be learned and used to improve reputation.
- Quality is meeting the customer requirements, and this is not restricted to the functional characteristics of the product or service.
- Reliability is the ability of the product or service to continue to meet the customer requirements over time.
- Organizations 'delight' the customer by consistently meeting customer requirements, and then achieve a reputation of 'excellence' and customer loyalty.

Understanding and building the quality chains

- Throughout all organizations there are a series of internal suppliers and customers. These form the so-called 'quality chains', the core of 'company-wide quality improvement'.
- The internal customer/supplier relationships must be managed by interrogation, i.e. using a set of questions at every interface. Measurement of capability is vital.
- There are two distinct but interrelated aspects of quality, design and conformance to design. *Quality of design* is a measure of how well the product or service is designed to achieve the agreed requirements. *Quality of conformance to design* is the extent to

Understanding quality

which the product or service achieves the design. Organizations should assess how much time they spend doing the right things right.

Managing quality

- Asking the question 'Have we done the job correctly?' should be replaced by asking 'Are we capable of doing the job correctly?' and 'Do we continue to do the job correctly?'
- Asking the questions in the right order replaces a strategy of *detection* with one of *prevention*.
- Everything we do is a process, which is the transformation of a set of inputs into the desired outputs.
- In every organization there are some core business processes that must be performed especially well if the mission and objectives are to be achieved. They are defined by SIPOC – suppliers-inputs-process-outputs-customers.
- Inspection is not *quality control*. The latter is the employment of activities and techniques to achieve and maintain the quality of a product, process or service.
- *Quality assurance* is the prevention of quality problems through planned and systematic activities.

Quality starts with understanding the needs

- Marketing processes establish the true requirements for the product or service. These must be communicated properly throughout the organization in the form of specifications.
- Excellent communications between customers and suppliers is the key to a total quality performance – the organization must establish feedback systems, such as CRM, to gather customer information.
- Appropriate research techniques should be used to understand the 'market' and keep close to customers and maintain the external perspective.

Quality in all functions

- All members of an organization need to work together on organization-wide quality improvement. The co-operation of everyone at every interface is necessary to achieve improvements in performance, which can only happen if the top management is really committed.

Chapter 2

Models and frameworks for Total Quality Management

EARLY TQM FRAMEWORKS

In the early 1980s when organizations in the West seriously became interested in quality and its management there were many attempts to construct lists and frameworks to help this process.

In the West the famous American 'gurus' of quality management, such as W. Edwards Deming, Joseph M. Juran and Philip B. Crosby started to try to make sense of the labyrinth of issues involved, including the tremendous competitive performance of Japan's manufacturing industry. Deming and Juran had contributed to building Japan's success in the 1950s and 1960s and it was appropriate that they should set down their ideas for how organizations could achieve success.

Deming had fourteen points to help management as follows:

1. Create constancy of purpose towards improvement of product and service.
2. Adopt the new philosophy. We can no longer live with commonly accepted levels of delays, mistakes, defective workmanship.
3. Cease dependence on mass inspection. Require, instead statistical evidence that quality is built in.
4. End the practice of awarding business on the basis of price tag.
5. Find problems. It is management's job to work continually on the system.
6. Institute modern methods of training on the job.
7. Institute modern methods of supervision of production workers. The responsibility of foremen must be changed from numbers to quality.
8. Drive out fear, so that everyone may work effectively for the company.
9. Break down barriers between departments.
10. Eliminate numerical goals, posters and slogans for the workforce asking for new levels of productivity without providing methods.
11. Eliminate work standards that prescribe numerical quotas.

12. Remove barriers that stand between the hourly worker and his right to pride of workmanship.
13. Institute a vigorous programme of education and retraining.
14. Create a structure in top management that will push every day on the above thirteen points.

Juran's ten steps to quality improvement were:

1. Build awareness of the need and opportunity for improvement.
2. Set goals for improvement.
3. Organize to reach the goals (establish a quality council, identify problems, select projects, appoint teams, designate facilitators).
4. Provide training.
5. Carry out projects to solve problems.
6. Report progress.
7. Give recognition.
8. Communicate results
9. Keep score.
10. Maintain momentum by making annual improvement part of the regular systems and processes of the company.

Phil Crosby, who spent time as Quality Director of ITT, had 'four absolutes:'

- Definition – conformance to requirements.
- System – prevention.
- Performance standard – zero defects.
- Measurement – price of non-conformance.

He also offered management fourteen steps to improvement:

1. Make it clear that management is committed to quality.
2. Form quality improvement teams with representatives from each department.
3. Determine where current and potential quality problems lie.
4. Evaluate the cost of quality and explain its use as a management tool.
5. Raise the quality awareness and personal concern of all employees.
6. Take actions to correct problems identified through previous steps.
7. Establish a committee for the zero defects programme.
8. Train supervisors to actively carry out their part of the quality improvement programme.
9. Hold a 'zero defects day' to let all employees realize that there has been a change.
10. Encourage individuals to establish improvement goals for themselves and their groups.
11. Encourage employees to communicate to management the obstacles they face in attaining their improvement goals.
12. Recognize and appreciate those who participate.
13. Establish quality councils to communicate on a regular basis.
14. Do it all over again to emphasize that the quality improvement programme never ends.

A comparison

One way to compare directly the various approaches of the three American gurus is in Table 2.1, which shows the differences and similarities clarified under 12 different factors.

Table 2.1 The American quality gurus compared

	Crosby	*Deming*	*Juran*
Definition of quality	Conformance to requirements	A predictable degree of uniformity and dependability at low cost and suited to the market	Fitness for use
Degree of senior-management responsibility	Responsible for quality	Responsible for 94% of quality problems	Less than 20% of quality problems are due to workers
Performance standard/motivation	Zero defects	Quality has many scales. Use statistics to measure performance in all areas. Critical of zero defects.	Avoid campaigns to do perfect work
General approach approach	Prevention, not inspection	Reduce variability by continuous improvement. Cease mass inspection	General management to quality – especially 'human' elements
Structure	Fourteen steps to quality improvement	Fourteen points for management	Ten steps to quality improvement
Statistical process control (SPC)	Rejects statistically acceptable levels of quality	Statistical methods of quality control must be used	Recommends SPLC but warns that it can lead to too-driven approach
Improvement basis.	A 'process', not a programme. Improvement goals.	Continuous to reduce variation. Eliminate goals without methods.	Project-by-project team approach. Set goals.
Teamwork	Quality improvement teams. Quality councils.	Employee participation in decision-making. Break down barriers between departments	Team and quality circle approach
Costs of quality	Cost of non-conformance. Quality is free	No optimum – continuous improvement	Quality is not free – there is an optimum
Purchasing and goods received	State requirements. Supplier is extension of business. Most faults due to purchasers themselves	Inspection too late – allows defects to enter system through AQLs. Statistical evidence and control charts required.	Problems are complex. Carry out formal surveys.
Vendor rating improve.	Yes *and* buyers. Quality audits useless.	No – critical of most systems	Yes, but help supplier
Single sources of supply		Yes.	No – can neglect to sharpen competitive edge.

Figure 2.1
Total quality
management model –
major features

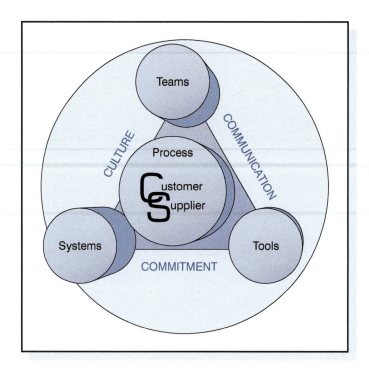

Our understanding of 'total quality management' developed through the 1980s and in earlier editions of this author's books on TQM, a broad perspective was given, linking the TQM approaches to the direction, policies and strategies of the business or organization. These ideas were captured in a basic framework – the TQM Model (Figure 2.1) which was widely promoted in the U.K. through the activities of the UK Department of Trade and Industry (DTI) 'Quality Campaign' and 'Managing into the 90s' programmes. These approaches brought together a number of components of the quality approach, including quality circles (teams), problem solving and statistical process control (*tools*) and quality systems, such as BS5750 and later ISO 9000 (systems). It was recognized that *culture* played an enormous role in whether organizations were successful or not with their TQM approaches. Good *communications*, of course, were seen to be vital to success but the most important of all was *commitment*, not only from the senior management but from everyone in the organization, particularly those operating directly at the customer interface. The customer/supplier or 'quality chains' were the core of this TQM model.

Many companies and organizations in the public sector found this simple framework useful and it helped groups of senior managers throughout the world get started with TQM. The key was to integrate the TQM activities, based on the framework, into the business or organization strategy and this has always been a key component of the author's approach.

Starting in Japan with the Deming Prize, companies started to get interested in quality frameworks that could be used essentially in three ways:

i. as the basis for awards
ii. as the basis for a form of 'self-assessment'
iii. as a descriptive 'what-needs-to-be-in-place' model.

The earliest approach to a *total* quality audit process is that established in the Japanese-based 'Deming Prize', which is based on a highly demanding and intrusive process. The categories of this award were established in 1950 when the Union of Japanese Scientists and Engineers (JUSE) instituted the prize(s) for 'contributions to quality and dependability of product' (www.juse.or.jp/e/deming).

This now defines TQM as

a set of systematic activities carried out by the entire organization to effectively and efficiently achieve the organization's objectives so as to provide products and services with a level of quality that satisfies customers, at the appropriate time and price.

As the Deming Award guidelines say, there is no easy success at this time of constant change and no organization can expect to build excellent quality management systems just by solving problems given by others:

They need to think on their own, set lofty goals and drive themselves to challenge for achieving those goals. For these organizations that introduce and implement TQM in this manner, the Deming Prize aims to be used as a tool for improving and transforming their business management.

The recognition that total quality management is a broad culture change vehicle with internal and external focus embracing behavioural and service issues, as well as quality assurance and process control, prompted the United States to develop in the late 1980s one of the most famous and now widely used frameworks, the Malcolm Baldrige National Quality Award (MBNQA). The award itself, which is composed of two solid crystal prisms 14 inches high, is presented annually to recognize companies in the USA that have excelled in quality management and quality achievement. But it is not the award itself, or even the fact that it is presented each year by the President of the USA which has attracted the attention of most organizations, it is the excellent framework for TQM and organizational self-assessments.

The Baldrige Performance Excellence Program, as it is now known, aims to:

- help improve organizational performance practices, capabilities and results
- facilitate communication and sharing of best practices information
- serve as a working tool for understanding and managing performance and for guiding, planning and opportunities for learning.

The award criteria are built upon a set of inter-related core values and concepts:

- visionary leadership
- customer-driven excellence

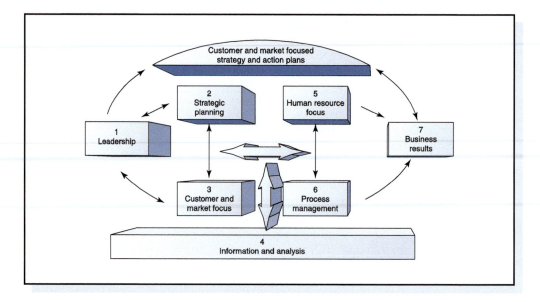

Figure 2.2
Baldrige criteria for performance excellence framework

- organizational and personal learning
- valuing employees and partners
- agility
- focus on the future
- managing for innovation
- management by fact
- public responsibility and citizenship
- focus on results and creating value
- systems developments.

These are embodied in a framework of seven categories which are used to assess organizations:

1. Leadership
   - organizational leadership
   - public responsibility and citizenship
2. Strategic planning
   - strategy development
   - strategy deployment
3. Customer focus
   - customer and market knowledge
   - customer relationships and satisfaction
4. Measurement, analysis and knowledge management
   - measurement and analysis of organizational performance
   - information management

5. Work force focus
   - work systems
   - employee education training and development
   - employee well-being and satisfaction
6. Operations focus
   - product and service processes
   - business processes
   - support processes
7. Results
   - customer focused results
   - financial and market results
   - human resource results
   - organizational effectiveness results.

Figure 2.2 shows how the framework's system connects and integrates the categories. The main driver is the senior executive leadership which creates the values, goals and systems, and guides the sustained pursuit of quality and performance objectives. The system includes a set of well-defined and designed processes for meeting the organization's direction and performance requirements. Measures of progress provide a results-oriented basis for channelling actions to deliver ever-improving customer values and organization performance. The overall goal is the delivery of customer satisfaction and market success leading, in turn, to excellent business results. The seven criteria categories are further divided into items and areas to address. These are described in some detail in the 'Criteria for Performance Excellence' available from the US National Institute of Standards and Technology (NIST), in Gaithesburg USA (www.nist.gov/baldrige).

The Baldrige Award led to a huge interest around the world in quality award frameworks that could be used to carry out self-assessment and to build an organization-wide approach to quality, which was truly integrated into the business strategy. It was followed in Europe in the early 1990s by the launch of the European Quality Award by the European Foundation for Quality Management (EFQM). This framework was the first one to include 'Business Results' and to really represent the whole business model.

Like the Baldrige, the 'EFQM Excellence Model', as it is now known, recognizes that processes are the means by which an organization harnesses and releases the talents of its people to produce results/performance. Moreover, improvement in performance can be achieved only by improving the processes by involving the people. This simple model is shown in Figure 2.3.

Figure 2.4 displays graphically the 'non-prescriptive' principles of the full Excellence Model. Essentially customer results, people (employee) results and favourable society results are achieved through leadership driving strategy, people, partnerships & resources and processes, products & systems, which lead ultimately to excellence in key results – the enablers deliver the results which in turn drive learning, creativity & innovation. The EFQM have provided a weighting for each of the criteria which may be used in scoring self-assessments and making awards (see Chapter 8).

Through usage and research, the Baldrige and EFQM Excellence models have continued to grow in stature since their inception. They were recognized as descriptive

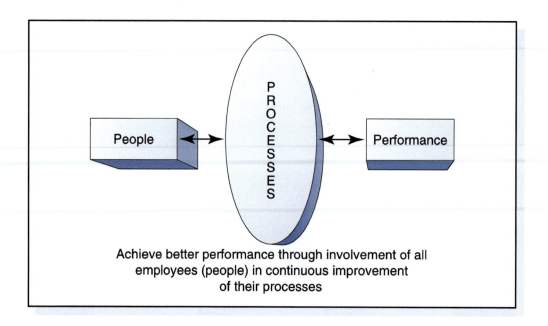

Figure 2.3
The simple model for improved performance

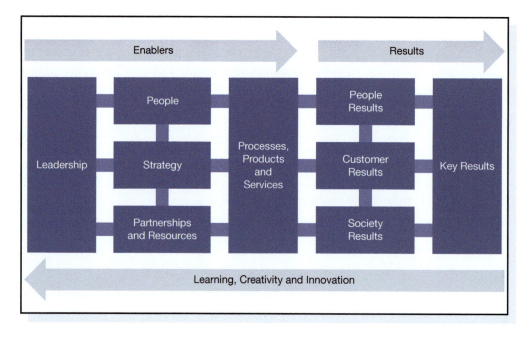

Figure 2.4
The EFQM Excellence Model

The foundations of TQM

holistic business models, rather than just quality models and mutated into frameworks for (Business) Excellence.

The NIST and EFQM have worked together well over recent years to learn from each other's experience in administering awards and supporting programmes, and from organizations which have used their frameworks 'in anger'.

The EFQM publications on the Excellence Model capture much of this learning and provide a framework which organizations can use to follow ten steps:

1. set direction through leadership
2. establish the results they want to achieve
3. establish and drive the strategy
4. set up and manage appropriately their approach to processes, people, partnerships and resources
5. deploy the approaches to ensure achievement of the strategies and thereby the results
6. assess the 'business' performance, in terms of customers, their own people and society results
7. assess the achievements of key performance results
8. review performance for strengths and areas for improvement
9. innovate to deliver performance improvements
10. learn more about the effects of the enablers on the results.

THE FOUR Ps AND THREE Cs OF TQM – A MODEL FOR TQM & OpEx

We have seen in Chapter 1 how *processes* are the key to delivering quality of products and services to customers. It is clear from Figure 2.4 that processes are a key linkage between the enablers of *planning* (leadership driving policy and strategy, partnerships and resources), through *people* into the *performance* of people, society, customers and key outcomes.

These 'four Ps' form the basis of a simple model for TQM and provide the 'hard management necessities' to take organizations successfully into the twenty-first century. These form the structure of the remainder of this book.

From the early TQM frameworks, however, we must not underestimate the importance of the three Cs – *Culture, Communication* and *Commitment*. The TQM model is complete when these 'soft outcomes' are integrated into the four Ps framework to move organizations successfully forward (Figure 2.5).

This TQM model, based on the extensive work done during the last century, provides a simple framework for excellent performance, covering all angles and aspects of an organization and its operation.

Performance is achieved, using a business excellence approach, and by planning the involvement of people in the improvement of processes. This has to include:

- *Planning* – the development and deployment of policies and strategies; setting up appropriate partnerships and resources; and designing in quality.

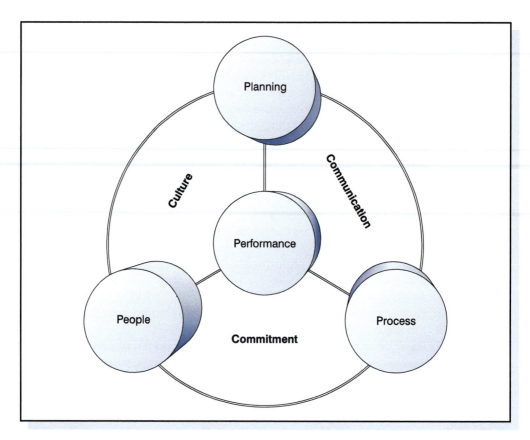

Figure 2.5
The framework for total quality management

- *Performance* – establishing a performance measure framework – a 'balanced scorecard' for the organization; carrying out self-assessment, audits, reviews and benchmarking.
- *Processes* – understanding, management, design and redesign; quality management systems; continuous improvement.
- *People* – managing the human resources; culture change; teamwork; communications; innovation and learning.

Wrapping around all this to ensure successful implementation is, of course, effective leadership and commitment, the subject of the next chapter.

BIBLIOGRAPHY

BQF (British Quality Foundation), *The Model in Practice* and *The Model in Practice 2*, London, 2000 and 2002.

EFQM (European Foundation for Quality Management), *The EFQM Excellence Model*, Brussels, 2013,

National Institute of Standard and Technology, *USA Malcolm Baldrige National Quality Award, Criteria for Performance Excellence*, NIST, Gaithesburg, 2013.

Pyzdek, T. and Keller. P., *The Handbook for Quality Management; A Complete Guide to Operational Excellence* (2nd edn), ASQ, Milwaukee, 2013.

Summers, D.C.S., *Quality Management* (2nd edn), Prentice Hall, London, 2008.

CHAPTER HIGHLIGHTS

Early TQM frameworks

- There have been many attempts to construct lists and frameworks to help organizations understand how to implement good quality management.
- The 'quality gurus' in America, Deming, Juran and Crosby, offered management fourteen points, ten steps and four absolutes (plus fourteen steps) respectively. These similar but different approaches may be compared using a number of factors, including definition of quality, degree of senior management responsibility and general approach.
- The understanding of quality developed and, in Europe and other parts of the world, the author's early TQM model, based on a customer/supplier chain core surrounded by systems, tools and teams, linked through culture, communications and commitment, gained wide usage.

Quality award models

- Quality frameworks may be used as the basis for awards for a form of 'self-assessment' or as a description of what should be in place.
- The Deming Prize in Japan was the first formal quality award framework established by JUSE in 1950. The examination viewpoints include: top management leadership and strategies; TQM frameworks, concepts and values; QA and management systems; human resources; utilization of information, scientific methods; organizational powers; realization of corporate objectives.
- The USA Baldrige Award aims to promote performance excellence and improvement in competitiveness through a framework of seven categories which are used to assess organizations: leadership; strategic planning; customer and market focus; information and analysis; human resource focus; process management; business results.
- The European (EFQM) Excellence Model operates through a simple framework of performance improvement through involvement of people in improving processes.
- The full Excellence Model is a non-prescriptive framework for achieving good results – customers, people, society, key performance – through the enablers – leadership,

strategy, people, processes, products, systems, partnerships and resources. The framework includes feedback loops of learning, innovation and creativity and proposed weightings for assessment.

The four Ps and three Cs – a model for TQM

- *Planning*, *People* and *Processes* are the keys to delivering quality products and services to customers and generally improving overall *Performance*. These four Ps form a structure of 'hard management necessities' for a simple TQM model which forms the structure of this book.
- The three Cs of culture, communication and commitment provide the glue or 'soft outcomes' of the model which will take organizations successfully into the twenty-first century.

Chapter 3

Leadership and commitment

THE TOTAL QUALITY MANAGEMENT APPROACH

'What is quality management?' Something that is best left to the experts is often the answer to this question. But this is avoiding the issue, because it allows executives and managers to opt out. Quality is too important to leave to the so called 'quality professionals'; it cannot be achieved on a company or organization-wide basis if it is left to the so-called experts. Equally dangerous, however, are the uninformed who try to follow their natural instincts because they 'know what quality is when they see it'. Usually this belies a narrow product quality focus and can only yield limited benefits. This type of intuitive approach can lead to serious attitude problems, which do no more than reflect the understanding and knowledge of quality that are present in an organization.

The organization that believes that the traditional quality control techniques and the way they have always been used will resolve their quality problems may be misguided. Employing more inspectors, tightening up standards, developing correction, repair and rework teams do not improve quality. Traditionally, quality has been regarded as the responsibility of a quality (assurance or control) department, and still it has not yet been recognized in some organizations that many quality problems originate in the commercial, engineering, service or administrative areas.

Total quality management is far more than shifting the responsibility of *detection* of problems from the customer to the producer. It requires a comprehensive approach that must first be recognized and then implemented if the rewards are to be realized. Today's business environment is such that managers must plan strategically to maintain a hold on market share, let alone increase it. In many companies quality problems have been found to seriously erode margins due to the cost of rectifying defective work both during the contract period and afterwards. We know that low cost operations have dominated life during the last decade or more but consumers still place a higher value on quality than on loyalty to suppliers, and price is often not the major determining factor in consumer choice. Costs in many areas have been slashed by improvements and changes in operations, including location, but quality is still of paramount importance in most industrial, service, hospitality and many other markets.

TQM is an approach to improving the competitiveness, effectiveness and flexibility of a whole organization. It is essentially a way of planning, organizing and understanding each activity, and depends on each individual at each level. For an organization to be truly effective, each part of it must work properly together towards the same goals, recognizing that each person and each activity affects and in turn is affected by others. TQM is also a way of ridding people's lives of wasted effort by bringing everyone into the processes of improvement, so that results are achieved in less time. The methods and techniques used in TQM can be applied throughout any organization. They are equally useful in manufacturing, public service, health care, education and hospitality industries.

The impact of TQM on an organization is first to ensure that the management adopts a strategic overview of quality. The approach must focus on developing a *problem-prevention* mentality; but it is easy to underestimate the effort that is required to change attitudes and approaches. Many people will need to undergo a complete change of 'mindset' to unscramble their intuition, which rushes into the detection/ inspection mode to solve quality problems – 'We have a quality problem, we had better double check every single item' – whether it is in electronics, plastics, heavy engineering or aerospace. Managers in the construction industries, where there are sometimes three tiers of quality inspection to make sure that everything is right, often report that no one takes responsibility, the first worker relies on the first inspector, the next worker on the second inspector and so on, with no one apparently accountable.

A better mindset may be achieved by looking at the sort of barriers that exist in key areas. Staff may need to be trained and shown how to reallocate their time and energy to studying their processes in teams, searching for causes of problems and correcting the causes, not the symptoms, once and for all. This often requires of management a positive, thrusting initiative to promote the right-first-time approach to work situations. Through *quality or process performance improvement teams*, these actions will reduce the inspection-rejection syndrome in due course. If things are done correctly first time round, the usual problems that create the need for inspection for failure should disappear.

The managements of many firms may think that their scale of operation is not sufficiently large, that their resources are too slim or that the need for action is not important enough to justify implementing TQM. Before arriving at such a conclusion, however, they should examine their existing performance by asking the following questions:

1. Is any attempt made to assess the costs arising from errors, defects, waste, customer complaints, lost sales, etc.? If so, are these costs minimal or insignificant?
2. Is the standard of management adequate and are attempts being made to ensure that quality is given proper consideration at the design stage?
3. Are the organization's quality management systems – documentation, processes, operations, etc. – in good order?
4. Have people been trained in how to prevent errors and problems? Do they anticipate and correct potential causes of problems, or do they find and reject?

5. Are subcontract suppliers being selected on the basis of the quality of their people and services as well as price?
6. Do job instructions contain the necessary quality elements, are they kept up to date and are employers doing their work in accordance with them?
7. What is being done to motivate and train employees to do work right first time?
8. How many errors and defects, and how much wastage occurred last year? Is this more or less than the previous year?

If satisfactory answers can be given to most of these questions, an organization can be reassured that it is already well on the way to using adequate quality management. Even so, it may find that the introduction of TQM causes it to reappraise activities throughout. If answers to the above questions indicate problem areas, it will be beneficial to review the top management's attitude to quality. Time and money spent on quality-related activities are *not* limitations of profitability; they make significant contributions towards greater efficiency and enhanced profits.

COMMITMENT AND POLICY

To be successful in promoting business effectiveness and efficiency, TQM must be truly organization-wide, it must include the supply chain and it must start at the top with the chief executive or equivalent. The most senior directors and management must all demonstrate that they are serious about quality. The middle management has a particularly important role to play, since they must not only grasp the principles of TQM, they must go on to explain them to the people for whom they are responsible, and ensure that their own commitment is communicated. Only then will TQM spread effectively throughout the organization. This level of management also needs to ensure that the efforts and achievements of their subordinates obtain the recognition, attention and reward that they deserve. Project managers in predominantly design organizations have a critical role as they often have responsibility for selecting the manufacturers, subcontractors and suppliers and have the challenge of creating a cohesive quality focused team on the project. They have to explain the quality strategy to all those involved and ensure that all parts of the team are committed to the shared values and goals.

The chief executive of an organization should accept the responsibility for and commitment to a quality policy in which he/she must really believe. This commitment is part of a broad approach extending well beyond the accepted formalities of the quality assurance function. It creates responsibilities for a chain of quality interactions between the marketing, design, production/operations, purchasing, distribution and service functions. Within each and every department of the organization at all levels, starting at the top, basic changes of attitude may be required to implement TQM approaches. If the owners or directors of the organization do not recognize and accept their responsibilities for the initiation and operation of TQM, then these changes will not happen. Controls, systems and techniques are very important in TQM, but they are not the primary requirement. It is more an attitude of mind, based on pride in the job and teamwork, and it requires from the management total commitment, which must then be extended to all employees at all levels and in all departments.

Leadership and commitment

Senior management commitment should be an obsession, not lip service. It is possible to detect real commitment; it shows in the factories, in the head offices, in the design offices – where the work is being done. Going into organizations sporting poster-campaigns of quality instead of belief, one is quickly able to detect the falseness. The people are told not to worry if problems arise, 'just do the best you can', 'the customer will never notice'. The opposite is an organization where total quality means something can be seen, heard, felt. Things happen at this operating interface as a result of *real* commitment. Material problems are corrected with suppliers, equipment difficulties are put right by improved maintenance programmes or replacement, people are trained, change takes place, partnerships are built and continuous improvement means just that.

One of the major challenges in project work, including consultancy, is the need to influence the values and behaviours of the entire delivery team. To be successful the project manager needs to select the project team on the basis of the fit in values as well as on their technical competence and costs. In construction, for example, it is the effectiveness of the inter-organizational team, working side by side, that will determine the success or failure of the project. In such a setting, the challenge of leadership is to create cohesion in values and behaviours across the entire project.

The quality policy

A sound quality policy, together with the organization and facilities to put it into effect, is a fundamental requirement, if an organization is to fully implement TQM. Every organization should develop and state its policy on quality, together with arrangements for its implementation. The content of the policy should be made known to all employees. The preparation and implementation of a properly thought out quality policy, together with continuous monitoring, make for smoother production or service operation, minimize errors and reduce waste.

Management should be dedicated to the regular improvement of quality, not simply a one-step improvement to an acceptable plateau. These ideas can be set out in a *quality policy* that requires top management to:

1. Identify the end customer's needs (including perception).
2. Assess the ability of the organization to meet these needs economically.
3. Ensure that any bought-in materials meet the required standards of performance and efficiency.
4. Ensure that subcontractors or suppliers share your values and process goals.
5. Concentrate on the prevention rather than detection philosophy.
6. Educate and train for quality improvement and ensure that your subcontractors do so as well.
7. Measure customer satisfaction at all levels, the end customer as well as customer satisfaction between the links of the supply chain.
8. Review the quality management systems to maintain progress.

The quality policy should be the concern of all employees, and the principles and objectives communicated as widely as possible so that it is understood at all levels of the organization and within the subcontract supply chain on construction projects. Practical assistance and training should be given, where necessary, to ensure the relevant knowledge and experience are acquired for successful implementation of the policy throughout the supply chain.

Examples of two company quality policies are given below.

Quality policy example 1 (a process industry company supplying the automotive sector)

- The company will concentrate on its customers and suppliers, both external and internal.
- The performance of our competitors will be communicated to all relevant units.
- Important suppliers and partners will be closely involved in our quality policy – this relates to both external and internal suppliers of goods, resources and services.
- Quality management systems will be designed, implemented, audited and reviewed to drive continuous improvement – they will be integrated into other management systems.
- Quality improvement is primarily the responsibility of management and will be tackled and followed up in a systematic and planned manner – this applies to every part of our organization.
- In order to involve everyone in the organization of quality improvement, management will enable all employees to participate in the preparation, implementation and evaluation of improvement activities.
- Quality improvement will be a continuous process and widespread attention will be given to education, training and skills development activities, which will be assessed with regard to their contribution to the quality policy.
- Publicity will be given to the quality policy in every part of the organization so that everyone may understand it – all available methods and media will be used for its internal and external promotion and communication.
- Reporting on the progress of the implementation of the policy will be a permanent agenda item in management meetings.

Quality policy example 2

This is shown in Figure 3.1.

CREATING OR CHANGING THE CULTURE

The culture within an organization is formed by a number of components:

1. Behaviours based on people interactions.
2. Norms resulting from working groups.
3. Dominant values adopted by the organization.
4. Rules of the game for 'getting on.'
5. The climate.

Culture in any 'business' may be defined then as how business is conducted and how employees behave and are treated. Any organization needs a *vision framework*

ABB Quality policy

There are many dimensions in which ABB can compete, but none of these are meaningful for our customers without a foundation of quality. The responsibility for quality is something that must be owned by every person, every business, and every location that ABB calls home.

Ulrich Spiesshofer

Ulrich Spiesshofer, CEO ABB

To ensure that we meet our responsibilities and obligations to our customers, our people, our partners, our suppliers and to our shareholders we are committed to the following Quality Objectives:

- Deliver on-time & on-quality products, systems and services that meet or exceed our customer's expectations.

- Identify and understand our customer's expectations, measure customer perceptions, and implement improvements to increase customer satisfaction.

- Enable and engage our people at all levels in a relentless drive to improve operational performance along the value chain from suppliers to customers.

- Increase the motivation and skills of our people to add value to our customers and our businesses, through continual training and development.

- Leverage our partners & suppliers strengths to improve our products and our businesses from product design through production, installation and operation.

- Embed social responsibility & company ethics policies in our business practices.

- Continually improve environmental, health and safety performance through all products, operations, systems and services.

Figure 3.1
Quality policy example 2

that includes its *guiding philosophy, core values and beliefs* and a *purpose*. These should be combined into a *mission*, which provides a vivid description of what things will be like when it has been achieved (Figure 3.2).

The *guiding philosophy* drives the organization and is shaped by the leaders through their thoughts and actions. It should reflect the vision of an organization rather than the vision of a single leader and should evolve with time, although organizations must hold on to the *core* elements.

The *core values and beliefs* represent the organization's basic principles about what is important in business, its conduct, its social responsibility and its response to changes in the environment. They should act as a guiding force, with clear and

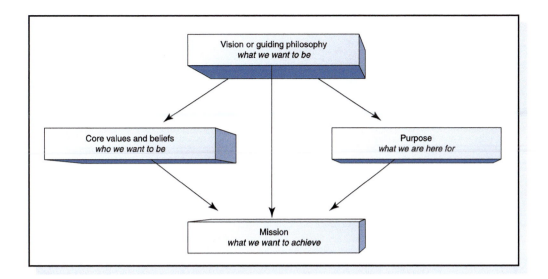

Figure 3.2
Vision framework for an organization

authentic values, which are focused on employees, suppliers, customers, society at large, safety, shareholders and generally stakeholders.

The *purpose* of the organization should be a development from the vision and core values and beliefs and should quickly and clearly convey how the organization is to fulfil its role.

The *mission* will translate the abstractness of philosophy into tangible goals that will move the organization forward and make it perform to its optimum. It should not be limited by the constraints of strategic analysis and should be proactive not reactive. Strategy is subservient to mission, the strategic analysis being done after, not during, the mission setting process.

Two examples of how leaders of organizations – one in the private sector and one in the public sector – develop their vision, mission and values and are role models of a culture of total quality excellence are given in Figure 3.3.

Control

The effectiveness of an organization and its people depends on the extent to which each person and function/department perform their role and move towards the common goals and objectives. Control is the process by which information or feedback is provided so as to keep all functions on track. It is the sum total of the activities that increase the probability of the planned results being achieved. Control mechanisms fall into three categories, depending upon their position in the managerial process (see Table 3.1).

Many organizations use after-the-fact controls, causing managers to take a reactive rather than a proactive position. Such 'crisis-orientation' needs to be replaced by a more anticipative one in which the focus is on preventive or before-the-fact controls.

Public Sector	Private Sector
The purpose and direction of the organisation – the Mission – is developed by a Task Team. Senior, middle and junior managers review and update the Mission, Vision and Values annually to ensure it supports Policy and Strategy.	To enable the company to set direction and achieve its Vision, the senior management team address priorities for improvement. These are driven by a business improvement process, which consists of: articulate a Vision, determine the actions to realize the Vision, define measures and set targets, then implement a rigorous review mechanism.
Leaders invite input from stakeholders via the Employee Involvement initiative, Monthly Update Meetings and Customer Service Seminars. The Values have been placed on help-cards for every employee and are continually re-emphasized at Monthly Update Meetings.	Each member of the team takes responsibility for one of the Excellence Model criteria. They develop improvement plans and personally ensure that these are properly resourced and implemented, and that progress is monitored. Improvements identified at local level are prioritized and resourced by local management against the organisation's annual business plan.
Leaders act as role models and have a list of Role Model Standards to follow, which they are measured against in their Performance Management System. All managers include TQM objectives in their Performance Agreements and Personal Development Plan, which are reviewed through the Review.	

Figure 3.3
Examples of vision framework statements from organizations in public and private sectors

Table 3.1 Three stages of control mechanisms

Before the fact	Operational	After the fact
Strategic plan	Observation	Annual reports
Action plans	Inspection and correction	Variance reports
Budgets	Progress review	Audits
Job descriptions	Staff meetings	Surveys
Individual performance objectives	Internal information and data systems	Performance review
Training and development	Training programmes	Evaluation of training

Attempting to control performance through systems, procedures or techniques *external* to the individual is not an effective approach, since it relies on 'controlling' others; individuals should be responsible for their own actions. An externally based control system can result in a high degree of concentrated effort in a specific area if the system is overly structured, but it can also cause negative consequences to surface:

1. Since all rewards are based on external measures, which are imposed, the 'team members' often focus all their effort on the measure itself, e.g. to

have it set lower (or higher) than possible, to manipulate the information which serves to monitor it, or to dismiss it as someone else's goal not theirs. In the budgeting process, for example, distorted figures are often submitted by those who have learned that their 'honest projections' will be automatically altered anyway.

2. When the rewards are dependent on only one or two limited targets, all efforts are directed at those, even at the expense of others. If short-term profitability is the sole criterion for bonus distribution or promotion, it is likely that investment for longer-term growth areas will be substantially reduced. Similarly, strong emphasis and reward for output or production may result in lowered quality.

3. The fear of not being rewarded, or even being criticized, for performance that is less than desirable may cause some to withhold information that is unfavourable but nevertheless should be flowing into the system.

4. When reward and punishment are used to motivate performance, the degree of risk-taking may lessen and be replaced by a more cautious and conservative approach. In essence, the fear of failure replaces the desire to achieve.

The following problem situations have been observed by the author and his colleagues within companies that have taken part in research and consultancy:

• The goals imposed are seen or known to be unrealistic. If the goals perceived by the subordinate are in fact accomplished, then the subordinate has proved himself wrong. This clearly has a negative effect on the effort expended, since few people are motivated to prove themselves wrong!

• Where individuals are stimulated to commit themselves to a goal and where their personal pride and self-esteem are at stake, then the level of motivation is at a peak. For most people the toughest critic and the hardest taskmaster they confront is not their immediate boss but themselves.

• Directors and managers are often afraid of allowing subordinates to set the goals for fear of them being set too low, or loss of control over subordinate behaviour. It is also true that many do not wish to set their own targets, but prefer to be told what is to be accomplished.

• Where external project managers are recruited to run projects and a reward is negotiated on the basis of a bonus package reflecting time and cost performance, all too often the company is left with the legacy of quality defects long after the project manager has finished his assignment and pocketed his/her bonuses.

• Some public sector client organizations, in moving towards the delivery of infrastructure projects through alliances, have developed very complex performance frameworks to attempt to drive project outcomes in non-cost areas such as safety, quality, community and legacy. The complexity of these can be such that the performance measures become an end in themselves and get in the way of management initiative.

TQM is concerned with moving the focus of control from outside the individual to within, the objective being to make everyone accountable for their own performance and to get them committed to attaining quality in a highly motivated fashion. The assumptions a director or manager must make in order to move in this direction are simply that people do not need to be coerced to perform well, and that people want to achieve, accomplish, influence activity and challenge their abilities. If there is belief in this, then only the techniques remain to be discussed.

Total Quality Management is user-driven – it cannot be imposed from outside the organization, as perhaps can a quality management system standard. This means that the ideas for improvement must come from those with knowledge and experience of the processes, activities and tasks; this has massive implications for training and follow-up. TQM is not a cost-cutting or productivity improvement device in the traditional sense, and it must not be used as such. Although the effects of a successful programme will certainly reduce costs and improve productivity, TQM is concerned chiefly with changing attitudes and skills so that culture of the organization becomes one of preventing failure – doing the right things, right first time, every time.

Effective leadership

Some management teams have broken away from the traditional style of management; they have made a 'managerial breakthrough'. Their approach puts their organization head and shoulders above others in the fight for sales, profits, resources, funding and jobs. Many public service organizations are beginning to move in the same way, and the successful quality-based strategy they are adopting depends very much on effective leadership.

Effective leadership starts with the chief executive's and his top team's vision, capitalizing on market or service opportunities, continues through a strategy that will give the organization competitive or other advantage, and leads to business or service success. It goes on to embrace all the beliefs and values held, the decisions taken and the plans made by anyone anywhere in the organization, and the focusing of them into effective, value-adding action.

Together, effective leadership and total quality management result in the company or organization doing the right things, right first time.

The five requirements for effective leadership are the following.

1. Developing and publishing clear documented corporate beliefs and purpose – a vision

Executives should express values and beliefs through a clear vision of what they want their company to be and its purpose – what they specifically want to achieve in line with the basic beliefs. Together, they define what the company or organization is all about. The senior management team will need to spend some time away from the 'coal face' to do this and develop their programme for implementation.

Clearly defined and properly communicated beliefs and objectives, which can be summarized in the form of vision and mission statements, are essential if the directors, managers and other employees are to work together as a winning team. The beliefs and objectives should address:

- The definition of the business, e.g. the needs that are satisfied or the benefits provided.
- A commitment to effective leadership and quality.

- Target sectors and relationships with customers, and market or service position.
- The role or contribution of the company, organization or unit, e.g. profit-generator, service department, opportunity-seeker.
- The distinctive competence – a brief statement which applies only to that organization, company or unit.
- Indications for future direction – a brief statement of the principal plans which would be considered.
- Commitment to monitoring performance against customers' needs and expectations, and continuous improvement.

These together with broad beliefs and objectives may then be used to communicate an inspiring vision of the organization's future. The top management must then show *TOTAL COMMITMENT* to it.

2. Develop clear and effective strategies and supporting plans for achieving the vision

The achievement of the company or service vision requires the development of business or service strategies, including the strategic positioning in the 'market place'. Plans can then be developed for implementing the strategies. Such strategies and plans can be developed by senior managers alone, but there is likely to be more commitment to them if employee participation in their development and implementation is encouraged.

3. Identify the critical success factors and critical processes (Figure 3.4)

The next step is the identification of the *critical success factors* (CSFs), a term used to mean the most important sub-goals of a business or organization. CSFs are what must be accomplished for the mission to be achieved. The CSFs are followed by the key, *core business processes* for the organization – the activities that must be done particularly well for the CSFs to be achieved. It is also critical for each of these to develop some performance measures that indicate progress towards each goal that has been set.

4. Review the management structure

Defining the corporate vision, strategies and CSFs might make it necessary to review the organizational structure. Directors, managers and other employees can be fully effective only if an effective structure based on process management exists. This includes both the definition of responsibilities for the organization's management and the operational procedures they will use. These must be the agreed best ways of carrying out the core processes.

The review of the management structure should also include the establishment of a process improvement team structure throughout the organization.

5. Empowerment – encouraging effective employee participation

For effective leadership it is necessary for management to get very close to the employees. They must develop effective communications – up, down and across the

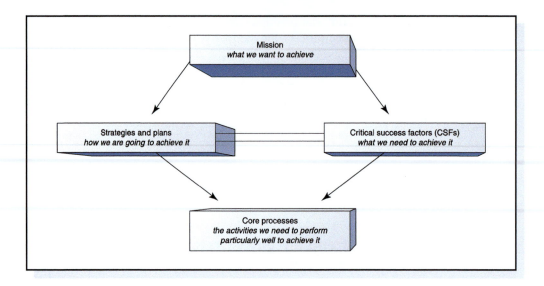

Figure 3.4
Mission into action through strategies, CSFs and core processes

organization – and take action on what is communicated; and they should encourage good communications between all suppliers and customers.

Particular attention must be paid to the following.

Attitudes

The key attitude for managing any winning company or organization may be expressed as follows: 'I will personally understand who my customers are and what are their needs and expectations and I will take whatever action is necessary to satisfy them fully. I will also understand and communicate my requirements to my suppliers, inform them of changes and provide feedback on their performance'. This attitude should start at the top – with the chairman or chief executive. It must then percolate down, to be adopted by each and every employee. That will happen only if managers lead by example. Words are cheap and will be meaningless if employees see from managers' actions that they do not actually believe or intend what they say.

Abilities

Every employee should be able to do what is needed and expected of him or her, but it is first necessary to decide what is really needed and expected. If it is not clear what the employees or subcontractors are required to do and what standards of performance are expected, how can managers expect them to do it? A good example of such confusion has been created on construction projects over the past few decades. Management has repeatedly said that they want the job done quickly, but rarely have they stressed that it should be correct the first time. This has resulted in a culture of 'we can fix it later, let's just get on with the job' and numerous defects have been incorporated into buildings; defects that are more expensive to rectify later.

Train, train, train and train again. Training is very important, but it can be expensive if the money is not spent wisely. The training should be related to needs, expectations and process improvement. It must be planned and *always* its effectiveness reviewed.

Participation

If all employees are to participate in making the company or organization successful (directors and managers included), then they must also be trained in the basics of disciplined management.

They must be trained to:

E Evaluate – the situation and define their *objectives*
P Plan – to achieve those objectives fully.
D Do – implement the plans.
C Check – that the objectives are being achieved.
A Amend – take corrective action if they are not.

The word 'disciplined' applied to people at all levels means that they will do what they say they will do. It also means that in whatever they do they will go through the full process of Evaluate, Plan, Do, Check and Amend, rather than the more traditional and easier option of starting by doing rather than evaluating. This will lead to a never-ending improvement helix (Figure 3.5).

This basic approach needs to be backed up with good management, planning techniques and problem-solving methods, which can be taught to anyone in a relatively short period of time. Management enables changes to be made successfully and the people to remove the obstacles in their way. Directors and managers need this training as much as other employees.

The 'operations management' has a very important and difficult leadership challenge – to create a cohesive team and galvanize it towards a set of shared goals. It is generally true that managers who do this well do so instinctively; it is not an area that has been identified and taught with any focus. Because many problems occur at the interfaces between departments or between suppliers and the organization, the challenge is that process innovation and problem solving has to be achieved by teams

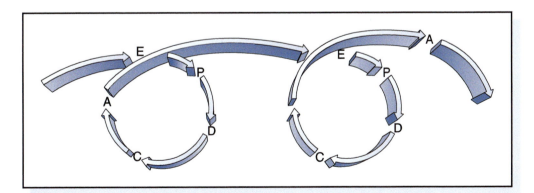

Figure 3.5
The helix of never-ending improvement

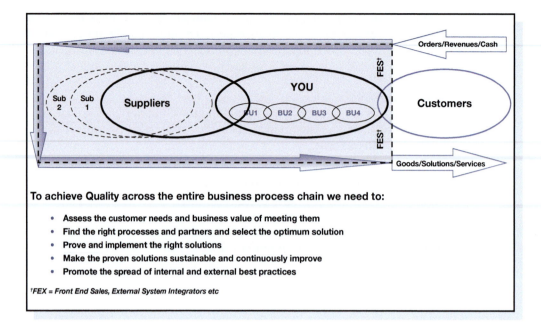

To achieve Quality across the entire business process chain we need to:

- Assess the customer needs and business value of meeting them
- Find the right processes and partners and select the optimum solution
- Prove and implement the right solutions
- Make the proven solutions sustainable and continuously improve
- Promote the spread of internal and external best practices

†FEX = Front End Sales, External System Integrators etc

Figure 3.6
Quality in the 21st century

of workers from the collaborating departments or organizations who are working side by side in any task. Hence, the challenge is to achieve shared goals and common action across the supply chain – '*Quality in the 21st Century.*' (Figure 3.6)

EXCELLENCE IN LEADERSHIP

The vehicle for achieving excellence in leadership is Total Quality Management. We have seen that its framework covers the entire organization, all the people and all the functions, including external organization and suppliers. Several requirements of TQM are becoming clear:

- Recognizing customers and discovering their needs (this refers to immediate and end user customers equally).
- Setting standards that are consistent with internal and end user customer requirements.
- Controlling processes, including systems, and improving their capability.
- Management's responsibility for setting the guiding philosophy, vision, quality policy, etc., and providing motivation through leadership and equipping people to achieve quality.
- Selecting the right employees and supply chain partners and empowering people at all levels in the organization and across the supply chain to act for quality improvement.

The task of implementing TQM can be daunting, and the chief executive and directors faced with it may become confused and irritated by the proliferation of theories and packages. A simplification is required. The *core* of TQM is the customer-supplier interfaces, both internally and externally, and the fact that at each interface there are processes to convert inputs to outputs. Clearly, there must be commitment to building-in quality through management of the inputs and processes.

How can senior managers and directors be helped in their understanding of what needs to be done to become committed to quality and implement the vision? The American and Japanese quality 'gurus' each set down a number of points or absolutes – words of wisdom in management and leadership – and many organizations have used these to establish a policy based on quality.

Similarly, the EFQM has defined the criterion of leadership and its sub-criteria as part of its model of Excellence. A fundamental principle behind all these approaches is that the behaviours of the leaders in an organization need to create clarity and constancy of purpose. This may be achieved through development of the vision, values and purpose needed for longer-term performance success.

Using as a construct of the 'Oakland TQM Model,' the four Ps and the three Cs plus a fourth C – Customers (which resides in 'performance') – the main items for attention to deliver excellence in leadership are given below.

Planning

- Develop the vision needed for constancy of purpose and for long-term success.
- Develop, deploy and update policy and strategy.
- Align organizational structure to support delivery of policy and strategy.

Performance

- Identify critical areas of performance.
- Develop measures to indicate levels of current performance.
- Set goals and measure progress towards their achievement.
- Provide feedback to people at all levels regarding their performance against agreed goals.

Processes

- Ensure a system for managing processes is developed and implemented.
- Ensure through personal involvement that the management system is developed, implemented and continuously improved.
- Prioritize improvement activities and ensure they are planned on an organization-wide basis.

People

- Train managers and team leaders at all levels in leadership skills and problem solving.

- Stimulate empowerment ('experts') and teamwork to encourage creativity and innovation.
- Encourage, support and act on results of training, education and learning activities.
- Motivate, support and recognize the organization's people – both individually and in teams.
- Help and support people to achieve plans, goals, objectives and targets.
- Respond to people and encourage them to participate in improvement activities.

Customers

- Be involved with customers and other stakeholders.
- Ensure customer (external and internal) needs are understood and responded to.
- Establish and participate in partnerships – as a customer demand continuous improvement in everything.

Commitment

- Be personally and actively involved in quality and improvement activities.
- Review and improve effectiveness of own leadership.

Culture

- Develop the values and ethics to support the creation of a total quality culture across the entire supply chain.
- Implement the values and ethics through actions and behaviours.
- Ensure creativity, innovation and learning activities are developed and implemented.

Communications

- Stimulate and encourage communication and collaboration.
- Personally communicate the vision, values, mission, policies and strategies.
- Be accessible and actively listen.

TQM should not be regarded as a woolly-minded approach to running an organization. It requires strong leadership with clear direction and a carefully planned and fully integrated strategy derived from the vision. One of the greatest tangible benefits of excellence in leadership is the improved overall performance of the organization. The evidence for this can be seen in some of the major consumer and industrial markets of the world. Moreover, effective leadership leads to improvements and superior quality which can be converted into premium prices. Research now shows that leadership and quality clearly correlate with profit but the less tangible benefit of greater employee participation is equally, if not more, important in the longer term. The pursuit of continual improvement must become a way of life for everyone in an organization if it is to succeed in today's competitive environment.

Bibliography

Adair, J., *Not Bosses but Leaders: How to Lead the Successful Way,* Talbot Adair Press, Guildford, 1987.

Adair, J., *The Action-Centred Leader,* Industrial Society, London, 1988.

Adair, J., *Effective Leadership* (2nd edn), Pan Books, London, 1988.

Davidson, H., *The Committed Enterprise,* Butterworth-Heinemann, Oxford, 2002.

Juran, J.M., *Juran on Leadership for Quality: An Executive Handbook,* The Free Press (Macmillan), New York, 1989.

Townsend, P.L. and Gebhardt, J.E., *Quality in Action – 93 Lessons in Leadership, Participation and Measurement,* J. Wiley Press, New York, 1992.

Chapter highlights

The Total Quality Management approach

- TQM is a comprehensive approach to improving competitiveness, effectiveness and flexibility through planning, organizing and understanding each activity, and involving each individual at each level. It is useful in all types of organization.
- TQM ensures that management adopts a strategic overview of quality and focuses on prevention, not detection, of problems.
- It often requires a mindset change to break down existing barriers. Managements that doubt the applicability of TQM should ask questions about the operation's costs, errors, wastes, standards, systems, training and job instructions.

Commitment and policy

- TQM starts at the top, where serious obsession and commitment to quality and leadership must be demonstrated. Middle management also has a key role to play in communicating the message.
- Every chief executive must accept the responsibility for commitment to a quality policy that deals with the organization for quality, the customer needs, the ability of the organization, supplied materials and services, education and training, and review of the management systems for never-ending improvement.

Creating or changing the culture

- The culture of an organization is formed by the beliefs, behaviours, norms, dominant values, rules and climate in the organization.
- Any organization needs a vision framework, comprising its guiding philosophy, core values and beliefs and purpose.
- The effectiveness of an organization depends on the extent to which people perform their roles and move towards the common goals and objectives.
- TQM is concerned with moving the focus of control from the outside to the inside of individuals, so that everyone is accountable for his/her own performance.

Leadership and commitment   47

Effective leadership

- Effective leadership starts with the chief executive's vision and develops into a strategy for implementation.
- Top management should develop the following for effective leadership: clear beliefs and objectives in the form of a vision; clear and effective strategies and supporting plans; the critical success factors and core processes; the appropriate management structure; employee participation through empowerment and the EPDCA helix; the challenge is to achieve shared goals and common action across the supply chain – *'Quality in the 21st Century'*.

Excellence in leadership

- The vehicle for achieving excellence in leadership is TQM. Using the construct of the Oakland TQM Model, the four Ps and four Cs provide a framework for this: Planning, Performance, Processes, People, Customers, Commitment, Culture and Communications.

Part I Discussion questions

1. You are planning to start an on-line retail business and have secured the necessary capital. Your aim is to attract customers who normally visit the high street to do their shopping. Discuss the key implications of this for the management of the business.
2. Explain the difference between quality and reliability; and between quality of design and quality of conformance.
3. Discuss the various facets of the 'quality control' function, paying particular attention to its interfaces with the other functional areas within the organization.
4. Explain what you understand by the term 'Total Quality Management', paying particular attention to the following terms: quality, supplier/customer interfaces, process.
5. Present a 'model' for total quality management, describing briefly the various elements of the model.
6. Select one of the so-called 'Gurus' of Quality Management, such as Juran, Deming, Crosby, Ishikawa, and explain their approach, with respect to the 'Oakland Model' of TQM. Discuss the strengths and weaknesses of their approach using this framework.
7. Compare and contrast the three models for total quality described by the Deming Prize in Japan, the Baldrige Award in the USA and the European Excellence Award.
8. In your new role as quality manager of the high-tech unit of a large national company, you identify a problem which is typified by the two internal memos shown below. Discuss in some detail the problems illustrated by this conflict, explaining how you would set about trying to make improvements.

From: Marketing Director
To: Managing Director
c.c.
Production Director
Works Manager
Date: 4 August

We have recently carried out a customer survey to examine how well we are doing in the market. With regard to our product range, the reactions were generally good, but the 24 byte microwinkle thrystor is a problem. Without exception everyone we interviewed said that its quality is not good enough. Although it is not yet apparent, we will inevitably lose our market share.

As a matter of urgency, therefore, will you please authorise a complete redesign of this product?

From: Works Manager
To: Production Director
Date: 6 August

This really is ridiculous!
I have all the QC records for the past three years on this product. How can there be anything seriously wrong with the quality when we only get 0.1% rejects at final inspection and less than 0.01% returns from customers?

9. Explain how the culture in an organization develops over time and describe the main components. How would you go about addressing negative cultural and behavioural aspects in a factory which are clearly leading to quality problems in the market place?

10. What are the aspects of leadership which are key to a successful total quality approach? Describe how you would go about helping a senior management team in a hospital gain the commitment of the medical, nursing and administration staff to deliver quality health services to the local community.

Part II

Planning

A mighty maze!
but not without a plan.

Alexander Pope, 1733,
from 'An Essay on Man'

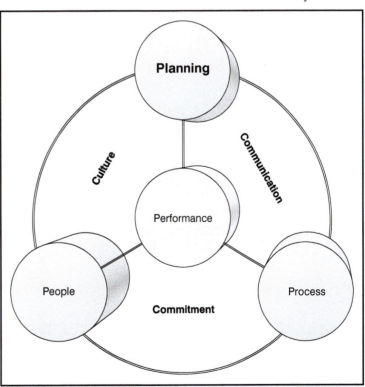

Chapter 4

Policy, strategy and goal deployment

INTEGRATING TQM INTO THE POLICY AND STRATEGY

One thing that all writers on strategy agree is that the leaders of any organization need a clear sense of direction and purpose, which they must communicate effectively throughout the organization – a clear message from the previous chapter on leadership. This typically involves the development of the vision, values and mission, which define the fundamental nature of the organization, and the strategic plan, which determines where resources will be invested to greatest net benefit.

For this to happen the vision and mission and their deployment must be based on the needs and expectations of the organization's stakeholders – present and future – and a thorough examination of the environment in which the organization exists. In today's dynamic environment, this requires information from research and learning activities and, even more importantly, accurate and timely measures of key performance criteria. Without this information, regularly reviewed and updated, it is difficult for managers to make the right strategic choices.

Included in the EFQM Excellence Model is a 'Strategy' criterion that is concerned with how the organization implements its vision and mission via a clear stakeholder-focused strategy, supported by relevant policies, plans, objectives, targets and processes.

The challenge for management is how to ensure that a gap doesn't appear between the chosen direction for the organization and the day-to-day operations. Any misalignment wastes resources, squanders opportunities and increases risk. There are six basic steps for achieving the alignment between the strategic choices, critical success factors, processes and people, and providing a foundation for the implementation of effective improvement. What is more, senior quality professionals can play an active role in facilitating this process.

Step 1 Develop a shared vision and mission for the business/organization

Once the top team is reasonably clear about the direction the organization should be taking it can develop vision and mission statements that will help to define the strategic choices and provide alignment with the processes, roles and responsibilities. This will lead to a co-ordinated flow of analysis of processes that cross the traditional functional areas at all levels of the organization, without changing formal structures, titles and systems which can create resistance. The vision framework was introduced in Chapter 3 (Figure 4.1).

The mission statement gives a purpose to the organization or unit. It should answer the questions 'what are we here for?' or 'what is our basic purpose?' and 'what have we got to achieve?' It therefore, defines the boundaries of the business in which the organization operates. This will help to focus on the 'distinctive competence' of the organization, and to orient everyone in the same direction of what has to be done. The mission must be documented, agreed by the top management team, sufficiently explicit to enable its eventual accomplishment to be verified and ideally be no more

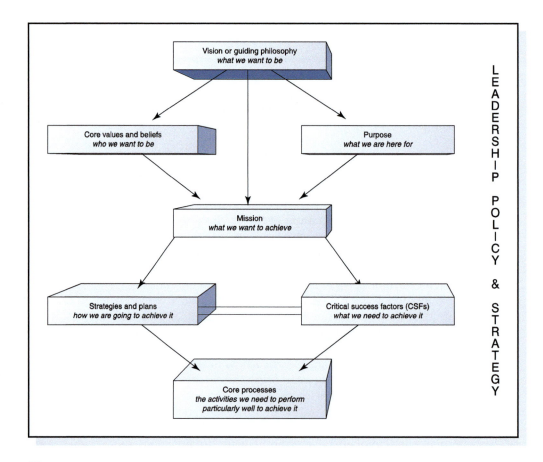

Figure 4.1
Vision framework for an organization

than four sentences. The statement must be understandable, communicable, believable and usable.

The mission statement is:

- an expression of the aspiration of the organization
- the touchstone against which all actions or proposed actions can be judged
- usually long term
- short term if the mission is survival.

Typical content includes a statement of:

- The role or contribution of the business or unit – for example, profit generator, service department, opportunity seeker.
- The definition of the business – for example, the needs you satisfy or the benefits you provide. Do not be too specific or too general.
- Your distinctive competence – this should be a brief statement that applies to only your specific unit. A statement which could apply equally to any organization is unsatisfactory.
- Indications for future direction – a brief statement of the principal things you would give serious consideration to.

Some questions that may be asked of a mission statement are, does it:

- define the organization's role?
- contain the need to be fulfilled:
  - is it worthwhile/admirable?
  - will employees identify with it?
  - how will it be viewed externally?
- take a long-term view, leading to, for example, commitment to a new product or service development, or training of personnel?
- take into account all the 'stakeholders' of the organization?
- ensure the purpose remains constant despite changes in top management?

It is important to establish in some organizations whether or not the mission is survival. This does not preclude a longer-term mission, but the short-term survival mission must be expressed, if it is relevant. The management team can then decide whether they wish to continue long-term strategic thinking. If survival is a real issue it is inadvisable to concentrate on long-term planning initially.

There must be open and spontaneous discussion during generation of the mission, but there must in the end be convergence on one statement. If the mission statement is wrong, everything that follows will be wrong too, so a clear understanding is vital.

Step 2 Develop the 'mission' into its critical success factors (CSFs) to coerce and move it forward

The development of the mission is clearly not enough to ensure its implementation. This is the 'danger gap' which many organizations fall into because they do not foster the skills needed to translate the mission through its CSFs into the core processes and people needs. Hence, they have 'goals without methods' and change is not integrated properly into the business.

Once the top managers begin to list the CSFs they will gain some understanding of what the mission or the change requires. The first step in going from mission to

CSFs is to brainstorm all the possible impacts on the mission. In this way 30 to 50 items ranging from politics to costs, from national cultures to regional market peculiarities may be derived.

The CSFs may now be defined – *what* the organization must accomplish to achieve the mission, by examination and categorization of the impacts. This should lead to a balanced set of deliverables for the organization in terms of:

- financial and non-financial performance
- customer/market satisfaction
- people/internal organization satisfaction
- environmental/societal satisfaction.

There should be no more than eight CSFs and no more than four if the mission is survival. They are the building blocks of the mission – minimum key factors or sub-goals that the organization **must have** or **needs** and which together will achieve the mission. They are the *whats* not the *hows*, and are not directly manageable – they may be in some case statements of hope or fear. But they provide direction and the success criteria, and are the end product of applying the processes. In CSF determination, a management team should follow the rule that each CSF is **necessary** and together they are **sufficient** for the mission to be achieved.

Some examples of CSFs may clarify their understanding:

- We must have right-first-time suppliers.
- We must have motivated, skilled workers.
- We need new products that satisfy market needs.
- We need new business opportunities.
- We must have best-in-the-field product quality.

The list of CSFs should be an agreed balance of strategic and tactical issues, each of which deals with a 'pure' factor, the use of 'and' being forbidden. It will be important to know when the CSFs have been achieved, but an equally important step is to use the CSFs to enable the identification of the processes.

Senior managers in large complex organizations may find it necessary or useful to show the interaction of divisional CSFs with the corporate CSFs in an impact matrix (see Figure 4.2 and discussion under Step 6).

Step 3 Define the key performance indicators as being the quantifiable indicators of success in terms of the mission and CSFs

The mission and CSFs provide the **what** of the organization, but they must be supported by measurable key performance indicators (KPIs) that are tightly and inarguably linked. These will help to translate the directional and sometimes 'loose' statements of the mission into clear **targets**, and in turn to simplify management's thinking. The KPIs will be used to monitor progress and as evidence of success for the organization, in every direction, internally and externally.

Each CSF should have an 'owner' who is a member of the management team that agreed the mission and CSFs. The task of an owner is to:

- define and agree the KPIs and associated *targets*
- ensure that appropriate data is collected and recorded

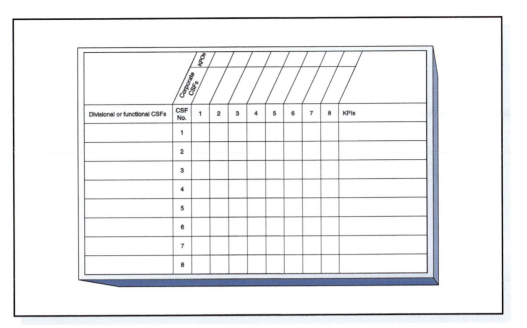

Figure 4.2
Interaction of corporate and divisional CSFs

- monitor and report progress towards achieving the CSF (KPIs and targets) on a regular basis
- review and modify the KPIs and targets where appropriate.

A typical CSF data sheet for completion by owners is shown in Figure 4.3.

The derivation of KPIs may follow the 'balanced scorecard' model, proposed by Kaplan, which divides measures into financial, customer, internal business and innovation and learning perspectives (see Chapter 7). In large complex organizations the skills and resources needed to achieve such demanding targets are rarely contained within one discrete 'business unit'. Therefore, it is vital that leaders are able to delegate responsibility for achieving contributory elements of the CSFs and KPIs to other members of the organization. It is through simple and effective performance management, allied with clarity of the processes, that this can be achieved.

Step 4 Understand the core processes and gain process sponsorship

This is the point when the top management team have to consider how to institutionalize the mission in the form of processes that will continue to be in place, until major changes are required.

The core business processes describe what actually is or needs to be done so that the organization meets its CSFs. As with the CSFs and the mission, each process which is **necessary** for a given CSF must be identified, and together the processes listed must

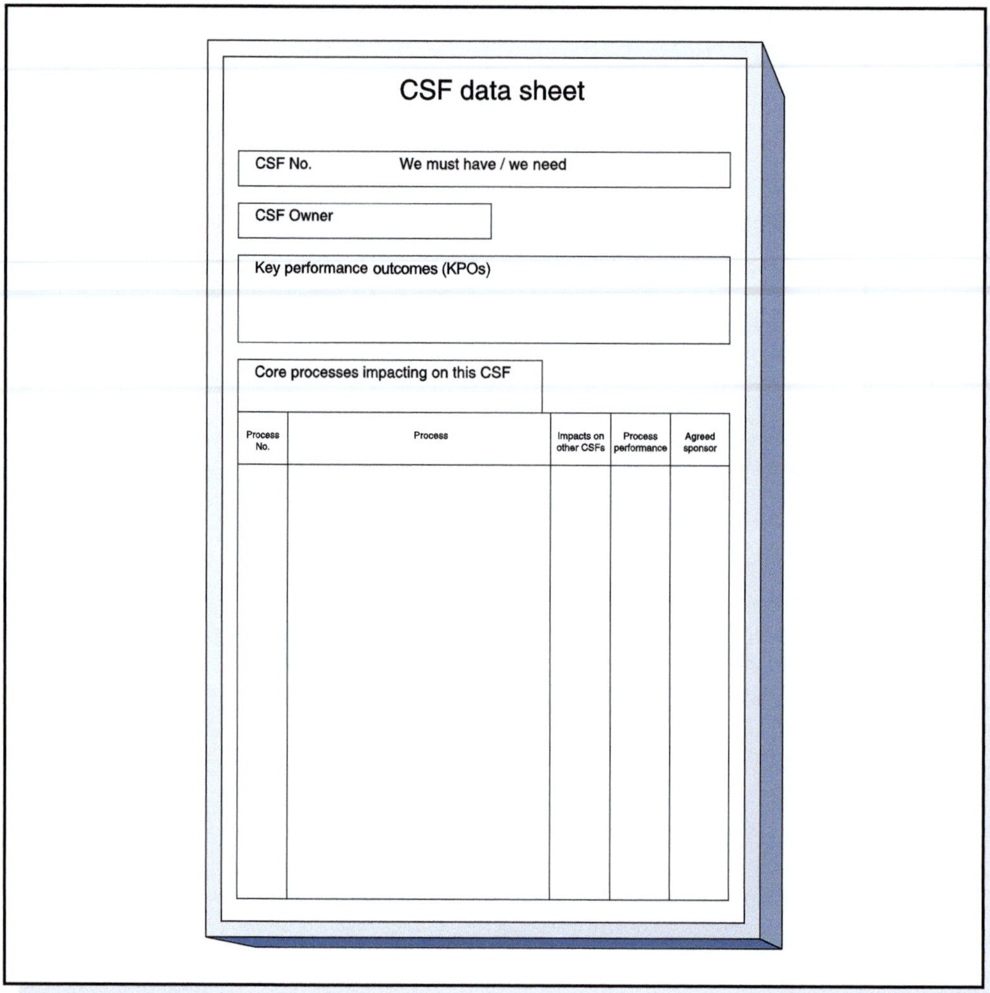

Figure 4.3
CSF data sheet

be **sufficient** for all the CSFs to be accomplished. To ensure that **processes** are listed, they should be in the form of verb plus object, such as research the market, recruit competent staff or manage supplier performance. The core processes identified frequently run across 'departments' or functions, yet they must be measurable.

Each core process should have a sponsor, preferably a member of the management team that agreed the CSFs.

The task of a sponsor is to:

• ensure that appropriate resources are made available to map, investigate and improve the process
• assist in selecting the process improvement team leader and members

- remove blocks to the teams' progress
- report progress to the senior management team.

The first stage in understanding the core processes is to produce a set of processes of a common order of magnitude. Some smaller processes identified may combine into core processes; others may be already at the appropriate level. This will ensure that the change becomes entrenched, the core processes are identified and that the right people are in place to sponsor or take responsibility for them. This will be the start of getting the process team organization up and running.

The questions will now come thick and fast; is the process currently carried out? By whom? When? How frequently? With what performance and how well compared with competitors? The answering of these will force process ownership into the business. The process sponsor may form a process team which takes quality and performance improvement into the next steps. Some form of prioritization using process performance measures is necessary at this stage to enable effort to be focused on the key areas for improvement. This may be carried out by a form of impact matrix analysis (see Figure 4.4). The outcome should be a set of 'most critical processes' (MCPs) which receive priority attention for improvement, based on the number of CSFs impacted by each process and its performance on a scale A to E.

This high-level picture as to how 'the business is wired up' provides a valuable forum for informed debate about where and how value is designed, created, delivered and communicated. Without these informed debates, executives can take a very functional approach to delivering the strategy, missing the huge opportunities created by taking an end-to-end view of the value chain. If they don't their customers certainly will.

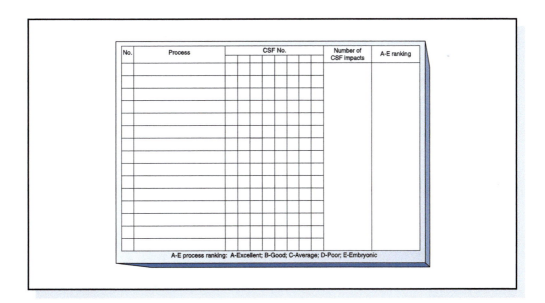

Figure 4.4
Process/CSF matrix

Step 5 Break down the core processes into sub-processes, activities and tasks and form improvement teams around these

Once an organization has defined and mapped out the high-level core processes, people need to understand what activities are required within these core processes and how they combine at operational levels. This 'top-down' approach is needed to ensure that the day-to-day activities of the organization are aligned with what is critical to achieving strategic success. If properly executed, this will develop the skills needed to understand how the new process structure will be analysed and made to work. It should then be possible to develop metrics for measuring the performance of the processes, sub-processes, activities, and tasks (Figure 4.5). The very existence of new process teams with new goals and responsibilities will force the organization into a learning phase. This will then provide the basis for change and the attitudes and behaviours necessary to meet the strategic goals.

An illustration of the breakdown from mission through CSFs and core processes, to individual tasks may assist in understanding the process required.

Individuals, tasks and teams

Having broken down the processes into sub-processes, activities and tasks in this way, it is now possible to link this with the Adair model of action centred leadership and teamwork (see Chapter 16).

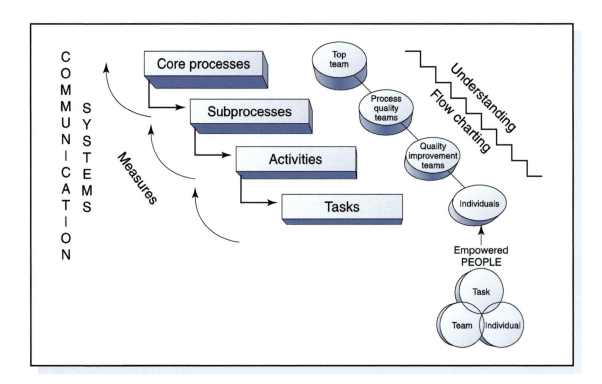

Figure 4.5
Breakdown of core processes into sub-processes, activities and tasks

The tasks are clearly performed, at least initially, by individuals. For example, somebody has to sit down and draft out the first version of a seminar leaflet. There has to be an understanding by the individual of the task and its position in the hierarchy of processes. Once the initial task has been performed, the results must be checked against the activity of co-ordinating the promotional booklet – say for TQM. This clearly brings in the team, and there must be interfaces between the needs of the *tasks*, the *individuals* who performed them and the *team* concerned with the *activities*.

Performance measurement and metrics

Once the processes have been analysed in this way, it should be possible to develop metrics for measuring the performance of the processes, sub-processes, activities and tasks. These must be meaningful in terms of the *inputs* and *outputs* of the processes, and in terms of the *customers* and of *suppliers* to the processes (Figure 4.5).

At first thought, this form of measurement can seem difficult for processes such as preparing a sales brochure or writing leaflets advertising seminars but, if we think carefully about the *customers* for the leaflet-writing tasks, these will include the *internal* ones, i.e. the consultants, and we can ask whether the output meets their requirements. Does it really say what the seminar is about, what its objectives are and what the programme will be? Clearly, one of the 'measures' of the seminar leaflet-writing task could be the number of typing errors in it, but is this a *key* measure of the performance of the process? Only in the context of office management is this an important measure. Elsewhere it is not.

The same goes for the *activity* of preparing the subject booklet. Does it tell the 'customer' what TQM or SPC is and how the consultancy can help? For the *sub-process* of preparing the company brochure, does it inform people about the company and does it bring in enquiries from which customers can be developed? Clearly, some of these measures require *external market research,* and some of them *internal research.* The main point is that metrics must be developed and used to reflect the *true performance* of the processes, sub-processes, activities and tasks. These must involve good contact with external and internal customers of the processes. The metrics may be quoted as *ratios*, e.g. numbers of customers derived per number of brochures mailed out. Good data-collection, record-keeping and analysis are clearly required.

It is hoped that this illustration will help the reader to:

* Understand the breakdown of processes into sub-processes, activities and tasks.
* Understand the links between the process breakdowns and the task, individual and team concepts.
* Link the hierarchy of processes with the hierarchy of quality teams.
* Begin to assemble a cascade of flowcharts representing the process breakdowns, which can form the basis of the quality management system and communicate what is going on throughout the business.
* Understand the way in which metrics must be developed to measure the true performance of the process, and their links with the customers, suppliers, inputs and outputs of the processes.

The changed patterns of co-ordination, driven by the process maps, should increase collaboration and information sharing. Clearly the senior and middle managers need to provide the right support. Once employees, at all levels, identify what kinds of new skill are needed, they will ask for the formal training programmes in order to develop those skills further. This is a key area, because teamwork around the processes will ask more of employees, so they will need increasing support from their managers.

This has been called 'just-in-time' training, which describes very well the nature of the training process required. This contrasts with the blanket or carpet-bombing training associated with many unsuccessful change programmes, which targets competencies or skills, but does not change the organization's patterns of collaboration and co-ordination.

Step 6 Ensure process and people alignment through a policy deployment or goal translation process

One of the keys to integrating excellence into the business strategy is a formal 'goal translation' or 'policy deployment' process. If the mission and measurable goals have

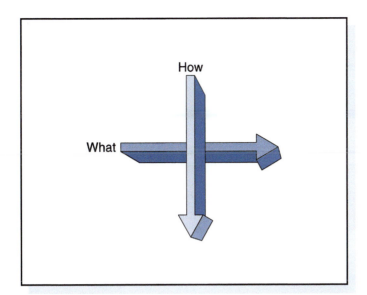

Figure 4.6
The goal translation process

been analysed in terms of critical success factors and core processes, then the organization has begun to understand how to achieve the mission. Goal translation ensures that the 'whats' are converted into 'hows', passing this right down through the organization, using a quality function deployment (QFD) type process, Figure 4.6 – Chapter 6). The method is best described by an example, as shown in Figure 4.6.

At the top of an organization in the chemical process industries, five measurable goals have been identified. These are listed under the heading 'What' in Figure 4.7. The top team listens to the 'voice of the customer' and tries to understand *how* these business goals will be achieved. They realize that product consistency, on-time delivery, and speed or quality of response are the keys. These CSFs are placed along the first row of the matrix and the relationships between the *what* and the *how* estimated as strong, medium or weak. A measurement target for the hows is then specified.

The *how* becomes the *what* for the next layer of management. The top team share their goals with their immediate reports and ask them to determine their *hows* indicate the relationship and set measurement targets. This continues down the organization through a 'catch-ball' process until the senior management goals have been translated through the *what/how* ‡ *what/how* ‡ *what/how* matrices to the individual tasks within the organization. This provides a good discipline to support the breakdown and understanding of the business process mapping described in Chapter 10.

A successful approach to policy/goal deployment and strategic planning in an organization with several business units or division is that mission, CSFs with KPIs and targets, and core processes, are determined at the corporate level, typically by the board. Whilst there needs to be some flexibility about exactly how this is translated into the business units, typically it would be expected that the process is repeated with the senior team in each business unit or division. Each business unit head should be part of the top team that did the work at the corporate level, and each of them would develop a version of the same process with which they feel comfortable.

Policy, strategy and goal deployment

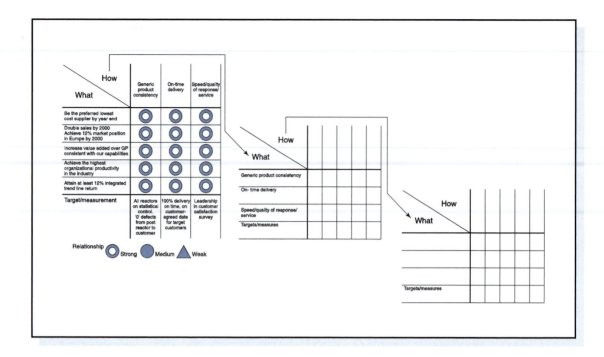

Figure 4.7
The goal translation process in practice

Each business unit would then follow a similar series of steps to develop their own mission (perhaps) and certainly their own CSFs and KPIs with targets. A matrix for each business unit showing the impact of achieving the business unit CSFs on the corporate CSFs would be developed. In other words, the first deployment of the corporate 'whats' CSFs is into the 'hows' – the business unit CSFs (Figure 4.2).

If each business unit follows the same pattern, the business unit teams will each identify unit CSFs, KPIs with targets and core processes, which are interlinked with the ones at corporate level. Indeed the core processes at corporate and business unit level may be the same, with any specific additional processes identified at business unit level to catch the flavour and business needs of the unit. It cannot be over-emphasized how much ownership there needs to be at the business unit management level for this to work properly.

With regard to core processes, each business unit or function will begin to map these at the top level. This will lead to an understanding of the purpose, scope, inputs, control and resources for each process and provide an understanding of how the sub-processes are linked together. Flow charting showing connections with procedures will then allow specific areas for improvement to be identified so that the continuous improvement, 'bottom up' activities can be deployed, and benefit derived from the process improvement training to be provided (Figure 4.8). Appropriate software is available to support all this analysis, of course.

It is important to get clarity at the corporate and business unit management levels about the whats/hows relationships, but the ethos of the whole process is one of

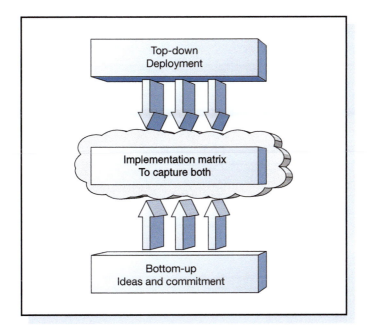

Figure 4.8
Implementation:
top-down and
bottom-up approach

involvement and participation in goal/target setting, based on good understanding of processes – so that it is known and agreed what can be achieved and what needs measuring and targeting at the business unit level.

Senior management may find it useful to monitor performance against the CSFs, KPIs and targets, and to keep track of processes using a reporting matrix, perhaps at their monthly meetings. A simplified version of this developed for use in a small company is shown in Figure 4.9. The frequency of reporting for each CSF, KPI and process can be determined in a business planning calendar.

As previously described, in a larger organization, this approach may be used to deploy the goals from the corporate level through divisions to site/departmental level (Figure 4.10). This form of implementation should ensure the top-down *and* bottom-up approach to the deployment of policies and goals.

Deliverables

The deliverables then after one planning cycle of this process in a business will be:

1. An agreed framework for policy/goal deployment through the business.
2. Agreed mission statement for the business and, if required, for the business units/division.
3. Agreed critical success factors (CSFs) with ownership at top team level for the business and business units/divisions.
4. Agreed key performance indicators (KPIs) with targets throughout the business.
5. Agreed core business processes, with sponsorship at top team level.
6. A corporate CSF/business unit CSF matrix showing the impacts and the first 'whats/hows' deployment.

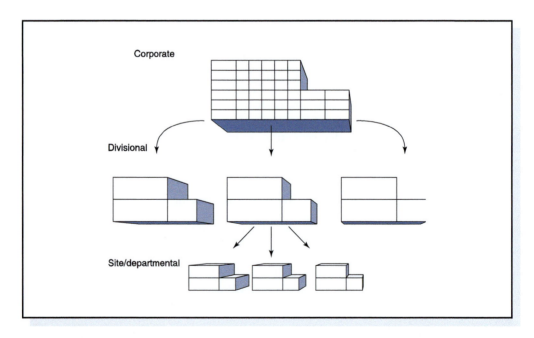

Figure 4.9
CSF/core process reporting matrix

Conduct research	Manage int. systems	Manage financials	Manage our accounts	Develop new business	Develop products	Manage people	Core processes	CSFs: We must have	Measures	Year targets	Target CSF owner
	×	×	×	×			×	Satisfactory financial and non-financial performance	Sales volume. Profit. Costs versus plan. Shareholder return Associate/employee utilisation figures	Turnover £2m. Profit £200k. Return for shareholders. Days/month per person	
×	×		×	×	×		×	A growing base of satisfied customers	Sales/customer Complaints/recommendations Customer satisfaction	>£200k = 1 client. £100k-£200k =5 clients. £50k-£100k= 6 clients <£50k=12 client	
	×	×					×	A sufficient number of committed and competent people	No. of employed staff/associates Gaps in competency matrix. Appraisal results Perceptions of associates and staff	15 employed staff 10 associates including 6 new by end of year	
×			×				×	Research projects properly completed and published	Proportion completed on time, in budget with customers satisfied. Number of publications per project	3 completed on time, in budget with satisfied customers	
		⁝	⁝				⁝	** = Priority for improvement			
								Process owner			
								Process performance			
								Measures and targets			

Figure 4.10
Deployment – what/how

7. At what/how (CSF/process) matrix approach for deploying the goals into the organization through process definition, understanding, and measured improvement at the business unit level.
8. Focused business improvement, linked back to the CSFs, with prioritized action plans and involvement of employees.

Strategic and operational planning

Changing the culture of an organization to incorporate a sustainable ethos of continuous improvement and responsive business planning will come about only as the result of a carefully planned and managed process. Clearly many factors are involved, including:

- identifying strategic issues to be considered by the senior management team
- balancing the present needs of the business against the vital needs of the future
- concentrating finite resources on important things
- providing awareness of impending changes in the business environment in order to adapt more rapidly, and more appropriately.

Strategic planning is the continuous process by which any organization will describe its destination, assess barriers standing in the way of reaching that destination, and select approaches for dealing with those barriers and moving forward. Of course the real contributors to a successful strategic plan are the participants.

The strategic and operational planning process described in this chapter will:

- Provide the senior management team with the means to manage the organization and strengths and weaknesses through the change process.
- Allow the senior management team members to have a clear understanding and to achieve agreement on the strategic direction, including vision and mission.
- Identify and document those factors critical to success (CSFs) in achieving the strategic direction and the means by which success will be measured (KPIs) and targeted.
- Identify and document and encourage ownership of the core processes that drive the business.
- Reach agreement on the priority processes for action by process improvement teams, incorporating current initiatives into an overall, cohesive framework.
- Provide a framework for successfully deploying all goals and objectives through all organizational levels through a two-way 'catch-ball' process.
- Provide a mechanism by which goals and objectives are monitored, reviewed and appropriate actions taken, at appropriate frequencies throughout the operational year.
- Transfer the skills and knowledge necessary to sustain the process.

The components outlined above will provide a means of effectively deploying a common vision and strategy throughout the organization. They will also allow for the incorporation of all change projects, as well as 'business as usual' activities, into a common framework which will form the basis of detailed operating plans.

Let us assume that a management team are to develop the policies and strategies based on stakeholder needs and the organization's capabilities, and that it wants to ensure these are communicated, implemented, reviewed and updated. Clearly a detailed review is required of the major stakeholders' needs, the performance of competitors, the state of the market and industry/sector conditions. This can then form the basis of top level goals, planning activities and setting of objectives and targets.

How individual organizations do this varies greatly, of course, and some of this variation can be seen in the case studies in this book. However some common themes emerge under six headings.

Customer/market

- Data collected, analysed and understood in terms of where the organization will operate.
- Customers' needs and expectations understood, now and in the future.
- Developments anticipated and understood, including those of competitors and their performance.
- The organization's performance in the market place known.
- Benchmarking against best in class organizations.

Shareholders/major stakeholders

- Shareholders/major stakeholders' needs and ideas understood.
- Appropriate economic trends/indicators and their impact analysed and understood.
- Policies and strategies appropriate to shareholder/stakeholder needs and expectations developed.
- Needs and expectations balanced.
- Various scenarios and plans to manage risks developed.

People

- The needs and expectations of the employees understood.
- Data collected, analysed and understood in terms of the internal performance of the organization.
- Output from learning activities understood.
- Everyone appropriately informed about the policies and strategies.
- Everyone appropriately trained and developed to provide the required competencies and skills the organization needs to deliver the policies and strategies.

Processes

- A key process framework to deliver the policies and strategies designed, understood and implemented.

- Key process owners identified.
- Each key process and its major stakeholders defined.
- Key process framework reviewed periodically in terms of its suitability to deliver to organization's requirements.

Partners/resources

- Appropriate technology understood.
- Impact of new technologies analysed.
- Needs and expectations of partners understood.
- Policies and strategies aligned with those of partners.
- Financial strategies developed.
- Appropriate buildings, equipment and materials identified/sourced.

Society

- Society, legal and environmental issues understood.
- Environment and corporate social (CSR) responsibility policies developed.

The whole field of business policy, strategy development and planning is huge and there are many excellent texts on the subject. It is outside the scope of this book to cover this area in detail, of course, but one of the most widely used and comprehensive texts is *Exploring Corporate Strategy – text and cases*, by Johnson, Scholes and Whittington. This covers strategic positioning and choices, and strategy implementation at all levels. The author is proud to have had a case study included in this text – see STMicroElectronics Case.

BIBLIOGRAPHY

Collins, J.C., *Good to Great*, Random House, New York, 2001.
Collins, J.C. and Porras, J.I., *Built to Last: successful habits of visionary companies*, Random House, New York, 1998.
Hardaker, M. and Ward, B.K., 'Getting things done – how to make a team work', *Harvard Business Review*, pp. 112–119, Nov/Dec., 1987.
Hutchins, D., *Hoshin Kanri – The Strategic Approach to Continuous Improvement*, Gower Publishing, Aldershot, 2008.
Johnson, G., Scholes, K. And Whittington, R., *Exploring Corporate Strategy, Text & Cases* (7th edn), Prentice-Hall, London, 2008.

CHAPTER HIGHLIGHTS

Integrating TQM into the policy and strategy

- Policy and strategy is concerned with how the organization implements its vision and mission in a clear stakeholder-focused strategy supported by relevant policies, plans, objectives, targets and processes.

- Senior management may begin the task of alignment through six steps:
  - develop a shared vision and mission
  - develop the critical success actors
  - define the key performance indicators (balanced scorecard)
  - understand the core process and gain ownership
  - break down the core processes into sub-processes, activities and tasks
  - ensure process and people alignment through a policy deployment or goal translation process.
- The deliverables after one planning cycle will include: an agreed policy/goal deployment framework; agreed mission statements; agreed CSFs and owners; agreed KPIs and targets; agreed core processes and sponsors; whats/hows deployment matrices; focussed business improvement plans.

The development of policies and strategies

- The development of policies that require a detailed review of the major stakeholders' needs, the performance of competitors, the market/industry/sector conditions to form the basis of top level goals, planning activities and setting of objectives and targets.
- The common themes for planning strategies may be considered under the headings of customers/market, shareholders/major stakeholders, people, processes, partners, resources and society.
- The field of policy and strategy development is huge and the text by Johnson, Scholes and Whittington is recommended reading.

Chapter 5

Partnerships and resources

PARTNERING AND COLLABORATION

Business, technologies and economies have developed in such a way that most organizations now recognize the increasing need to establish mutually beneficial relationships with other organizations, often called 'partners'. The philosophies behind the various TQM and Excellence models support the establishment of partnerships and lay down principles and guidelines for them.

Even in the twenty-first century, however, some organizations do not fully appreciate the important role supply chains play in delivering on-quality, on-time, on-cost products and services in their organization. Yet the 'supply chain' is often responsible for most of the labour, materials and equipment involved. Because *joint design and manufacturing* (JDM) partnering has become commonplace in some industries, the supply chain now includes the designers and developers. Partner organization employees can constitute upwards of 80 per cent of workers in the product or service chain, so a very important part of each organization's strategy must be the consideration of the values, skills, knowledge and attitudes of partner management and employees.

How companies in the private sector plan and manage their partnerships can mean the difference between success and failure for it is now extremely rare to find companies which can sustain a credible business operation without a network of co-operation between individuals and organizations or parts of them. This extends the internal customer–supplier relationship ideas into the supply chain of an organization, making sure that all the necessary materials, services, equipment, information skills and experience are available in totality to deliver the right products or services to the end customer. Gone are the days, hopefully, of conflict and dispute between a customer and their suppliers. An efficient supply chain process, built on strong confident partnerships, will create high levels of people satisfaction, customer satisfaction and support and, in turn, good business results.

Similarly in public sector organizations, where the involvement of the private sector has been a key development feature in recent times, there is the need to recognize and build strong external partnerships. The bidding and tendering processes of such organizations by necessity will always be different to those in the private sector. Nevertheless, government departments, the health service, education, police, the armed forces, tax collection bodies and local authorities need to understand, develop and deliver strong external partnerships if they are to achieve the performance levels and targets that the general public in any country desires.

How an organization plans and manages the external partnerships must be in line with its overall policies and strategies, being designed and developed to support the effective operation of its processes. A key part of this, of course, is identifying with whom those key strategic partnerships will be formed. Whether it is working with key suppliers to deliver materials or components to the required quality, plan (lead-times) and costs, or the supply of information technology, transport, broadcasting or consultancy services, the quality of partnerships has been recognized throughout the world as a key success criterion.

Understanding and applying a sound approach to improving strategic partnerships and supplier relationships will have a significant return on investment by:

- Significantly reducing the management overhead through increasing the level of collaboration on a number of key dimensions that research shows to be critical in effective partnerships.
- Establishing the right partnerships from the outset through understanding what makes partnerships work and building that into the selection, review and 're-contracting' process.

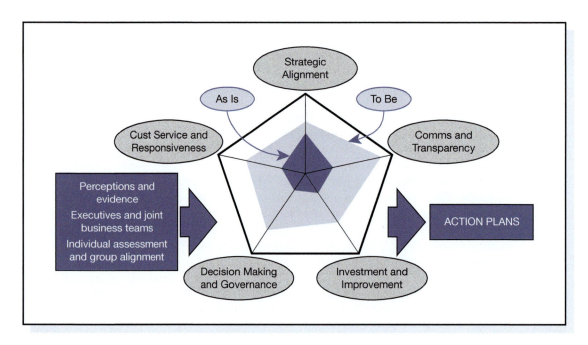

Figure 5.1
Partnering collaboration radar

Once the goals for the partnership are established, partners may carry out a 'health check' on the current operations to assess strengths and areas for improvement.

The approach recommended looks at five key dimensions of good collaborative working (Figure 5.1):

- *Strategic alignment*: how well aligned are the partners and how do they achieve this alignment at all key levels of management?
- *Customer focus*: to what extent do both parties develop and deliver the desired standards of service and experience across the whole service chain?
- *Decision-making and governance*: how the partnership is managed to best effect and efficiency
- *Communications and transparency*: how well data and knowledge is captured, shared and disseminated in a way that builds value and not cost
- *Investment and improvement*: the extent to which the partnership jointly invests in and improves the partnership operations and outcome measures.

There are various ways of ensuring the partnership processes work well for an organization. These range from the use of quality management system audits and reviews, through certificates of competence, to performance reviews and joint action plans. A key aspect of successful partnerships is good communications and exchange of information which supports learning between two organizations and often leads to innovative solutions to problems that have remained unsolved in the separate organizations, prior to their close collaboration.

One example of this in a medium sized company involved a close working relationship with a partner who had been responsible for producing the written content of the company's work-based learning materials. The customer of this service had shared their ideas for the future in order to exploit areas of possible mutual benefit. Arising from this and other similar partnership relationships in the company was improved customer satisfaction and refined processes.

When establishing partnerships attention should be given to:

- maximizing the understanding of what is to be delivered by the partnership – the needs of the customer and the capability of the supplier must match perfectly if satisfaction and loyalty are to be the result
- understanding what represents value for money – getting the commercial relationship right
- understanding the respective roles and ensuring an appropriate allocation of responsibilities – to the party best able to manage them
- working in a supportive, constructive and a team-based relationship
- having solid programmes of work, comprising agreed plans, timetables, targets, key milestones and decision points
- structuring the resolution of complaints, concerns or disputes rapidly and at the lowest practical level
- enabling the incorporation of knowledge transfer and making sure this adds value
- developing a stronger and stronger working relationship geared to delivering better and better products or services to the end customer – based on continuous improvement principles.

Research by *The Economist* has shown that the top two reasons for the selection of 'global suppliers' were the pursuit of new markets and reducing costs. Product/service quality and customer service improvements did feature but they were down the list of priorities. Yet, a mountain of evidence now points to the fact that poor quality can quickly, quietly and devastatingly demolish the benefits of any new market gains or cost reductions. Moreover, the difficulty of managing quality increases in direct relation to the distance between an organization and its partners, hence the need for a good 'assurance' model.

From direct experience of working with global manufacturers and financial service providers and within supplier quality management, the author and his colleagues have identified three key areas of opportunity and risk that are common across both global in-sourcing and outsourcing.

Managing performance and exposure to risk

We know that quality cannot be inspected into a product or service and that it must be built in, which makes it surprising that so many companies that rule out inspection as a quality control methodology in developed economies, on grounds of prohibitive cost and ineffectiveness, decide that it is appropriate when sourcing from developing economies. Inspection and trying to control at the end do not prevent poor quality happening, nor are they totally effective in preventing poor quality escaping to customers.

The real problem, all too often, is that there is no clear view of risk across a global manufacturing or operations footprint: there may be a lot of data and 'noise', but little in the way of real knowledge in a form that executives can digest and act upon – we can only manage what we can measure.

Getting organized for success

Building a global quality management capability needs leadership and a clear policy for managing quality on a global scale. The approach must integrate quality assurance, quality control, and quality improvement techniques, such as Lean Six Sigma, into an integrated approach to quality management. Pure reliance on inspection and 'quality control' will not by itself eliminate the risk of producing poor quality products or services.

Without the right organization in place, chief operating offices of global operations can find themselves exposed to a cost and quality 'killer' ... VARIATION. Research indicates that variation can add up to 30 per cent to manufacturing and support costs. The challenge is to move from inspection to 'assurance' and relatively small improvements here can multiply into big benefits.

Delivering better value from the global supply chain

The ongoing management of quality on a global scale requires an appropriate balance of stewarding global quality performance, facilitating global improvement, and managing risk. It is also vital to ensure that quality is not solely seen as the responsibility of the 'Quality Department'. As quality is everyone's responsibility it must rank with equal importance with other critical areas. Part of the solution is to create a central team that will genuinely be regarded as a 'value-centre' rather than a cost-centre. It is all a matter of getting value, not just cutting costs.

SUPPLY CHAIN EFFECTIVENESS IN THE GLOBAL ECONOMIES

The urgent pursuit of greater security of supply or improved efficiencies can necessitate the transfer of manufacturing or re-sourcing to alternative supply routes. These strategies represent great opportunities but also carry big risks. In turn, supply chains can represent the source of real savings if better managed.

Reliance upon a single source of supply can be risky at any time, but in unpredictable economic conditions it can be downright dangerous. An organization must establish whether it is at risk. Below is a brief, high-level overview of a process to strengthen sourcing security in terms of quality, timeliness and cost, by sourcing a second supplier.

1 Decide the selection criteria that are right for the business

Consider sourcing security needs by reference to a set of sourcing criteria. The list will be unique to a business or organization and its needs, but a sample to show the idea is given in Figure 5.2.

The list might include, for example, sourcing and supply chain capability, technical capability, product cost profile, and so on. The important thing is to create a tailored list and time spent on this exercise will repay itself many times.

2 Make an initial selection

Using the selection criteria make an initial selection of candidates.

3 Toughen up the selection criteria and make a shortlist

Using a more rigorous set of selection criteria reduce the list to – say – five candidates.

4 Conduct site visits

At this stage, really put the candidates under the microscope by scrupulously examining all the factors around the potential deal, including:

- Approach to managing a large production and assembly account
- Level of interest in the commercial opportunity
- Professionalism and ability to quickly accommodate site visits and assessments.

Factor	Spectrum		
Business size and scale		Small	Upper tier
Geography	Single plant locations		Global footprint
Business relationship	Simple contract		Business partnership
Business maturity and capability	Capability		Capability + maturity

Figure 5.2
Sample set of sourcing criteria

5 Make a final evaluation . . . and choice

When the site visits are complete, re-evaluate the key business criteria scores, moving them up and down as appropriate. Also, determine the candidates' key strengths and weaknesses, together with associated risks, and draw the conclusions.

In one company, a potential list of fourteen new suppliers was reduced to a shortlist of five, with one finally being selected as an additional supplier to the original single source supplier. The benefits included manufacturing costs halved compared to the existing source and expectations built of a step change in quality with indications that the defect rate would drop over the first twelve months from the initial level of 2 per cent to less than 0.5 per cent. This is a typical return from a more secure supply chain.

THE ROLE OF PROCUREMENT/PURCHASING IN PARTNERSHIPS

As we have seen, very few organizations are self-contained to the extent that their products and services are all generated at one location, from basic materials. Some materials or services are usually purchased from outside organizations, and the primary objective of a 'Purchasing' or 'Procurement' function is to obtain the correct equipment, materials and services in the right quantity, of the right quality, from the right origin, at the right time and cost. Procurement or Purchasing can also play a vital role as the organization's 'window-on-the-world', providing information on any new products, processes, materials and services that become available. It can also advise on probable prices, deliveries and performance of products under consideration by the research, design and development functions. In other words it should support any partnership in the supply chain.

The purchasing or procurement system should be documented and include:

1. Assigning responsibilities for and within the purchasing/procurement function.
2. Defining the manner in which suppliers are selected, to ensure that they are continually capable of supplying the requirements.
3. Specifying the purchasing documentation – written orders, specifications, etc. – required in any modern procurement activity.

Most leading organizations now have supply chain or procurement specialists working in them. Their role is to work across the business functions exploring ways to optimize the supply process through strategies such as outsourcing, early supplier involvement and off-shoring. In large organizations the purchase of goods and services tends to be negotiated centrally with the detailed management of their requirements organized locally. Purchasing/procurement should bring value to their organizations by improved contract management and fostering supplier compliance across the entire lifecycle of contracts resulting in continued overall cost reduction. Nevertheless, according to the UK Chartered Institute of Purchasing and Supply (CIPS), much work remains to be done in understanding how to capture the breadth and depth of the role of supplier management in achieving cost reductions and other benefits – the traditional measures of cost saving are no longer appropriate. Clearly, the link between procurement strategy and business performance and the resulting shareholder value need to be considered.

Procurement needs to expand its remit into managing risk and vulnerability within the supply chain, particularly in the context of geographically dispersed and distant suppliers. Additional complexity also arises as a direct consequence of the volatility of commodities, currencies and interest rates. Senior management should demand that procurement be able to avoid or reduce the vulnerability of supply chains to disruptions, and increase their resilience should problems occur. Risk-related issues include managing exposure to commercial and reputational risks, as well as providing improved supply market intelligence, such as forecasting future shortages. Collaborative relationships are a prerequisite for tapping into innovations available from suppliers globally and suppliers are becoming increasingly integrated into new product development efforts. In some organizations this requires the development of new remuneration and risk sharing models to share the benefits of technological development with suppliers. Most large organizations and governments worldwide are increasingly focusing on the issue of sustainability, both from the societal and the environmental perspectives. This has the potential to be the single most significant influence on organizational strategy, with procurement playing an important role in this effort.

Commitment and involvement

The process of improving supplier performance is complex and clearly relies very heavily on securing real commitment from the senior management of both organizations to a partnership. This may be aided by presentations made to groups to directors of the suppliers brought together to share the realization of the importance of their organizations' performance in the quality chains. The synergy derived from members of the partnership meeting together, being educated and discussing mutual problems will be tremendous. If this can be achieved, within the constraints of business and technical confidentiality, it is always a better approach than the arms-length approach to purchasing still used by many companies.

The author recalls the benefits that accrued from bringing together suppliers of a photocopier, paper and ring binders to explain to them the way their inputs were used to generate training-course materials and how they in turn were used during the courses themselves. The suppliers were able to understand the business in which their customers were engaged and play their part in the whole process. A supplier of goods *or* services that has received such attention, education and training, and understands the role its inputs play, is less likely knowingly to offer nonconforming materials and services, and more likely to alert customers to potential problems.

Policy

One of the first things to communicate to any external supplier is the purchasing organization's policy on quality of incoming goods and services. This can include such statements as:

- It is the policy of this company to ensure that the quality of all purchased materials and services meets its requirements.
- Suppliers who incorporate a quality management system into their operations will be selected. This system should be designed, implemented and operated accordingly to the International Standards Organization (ISO) 9000 series (see Chapter 11).
- Suppliers who incorporate statistical process control (SPC) and continuous improvement methods into their operations (see Chapter 13) will be selected.
- Routine inspection, checking, measurement and testing of incoming goods and services will *not* be carried out by this company on receipt.
- Suppliers will be audited and their operating procedures, systems and SPC methods will be reviewed periodically to ensure a never-ending improvement approach.
- It is the policy of this company to pursue uniformity of supply, and to encourage suppliers to strive for continual reduction in variability. (This may well lead to the narrowing of specification ranges.)

Quality management system assessment certification

Many customers examine their suppliers' quality management systems themselves, operating a second party assessment scheme (see Chapters 8 and 11). Inevitably this leads to high costs and duplication of activity, for both the customer and supplier. If a qualified, independent third party is used instead to carry out the assessment, attention may be focused by the customer on any special needs and in developing closer partnerships with suppliers. Visits and dialogue across the customer/supplier interface are a necessity for the true requirements to be met and for future growth of the whole business chain. Visits should be concentrated, however, on improving understanding and capability, rather than on close scrutiny of operating procedures, which is best left to experts, including those within the supplier organizations charged with carrying out internal system audits and reviews.

JUST-IN-TIME (JIT) MANAGEMENT

The prohibitive cost of holding large stocks of components and raw materials has pushed forward the 'just-in-time' (JIT) concept. As this requires that suppliers make frequent, on time, deliveries of small quantities of material, parts, components, etc., often straight to the point of use, in order that stocks can be kept to a minimum, the approach requires an effective supplier network – one producing goods and services that can be trusted to conform to the real requirements with a high degree of confidence.

JIT, like many modern management concepts, is credited to the Japanese, who developed and began to use it in the late 1950s. It took approximately 20 years for JIT methods to reach Western hard goods industries and a further 10 years before businesses realized the generality of the concepts.

Basically JIT is a programme directed towards ensuring that the right quantities are purchased or produced at the right time, and that there is no waste. Anyone who perceives it purely as a material-control system, however, is bound to fail with JIT. JIT fits well under the TQM umbrella, for many of the ideas and techniques are very similar and, moreover, JIT will not work without TQM in operation. Writing down a definition of JIT for all types of organization is extremely difficult, because the range of products, services and organization structures leads to different impressions of the nature and scope of JIT. It is essentially:

- A series of operating concepts that allows systematic identification of operational problems.
- A series of technology-based tools for correcting problems following their identification.

An important outcome of JIT is a disciplined programme for improving productivity and reducing waste. This programme leads to cost-effective production or operation and delivery of only the required goods or services, in the correct quantity, at the right time and place. This is achieved with the minimum amount of resources – facilities, equipment, materials, and people. The successful operation of JIT is dependent upon a balance between the suppliers' flexibility and the users' stability, and of course requires total management and employee commitment and teamwork.

Aims of JIT

The fundamental aims of JIT are to produce or operate to meet the requirements of the customer exactly, without waste, immediately on demand. In some manufacturing companies JIT has been introduced as 'continuous flow production', which describes very well the objective of achieving conversion of purchased material or service receipt to delivery, i.e. from supplier to customer. If this extends into the supplier and customer chains, all operating with JIT a perfectly continuous flow of material, information or service will be achieved. JIT may be used in non-manufacturing, in administration areas, for example, by using external standard as reference points.

The JIT concepts identify operational problems by tracking the following:

1. *Material movements* – when material stops, diverts or turns backwards, these always correlate with an aberration in the 'process.
2. *Material accumulations* – these are there as a buffer for problems, excessive variability, etc., like water covering up 'rocks'.
3. *Process flexibility* – an absolute necessity for flexible operation and design.
4. *Value-added efforts* – much of what is done does not add value and the customer will not pay for it.

The operation of JIT

The tools to carry out the monitoring required are familiar quality and operations management methods, such as:

- Process study and analysis
- Preventive maintenance
- Plant layout methods
- Standardized design
- Statistical process control
- Value analysis and value engineering.

But some techniques are more directly associated with the operation of JIT systems:

1. Batch or lot size reduction
2. Flexible workforce
3. Kanban or cards with material visibility
4. Mistake-proofing
5. Pull-scheduling
6. Set-up time reduction
7. Standardized containers.

In addition, joint development programmes with suppliers and customers will be required to establish long-term relationships and develop single sourcing arrangements that provide frequent deliveries in small quantities. These can only be achieved through close communications and meaningful certified quality.

There is clear evidence that JIT has been an important component of business success in the Far East and that it is used by Japanese companies operating in the West. Many European and American companies that have adopted JIT have made spectacular improvements in performance. These include:

- Increased flexibility (particularly of the workforce)
- Reduction in stock and work-in-progress, and the space it occupies
- Simplification of products and processes.

These programmes are *always* characterized by a real commitment to continuous improvement. Organizations have been rewarded, however, by the low cost, low risk aspects of implementation provided a sensible attitude prevails. The golden rule is to never remove resources – such as stock – before the organization is ready and able to correct the problems that will be exposed by doing so. Reduction of the water level to reveal the rocks, so that they may be demolished, is fine, provided that we can quickly get our hands back on the stock while the problem is being corrected.

The Kanban system

Kanban is a Japanese word meaning 'visible record', but in the West it is generally taken to mean a 'card' that signals the need to deliver or produce more parts or components. In manufacturing, various types of records, e.g. job orders or route information, are used for ordering more parts in a *push* type, schedule-based system. In a push system a multi-period master production schedule of future demands is prepared, and a computer explodes this into detailed schedules for producing or purchasing the appropriate parts or materials. The schedules then *push* the production of the part or components, out and onward. These systems, when computer-based, were originally called Material Requirements Planning (MRP) but have been extended in many organizations to 'Enterprise Resource Planning' (ERP) systems. The main feature of the Kanban system is that it *pulls* parts and components through the production processes when they are needed. Each material, component or part has its own special container designed to hold a precise, preferably small, quantity. The number of containers for each part is a carefully considered management decision. Only standard containers are used, and they are always filled with the prescribed quantity.

A Kanban system provides parts when they are needed but without guesswork, and therefore without the excess inventory that results from bad guesses. The system will only work well, however, within the context of a JIT system in general, and the reduction of set-up times and lot sizes in particular. A JIT programme can succeed without a Kanban-based operation, but Kanban will not function effectively independently of JIT.

Just-in-time in partnerships and the supply chain

The development of long-term partnerships with a few suppliers, rather than short-term ones with many, leads to the concept of *co-producers* in networks of trust providing dependable quality and delivery of goods and services. Each organization in the chain of supply is often encouraged to extend JIT methods to its suppliers. The requirements of JIT mean that suppliers are usually located near the purchaser's premises, delivering small quantities, often several times per day, to match the usage rate. Administration is kept to a minimum and standard quantities in standard containers are usual. The requirement for suppliers to be located near the buying organization, which places those at some distance at a competitive disadvantage, causes lead times to be shorter and deliveries to be more reliable.

It can be argued that JIT purchasing and delivery are suitable mainly for assembly line operations, and less so for certain process and service industries, but the reduction in the inventory and transport costs that it brings should encourage innovations to lead to its widespread adoption. The main point is that there must be recognition of the need to develop closer relationships and to begin the dialogue – the sharing of information and problems – that leads to the product or service of the right quality, being delivered in the right quantity, at the right time.

RESOURCES

All organizations assemble resources, other than human, to support the effective operation of the processes that hopefully will deliver the strategy. These come in many forms but certainly include financial resources, buildings, equipment, materials, technology, information and knowledge. How these are managed will have a serious effect on the effectiveness and efficiency of any establishment, whether it be in manufacturing, service provision or the public sector.

Financial resources

Investment is key for the future development and growth of business. The ability to attract investment often determines the strategic direction of commercial enterprises. Similarly the acquisition of funding will affect the ability of public sector organizations in health, education or law establishments to function effectively. The development and implementation of appropriate financial strategies and processes will, therefore, be driven by the financial goals and performance of the business. Focus on, for example, improving earnings before interest and tax (EBIT – a measure of profitability) and economic value added (EVA – a measure of the degree to which the returns generated exceed the costs of financing the assets used) can in a private company be the drivers for linking the strategy to action. The construction of plans for the allocation of financial resources in support of the policies and strategies should lead to the appropriate and significant activities being carried out within the business to deliver the strategy.

Consolidation of these plans, coupled with an iterative review and approval, provides a mechanism of providing the best possible chance for success. Use of a 'balanced scorecard' approach (see Chapters 2, 7 and 8) can help in ensuring that the long-term impact of financial decisions on processes, innovation and customer satisfaction is understood and taken into account. The extent to which financial resources are being used to support strategy needs to be subject to continuous appraisal – this will include evaluating investment in the tangible and non-tangible assets, such as knowledge.

In the public sector, of course, financial policies are often derived from legislation and pubic accountability. In such situations a 'Director of Finance' often supports the organization's financial system, which is subject to independent review by appropriate authorities. Whatever the system, it is important to ensure alignment with policy and strategy, and that objectives are agreed through incorporation of targets, budgets and accounts, and that the risks to financial resources are managed.

In small and medium sized enterprises it is even more important that the financial strategy forms a key part of the strategic planning system, key financial goals are identified, deployed, and regularly scrutinized.

Other resources

Many different types of resources are deployed by different types of organizations. Most organizations are established in some sort of building, use equipment and consume materials. In these areas directors and managers must pay attention to:

- utilization of these resources
- security of the assets
- maintenance of building and equipment
- managing material inventories and consumption
- waste reduction and recycling
- environmental aspects, including conservation of non-renewable resources and adverse impact of products and processes.

Technology is a splendid and vital resource in the modern age. Exciting alternative and emerging technologies need to be identified, evaluated and appropriately deployed in the drive towards achieving organizational goals. This will include managing the replacement of 'old technologies' and the innovations which will lead to the adoption of new ones. There are clear links here, of course, with process re-design and re-engineering (see Chapter 11). It is not possible to create the 'paper-less' courtroom, for example, without consideration of the processes involved. A murder trial typically involves a million pieces of paper, which are traditionally wheeled into courtrooms on trolleys. To replace this with computer systems and files on disks requires more than just a flick of a switch. The whole end-to-end process of the criminal justice system may come under scrutiny in order to deliver the paper-free trial, and this will involve many agencies in the process – police, prosecution service, courts, probation services and the legal profession. Their involvement in the end-to-end process design will be vital if technology solutions are to add value and deliver the improvement in justice and reductions in costs that the systems in most countries clearly need.

Most organizations' strategies these days have considerable focus on technology and information systems, as these play significant roles in how they supply products and services to and communicate with customers. They need to identify technology requirements through business planning processes and work with technology partners and IT system providers to exploit technology to best advantage, improve processes and meet business objectives. Whether this requires a dedicated IT team to develop the strategy will depend on the size and nature of the business but it will always be necessary to assess information resource requirements, provide the right balance and ensure value for money is provided. This is often a tall order it seems in the provision of IT services! Close effective partnerships that deliver in this area are often essential.

In the piloting and evaluation of new technology, the impact on customers and the business itself should be determined. The roll out of any new systems involves people across the organization and communication cycles need to be used to identify any IT issues and feedback to partners. IT support should be designed in collaboration with users to confirm business processes, functionality and the expected utilization and availability. Responsibilities and accountabilities are important here, of course, and in smaller organizations this usually falls on line management.

Like any other resource, knowledge and information need managing and this requires careful consideration in its own right. Chapter 16 on communication, innovation and learning covers this in some detail.

In the design of quality management systems, resource management is an important consideration and is covered by the detail to be found in the ISO 9000 family of standards (see Chapter 11).

COLLABORATIVE BUSINESS RELATIONSHIPS (BS 11000)

The British Standard BS 11000, first published in 2010, provides a strategic framework to establish and improve collaborative relationships in organizations of all sizes, with a view to ensuring that they are effective, optimized and deliver enhanced benefits to the stakeholders.

BS 11000 outlines different approaches to collaborative working that have proven to be successful in businesses of all sizes and sectors. It shows how to eliminate the known pitfalls of poor communication by defining roles and responsibilities, and creating partnerships that do nothing but add value to a business. The benefits of BS 11000 Collaborative Business Relationships are claimed to be:

- Collaborating successfully with chosen partners
- Creating a neutral platform for mutual benefit with business partners
- Defining roles and responsibilities to improve decision making processes
- Sharing cost, risks, resources and responsibilities
- Providing staff with wider training opportunities
- Building better relationships that lead to quicker results.

The standard gives 'ten top tips' for implementing BS 11000:

1. Get commitment and support from senior management.
2. Engage the whole business with good internal communication.
3. Compare existing business relationships with BS 11000 requirements.
4. Get partner and stakeholder feedback on current collaborative working.
5. Establish an implementation team to get the best results.
6. Map out and share roles, responsibilities and timescales.
7. Adapt the basic principles of the BS 11000 standard to the business.
8. Motivate staff involvement with training and incentives.
9. Share BS 11000 knowledge and encourage staff to train as internal auditors.
10. Regularly review the BS 11000 system to make sure it remains effective and is being continually improved.

BS 11000 based collaborative business relationship management systems can be integrated with other management systems in place. This allows organizations to combine and streamline the way they manage the processes that apply to more than one system.

BIBLIOGRAPHY

Ansari, A. and Modarress, B., *Just-in-time Purchasing,* The Free Press (Macmillan), New York, 1990.
Bineno, J., *Implementing JIT,* IFS, Bedford, 1991.
British Standards Institution, *Partnering*, BSI, Milton Keynes, 2011.
Harrison, A., *Just-in-Time Manufacturing in Perspective*, Prentice-Hall, Englewood Cliffs, NJ, 1992.
Lysons, K., *Purchasing and Supply Chain Management*, Prentice-Hall, New York, 2000.
Muhlemann, A.P., Oakland, J.S. and Lockyer, K.G., *Production and Operations Management* (6th edn), Pitman, London, 1992.
Voss, C.A. (ed.), *Just-in-Time Manufacture*, IFS Publications, Bedford, 1989.

CHAPTER HIGHLIGHTS

Partnering and collaboration

- Organizations increasingly recognize the need to establish mutually beneficial relationships in partnerships. The philosophies behind TQM and 'Excellence' lay down principles and guidelines to support them.
- How partnerships are planned and managed must be in line with overall policies and strategies and support the operation of the processes.
- Establishing effective partnerships requires attention to: strategic alignment, customer focus, decision making and governance, communications and transparency, and investment and improvement.

Global outsourcing

- The top two reasons organizations select global suppliers are the pursuit of new markets and reducing cost, yet poor quality quickly, quietly and devastatingly demolishes the benefits of any new market gains or cost reductions.
- The difficulty of managing quality increases in direct relation to the distance between an organization and its partners, so a good 'assurance' model is needed.
- Three key areas of opportunity and risk that are common across both global in-sourcing and outsourcing are: managing performance and exposure to risk, getting organized for success and delivering better value from the supply chain.

Supply chain effectiveness in global economies – the secret of safe sourcing

Reliance upon a single source of supply can be risky so an organization must establish whether it is at risk.

- To strengthen sourcing security in terms of quality, timeliness and cost, by sourcing a second supplier: decide the selection criteria; make an initial selection; toughen up the criteria and make a shortlist; conduct site visits, make a final selection and choice.

Partnerships and resources

Role of purchasing in partnerships

- The prime objective of purchasing is to obtain the correct equipment, materials and services in the right quantity, of the right quality, from the right origin, at the right time and cost. Purchasing also acts as a 'window-on-the-world'.
- The purchasing system should be documented and assign responsibilities, define the means of selecting suppliers and specify the documentation to be used.
- Improving supplier performance requires from the suppliers' senior management commitment, education, a policy, an assessed quality system and supplier approval.

Just-in-time (JIT) management

- JIT fits well under the TQM umbrella and is essentially a series of operating concepts that allow the systematic identification of problems *and* tools for correcting them.
- JIT aims to produce or operate, in accordance with customer requirements, without waste, immediately on demand. Some of the direct techniques associated with JIT are batch or lot size reduction, flexible workforce, Kanban cards, mistake proofing, set up time reduction, standardized containers.
- The development of long-term relationships with a few suppliers or 'co-producers' is an important feature of JIT. These exist in a network of trust to provide quality goods and services.

Resources

- All organizations assemble resources to support operation of the processes and deliver the strategy. These include finance, buildings, equipment, materials, technology, information and knowledge.
- Investment and/or funding is key for future development of all organizations and often determines strategic direction. Financial goals and performance will, therefore, drive strategies and processes. Use of a 'balanced scorecard' approach with continuous appraisal helps in understanding the long-term impact of financial decisions.
- In the management of buildings, equipment and materials, attention must be given to utilization, security, maintenance, inventory, consumption, waste and environmental aspects.
- Technology plays a key role in most organizations and management of existing alternative and emerging technologies need to be identified, evaluated and deployed to achieve organizational goals.
- There are clear links between the introduction of new or the replacement of old technologies and process redesign/engineering (see Chapter 10). The roll out of any new systems also involves people across the organization and good communications are vital.

Collaborative business relationships (BS 11000)

- BS 11000 provides a strategic framework to establish and improve collaborative relationships in organizations of all sizes to ensure they are effective, optimized and deliver enhanced benefits to the stakeholders.

- BS 11000 outlines different approaches to collaborative working that can be successful in businesses of all sizes and sectors. It shows how to eliminate the known pitfalls of poor communication by defining roles and responsibilities, and creating partnerships that add value to a business.
- The standard lists the benefits of using the approach, including defining roles and responsibilities to improve decision making processes, sharing cost, risks, resources and responsibilities, providing staff with wider training opportunities and building better relationships that lead to quicker results. BS 11000 also gives 'ten top tips' for implementation.

Design for quality

DESIGN, INNOVATION AND IMPROVEMENT

Products, services and processes are designed both to add value to customers and to generate profit. But leadership and management style is also designed through the creation of symbols and processes which are reflected in internal communication methods, materials and behaviour. Almost all areas of all organizations have design aspects inherent within them.

Design can be used to gain and hold on to competitive edge, save time and effort, deliver innovation, stimulate and motivate staff, simplify complex tasks, delight clients and stakeholders, dishearten competitors, achieve impact in a crowded market and justify a premium price. Design can be used to take the drudgery out of the mundane and turn it into something inspiring, or simply make money. Design can be considered as a management function, a cultural phenomenon, an art form, a process, a discrete activity, an end-product, a service or, often, a combination of several of these.

In the Collins *Cobuild English Language Dictionary*, design is defined as: 'the way in which something has been planned and made, including what it looks like and how well it works.' Using this definition, there is very little of an organization's activities that are not covered by 'planning' or 'making'. Clearly the consideration of what it looks like and how well it works in the eyes of the customer determines the success of products or services in the market place.

All organizations need to update their products, services and processes periodically. In markets such as electronics, audio and visual goods, and office automation, new variants of products are offered frequently – almost like fashion goods. While in other markets the pace of innovation may not be as fast and furious there is no doubt that the rate of change for product, service, technology and process design has accelerated on a broad front.

Innovation entails both the invention and design of radically new products and services, embodying novel ideas, discoveries and advanced technologies, *and* the continuous development and improvement of existing products, services and processes to enhance their performance and quality. It may also be directed at reducing costs of production or operations throughout the life cycle of the product or service system.

Within all industries rapid innovation is changing every aspect of the business, including the products and the services offered. These include an increase in the use of IT based technologies in design, communication, management, manufacturing and service delivery. In addition there are numerous examples of new technologies such as new equipment for materials handling and assembly, new products and materials, new financing arrangements and new procurement processes which involve the sharing of risk.

In many organizations innovation is predominantly either technology-led e.g. in some information and communications industries, or marketing-led e.g. in fmcg (fast moving consumer goods). What is always striking about leading product or service innovators is that their developments are market-led, which is different from marketing-led. The latter means that the marketing function takes the lead in product and service developments. But most leading innovators identify and set out to meet the existing and potential demands profitably and, therefore, are market-led constantly striving to meet the requirements even more effectively through appropriate experimentation.

Everything we experience in or from an organization is the result of a design decision, or lack of one. This applies not just to the tangible things like products and services, but the intangibles too: the systems and processes which affect the generation of products and delivery of services. Design is about combining function and form to achieve fitness for purpose: be it an improvement to a supersonic aircraft, the synthesis of a new drug, the development of a new building material or product, a new management process, a staff incentive scheme or a hand-held media device.

Once fitness for purpose has been achieved, of course, the goal posts change. Events force a reassessment of needs and expectations and customers want something different. In such a changing world, design is an on-going activity, dynamic not static, a verb not a noun – *design is a process.*

THE DESIGN PROCESS

Commitment from the most senior management helps to build quality throughout the *design process* and to ensure good relationships and communication between various groups and functional areas, both within the organization and across the supply chain. Designing customer satisfaction and loyalty into products and services contributes greatly to competitive success. Clearly, it does not guarantee it, because the conformance aspect of quality must be present and the operational processes must be capable of producing to the design. As in the marketing/operations interfaces, it is never acceptable to design a product, service, system or process that the customer wants but the organization is incapable of achieving.

The design process often concerns technological innovation in response to, or in anticipation of, changing market requirements and trends in technology. Those companies with impressive records of product- or service-led growth have demonstrated a state-of-the-art approach to innovation based on three principles:

- *Strategic balance* between product/service and process development to ensure that product and service innovation maintains market position, while process innovation ensures that production risks in safety, quality and productivity are effectively controlled and reduced.
- *Top management approach* to design and product/service creation to set the tone and ensure that commitment is the common objective by visibly supporting the design effort. Direct control should be concentrated on critical decision points, since over-meddling by very senior people in day-to-day project management can delay and demotivate staff.
- *Teamwork*, to ensure that once projects are under way, specialist inputs, e.g. from marketing and technical experts, are fused and problems are tackled simultaneously. The teamwork should be urgent yet informal, for too much formality can stifle initiative, flair and the 'fun' of design.

The extent of the product/service creation process (PSCP) should not be underestimated, but it often is. Many people associate design with *styling* of products, and this is certainly an important aspect. But for certain products and many service operations the *secondary design* considerations are vital. Anyone who has bought an 'assemble-it-yourself' kitchen unit will know the importance of the design of the assembly instructions, for example. Aspects of design that affect quality in this way are packaging, customer-service arrangements, maintenance routines, warranty details and their fulfilment, spare-part availability, etc.

An industry that has learned much about the secondary design features of its products is personal computers. Many of the problems of customer dissatisfaction experienced in this market have not been product design features but problems with user support, availability and loading of software, and applications. For technically complex products or service systems, the design and marketing of after-sales arrangements are an essential component of the design activity. The design of production equipment and its layout to allow ease of access for repair and essential maintenance, or simple use as intended, widens the management of design quality into the supply chain and contractors and requires their total commitment.

In the construction industry, design has a much larger role than is generally recognized; for example, on site, in the process of assembling a building, the selection and positioning of equipment is a design task. Similarly the selection of technology – the decision between using cast in place or prefabricated components – is also essentially a design task. Proper design of plant, equipment, buildings and surrounding environments plays a major role in good process operation, the elimination of errors, defectives and waste. Correct initial design also obviates the need for costly and wasteful modifications to be carried out. It is at the design stage that such important matters as variability of details, reproducibility, technical risk of failure due to workmanship, ease of use in operation, maintainability, etc. should receive detailed consideration.

Designing

If design quality is taking care of all aspects of the customer's requirements, including cost, production, safe and easy use, and maintainability of products and services, then *designing* must take place in all aspects of:

- Identifying the need (including need for change).
- Developing that which satisfies the need.
- Checking the conformance to the need.
- Ensuring that the need is satisfied.

Designing covers every aspect, from the identification of a problem to be solved, usually a market need, through the development of design concepts and prototypes to the generation of detailed specifications or instructions required to produce the artefact or provide the service. It is the process of presenting needs in some physical form, initially as a solution, and then as a specific configuration or arrangement of materials resources, equipment and people. Design permeates strategically and operationally many areas of an organization and, while design professionals may control detailed product styling, decisions on design involve many people from other functions. Total quality management supports such a cross-functional interpretation of design.

In the construction environment, this broad conceptualization of the design function is essential as design impacts on every stage of the production process: safety during construction, constructability, the cost and ease of prefabrication of engineered products, and the reliable achievement of product quality on site.

Design like any other activity, must be carefully managed. A flowchart of the various stages and activities involved in the design and development process appears in Figure 6.1.

By structuring the design process in this way, it is possible to:

- Control the various stages.
- Check that they have been completed.
- Decide which functions need to be brought in and at what stage.
- Estimate the level of resources needed.

The control of the design process must be carefully handled to avoid stifling the creativity of the designer(s), which is crucial in making design solutions a reality. It is clear that the design process requires a range of specialized skills, and the way in which these skills are managed, the way they interact and the amount of effort devoted to the different stages of the design and development process is fundamental to the quality, producibility and price of the service or final product. A team approach to the management of design is critical to the success of a project. The input of manufacturers, engineering fabricators and site assemblers is as crucial as the input from the end-user market.

It is never possible to exert the same tight control on the design effort as on other operational efforts, yet the cost and the time used are often substantial, and both must appear somewhere within the organization's budget.

Certain features make control of the design process difficult:

1. No design will ever be 'complete' in the sense that, with effort, some modification or improvement cannot be made.
2. Few designs are entirely novel. An examination of most 'new' products, services or processes will show that they employ existing techniques, components or systems to which have been added novel elements.
3. The longer the time spent on a design, the less the increase in the value of the design tends to be unless a technological breakthrough is achieved.

Design for quality

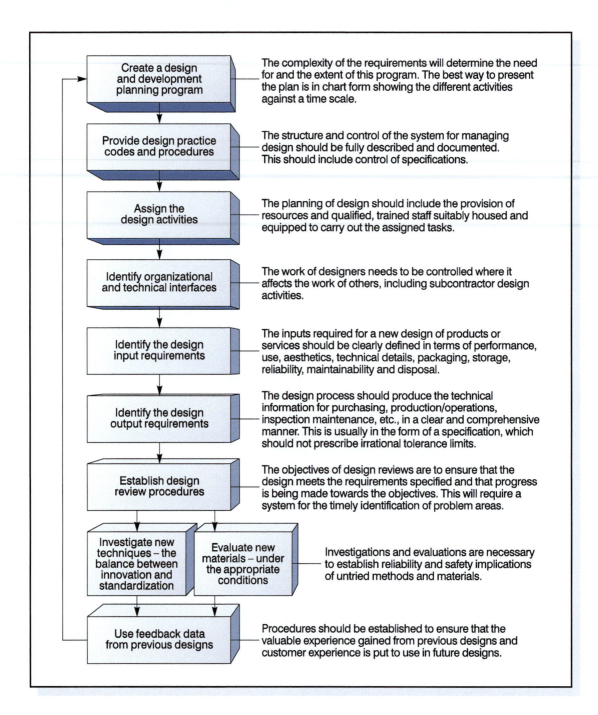

Figure 6.1
The design and development process

This diminishing return from the design effort must be carefully managed but this has to be balanced with the need for adequate design resolution and sound documentation, because production risk increases when the design is not properly resolved and effectively communicated.

4.   The design process is information intensive and the timing of decision-making, both by the clients and the design team, is critical to the efficiency of the entire process. It is not practical to manage the design process in the same manner as we do the production process – on the basis of tasks. For every task there may be up to ten information flows and the ratio of information flows to tasks is highly variable. Also there are a great number of concurrent and interdependent activities which need skill and experience in their effective resolution.

5.   External and/or internal customers will often impose limitations on design time and cost. It is as difficult to imagine a design project whose completion date is not implicitly fixed, either by a promise to a customer, the opening of a trade show or exhibition, a seasonal 'deadline', a production schedule or some other constraint, as it is to imagine an organization whose funds are unlimited, or a product/service whose price has no ceiling.

Total design processes

Quality of design, then, concerns far more than the product or service design and its ability to meet the customer requirements. It is also about the activities of design and development. The appropriateness of the actual *design process* has a profound influence on the performance of any organization, and much can be learned by examining successful companies and how their strategies for research, design and development are linked to the efforts of marketing and operations. In some quarters this is referred to as 'total design', and the term 'simultaneous engineering' has been used. This is an integrated approach to a new product or service introduction, similar in many ways to Quality Function Deployment (QFD – see next section) in using multifunction teams or task forces to ensure that research, design, development, manufacturing, purchasing, supply and marketing all work in parallel from concept through to the final launch of the product or service into the market place, including servicing and maintenance.

Most companies now recognize the need to develop and successfully deploy an end-to-end Product or Service Creation Process (PCP/SCP). This is built to ensure that the product or service requirements are translated from identification of need (or potential need) into the reality of a tangible new/revised product/service. Such a PCP/SCP will transgress the whole organization and involve the engagement of all functional areas. An example of such a process is shown in Figure 6.2.

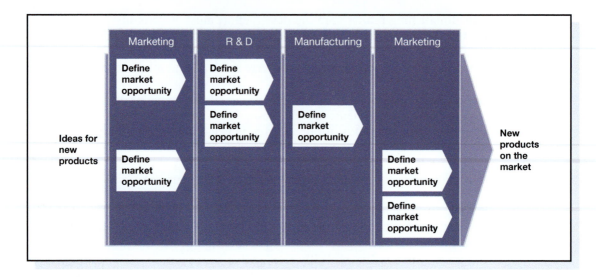

Figure 6.2
Cross-functional new product creation process

QUALITY FUNCTION DEPLOYMENT (QFD) – THE HOUSE OF QUALITY

The 'house of quality' is the framework of an approach to design management known as quality function deployment (QFD). Dr Yoji Akao originally developed QFD in Japan in 1966, combining his work on quality assurance and quality control with function deployment as used in value engineering.

It originated in Japan in 1972 at Mitsubishi's Kobe shipyard, but it has been developed in numerous ways by Toyota and its suppliers, and many other organizations. The house of quality (HoQ) concept, initially referred to as quality tables, has been used successfully by manufacturers of integrated circuits, synthetic rubber, construction equipment, engines, home appliances, clothing and electronics, mostly Japanese. Ford and General Motors use it, and other organizations, including AT&T, Bell Laboratories, Digital Equipment, Hewlett-Packard, Procter & Gamble, ITT, Rank Xerox and Jaguar have applications. In Japan, its design applications include public services and retail outlets. In the construction sector the application of QFD is limited to companies that have specialized in a specific market sector; for example, in Japan and Brazil it has been used to design apartment layouts, and it has application for areas of mass production of products and materials.

Quality function deployment (QFD) is a 'system' for designing a product or service, based on customer requirements, with the participation of members of all functions of the supplier organization. It translates the customer's requirements into the appropriate technical requirements for each stage. The activities included in QFD are:

1. Market research.
2. Basic research.
3. Innovation.
4. Concept design.
5. Prototype testing.
6. Final-product or service testing.
7. After sales service and trouble-shooting.

These are performed by people with different skills in a team whose composition depends on many factors, including the products or services being developed and the size of the operation. In many industries, such as cars, electronic equipment and computers, 'engineering' designers are seen to be heavily into 'designing'. But in other industries and service operations designing is carried out by people who do not carry the word 'designer' in their job title. The failure to recognize the design inputs they make, and to provide appropriate training and support, will limit the success of the design activities and result in some offering that does not necessarily satisfy the customer. This is particularly true of services generally.

The QFD team in operation

The first step of a QFD exercise is to form a cross-functional QFD team. Its purpose is to take the needs of the market and translate them into such a form that they can be satisfied within the operating unit and delivered to the customers.

As with all organizational problems, the structure of the QFD team must be decided on the basis of the detailed requirements of each organization. One thing, however, is clear – close liaison must be maintained at all times between the design, marketing and operational functions represented in the team.

The QFD team must answer three questions – **WHO, WHAT** and **HOW**, i.e.

WHO are the customers?
WHAT does the customer need?
HOW will the needs be satisfied?

WHO may be decided by asking 'Who will benefit from the successful introduction of this product, service or process? Once the customers have been identified, WHAT can be ascertained through interview/questionnaire/focus group processes, or from the knowledge and judgement of the QFD team members. HOW is more difficult to determine, and will depend on the attributes of the product, service or process under development. This will constitute many of the action steps in a 'QFD strategic plan'.

WHO, WHAT and HOW are entered into a QFD matrix or grid of 'house of quality' (HoQ), which is a simple 'quality table'. The *WHATs* are recorded in rows and the *HOWs* are placed in the columns.

The house of quality provides structure to the design and development cycle, often likened to the construction of a house, because of the shape of matrices when they are fitted together. The key to building the house is the focus on the customer requirements, so that the design and development processes are driven by what the customer needs as well as innovations in technology. This ensures that more effort is used to obtain vital customer information. It may increase the initial planning time in a particular development project, but the overall time, including design and redesign, taken to bringing a product or service to the market will be reduced.

This requires that marketing people, design staff (including architects, engineers, physicists, chemists, etc), and production/operations personnel work closely together from the time the new service, process or product is conceived. It will need to replace in many organizations the 'throwing it over the wall' approach, where a solid wall exists between each pair of functions (Figure 6.3).

The HoQ provides an organization with the means for inter-departmental or inter-functional planning and communications, starting with the so-called customer attributes (CAs). These are phrases customers use to describe product, process and service characteristics.

A complete QFD project will lead to the construction of a sequence of house of quality diagrams, which translate the customer requirements into specific operational process steps. For example, the 'feel' that customers like on the steering wheel of a motor car may translate into a specification for 45 standard degrees of synthetic polymer hardness, which in turn translates into specific manufacturing process steps, including the use of certain catalysts, temperatures, processes and additives. Similarly in construction, the acoustic privacy that home-owners demand is translated into a measurable decibel transfer rate and specific construction systems to achieve it. The first steps in QFD lead to a consideration of the product as a whole and subsequent steps to consideration of the individual components. For example, a complete hotel service would be considered at the first level, but subsequent QFD exercises would tackle the restaurant, bedrooms and reception. Each of the sub-services would have customer requirements, but they all would need to be compatible with the general service concept.

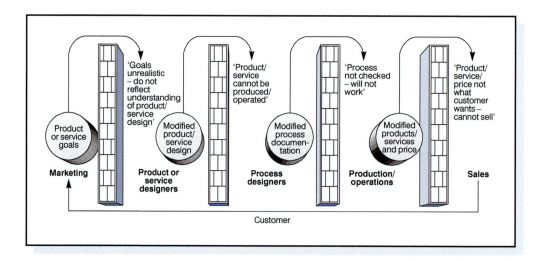

Figure 6.3
'Throw it over the wall.' The design and development process is sequential and walled into separate functions

The QFD or house of quality tables

Figure 6.4 shows the essential components of the quality table or HoQ diagram. The construction begins with the *customer requirements*, which are determined through the 'voice of the customer' – the marketing and market research activities. These are entered into the blocks to the left of the central relationship matrix. Understanding and prioritizing the customer requirements by the QFD team may require the use of competitive and complaint analysis, focus groups, and the analysis of market potential. The prime or broad requirements should lead to the detailed WHATs.

Once the customer requirements have been determined and entered into the table, the *importance* of each is rated and rankings are added. The use of the 'emphasis technique' or paired comparison may be helpful here (see Chapter 13).

Each customer requirement should then be examined in terms of customer rating; a group of customers may be asked how they perceive the performance of the organization's product or service versus those of competitors. These results are placed to the right of the central matrix. Hence the customer requirements' importance rankings and competition ratings appear from left to right across the house.

The WHATs must now be converted into the HOWs. These are called the *technical design requirements* and appear on the diagram from top to bottom in terms of requirements, rankings (or costs) and ratings against competition (technical benchmarking, see Chapter 9). These will provide the 'voice of the process'.

The technical design requirements themselves are placed immediately above the central matrix and may also be given a hierarchy of prime and detailed requirements. Immediately below the customer requirements appear the rankings of technical difficulty, development time or costs. These will enable the QFD team to discuss the

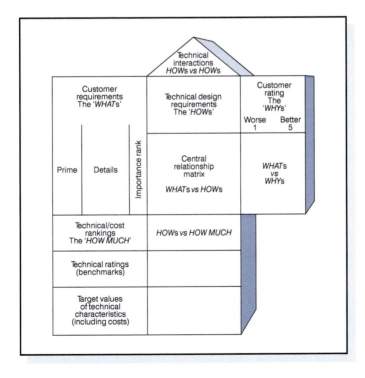

Figure 6.4
The house of quality

efficiency of the various technical solutions. Below the technical rankings on the diagram comes the benchmark data, which compares the technical processes of the organization against its competitors.

The *central relationship matrix* is the working core of the house of quality diagram. Here the WHATs are matched with the HOWs, and each customer requirement is systematically assessed against each technical design requirement. The nature of any relationship – strong positive, positive, neutral, negative, strong negative – is shown by symbols in the matrix. The QFD team carries out the relationship estimation, using experience and judgement, the aim being to identify HOW the WHATs may be achieved. All the HOWs listed must be necessary and together sufficient to achieve the WHATs. Blank rows (customer requirement not met) and columns (redundant technical characteristics) should not exist.

The roof of the house shows the interactions between the technical design requirements. Each characteristic is matched against the others, and the diagonal format allows the nature of relationships to be displayed. The symbols used are the same as those in the central matrix.

The complete QFD process is time-consuming, because each cell in the central and roof matrices must be examined by the whole team. The team must examine the matrix to determine which technical requirement will need design attention, and the costs of that attention will be given in the bottom row. If certain technical costs become a major issue, the priorities may then be changed. It will be clear from the central matrix if there is more than one way to achieve a particular customer requirement, and the roof matrix will show if the technical requirements to achieve one customer requirement will have a negative effect on another technical issue.

The very bottom of the house of quality diagram shows the *target* values of the *technical characteristics*, which are expressed in physical terms. They can only be decided by the team after discussion of the complete house contents. While these targets are the physical output of the QFD exercise, the whole process of information gathering, structuring and ranking generates a tremendous improvement in the team's cross-functional understanding of the product/service design delivery system. The target technical characteristics may be used to generate the next level house of quality diagram, where they become the WHATs, and the QFD process determines the further details of HOW they are to be achieved. In this way the process 'deploys' the customer requirements all the way to the final operational stages. Figure 6.5 shows how the target technical characteristics, at each level, become the input to the next level matrix.

QFD progresses now through the use of the 'seven new planning tools' and other standard techniques such as value analysis, experimental design, statistical process control and so on.

SPECIFICATIONS AND STANDARDS

There is a strong relationship between standardization and specification. To ensure that a product or a service is *standardized* and may be repeated a large number of times in exactly the manner required, *specifications* must be written so that they are open to only one interpretation. The requirements, and therefore the quality, must be built into the design specification. There are national and international standards which, if

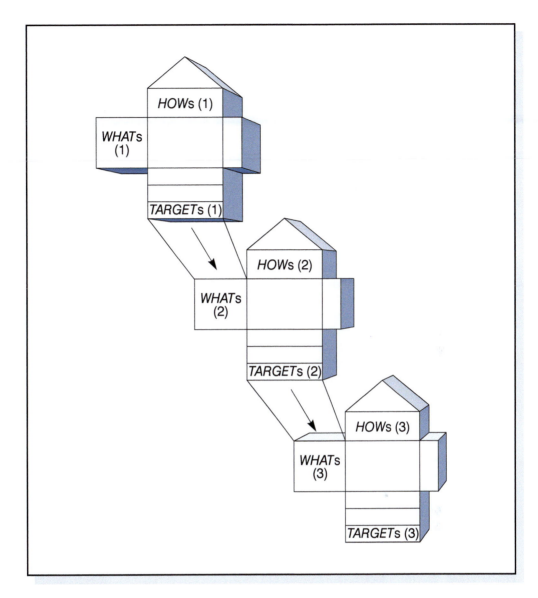

Figure 6.5
The deployment of the 'voice of the customer' through quality tables

used, help to ensure that specifications will meet certain accepted criteria of technical or managerial performance, safety, etc.

Standardization does not guarantee that the best design or specification is selected. It may be argued that the whole process of standardization slows down the rate and direction of technological development, and affects what is produced. If standards are used correctly, however, the process of drawing up specifications should provide opportunities to learn more about particular innovations and to change the standards accordingly.

Design for quality

These ideas are well illustrated by the construction sector's approach worldwide to the adoption of performance-based specifications wherever possible. Performance-based standards encourage innovation against measurable and transparent technical requirements. This allows the opportunity for manufacturers with new products and innovative solutions to have their ideas accredited and gain market entry. In areas like waterproofing, however, a building contractor might prefer to be very prescriptive in specifying the precise technical solution he wants. This is a particularly important area of construction where, based on everyday experience, we know that the risk of failure is high and its consequences of water leaking through the roof or out of a bathroom into adjoining rooms is simply unacceptable. In such areas, correct design and implementation is critical to managing an important area of risk for the general contractor.

It is possible to strike a balance between innovation and standardization; however, a sound approach to innovation clearly recognizes areas of design innovation that add value for the customer and areas of standardization that reduce risk in the production process. Clearly, it is desirable for designers to adhere where possible to past-proven materials and methods, in the interests of reliability, maintainability and variety control. Hindering designers from using recent developed materials, components or techniques, however, can cause the design process to stagnate technologically. A balance must be achieved by analysis of materials, products and processes proposed in the design, against the background of their known reproducibility and reliability. If breakthrough innovations are proposed, then analysis or testing should be indicated objectively, justifying their adoption in preference to the established alternatives.

It is useful to define a specification. The International Standards Organization (ISO) defines it in ISO 8402 (1986) as 'The document that prescribes the requirements with which the product or service has to conform'. A document not giving a detailed statement or description of the requirements to which the product, service or process must comply cannot be regarded as a specification, and this is true of much sales literature.

The specification conveys the customer requirements to the supplier to allow the product or service to be designed, engineered, produced or operated by means of conventional or stipulated equipment, techniques and technology. The basic requirements of a specification are that it gives the:

- Performance requirements of the product or service in measurable terms.
- Parameters – such as dimensions, concentration, turn-round time – which describe the product or service adequately (these should be quantified and include the units of measurement).
- Materials to be used by stipulating properties or referring to other specifications.
- Method of production or delivery of the service.
- Inspection/testing/checking requirements.
- References to other applicable specifications or documents.

To fulfil the purpose specifications must be written in terminology that is readily understood, and in a manner that is unambiguous and so cannot be subject to differing interpretation. This is not an easy task, and one which requires all the expertise and knowledge available.

It is in relation to the clear communication of process specifications that the use of 3D and virtual reality (VR) technologies are showing great potential. At many stages of the design and construction process, complex information has to be communicated to the partners in the supply chain or to customers and their design teams. Often, end-clients and other stakeholders are not able to conceptualize the design elements of a project; however, through the use of visualization tools their ability to interact with the design team is greatly enhanced. In other instances, the process design and detailing of parts of a structure can be very complex and VR simulation can assist both in optimizing the process through virtual prototyping and then in communicating the process to the people executing the work.

Good specifications are usually the product of much discussion, deliberation and sifting of information and data, and represent tangible output from a QFD team.

DESIGN QUALITY IN THE SERVICE SECTOR

The emergence of the services sector has been suggested by economists to be part of the natural progression in which economic dominance changes first from agriculture to manufacturing and then to services. It is argued that if income elasticity of demand is higher for services than it is for goods, then as incomes rise, resources will shift toward services. The continuing growth of services verifies this, and is further explained by changes in culture, fitness, safety, demography and life styles.

In considering the design of services it is important to consider the differences between goods and services. Some authors argue that the marketing and design of goods and services should conform to the same fundamental rules, whereas others claim that there is a need for a different approach to service because of the recognizable differences between the goods and services themselves.

In terms of design, it is possible to recognize three distinct elements in the service package – the physical elements or facilitating goods, the explicit service or sensual benefits, and implicit service or psychological benefits. In addition, the particular characteristics of service delivery systems may be itemised:

- Intangibility
- Perishability
- Simultaneity
- Heterogeneity.

It is difficult, if not impossible, to design the intangible aspects of a service, since consumers often must use experience or the reputation of a service organization and its representatives to judge quality.

Perishability is often an important issue in services, since it is frequently undesirable or impossible to hold stocks of the explicit service element of the service package. This aspect often requires that service operation and service delivery must exist simultaneously, such as in a restaurant business.

Simultaneity occurs because the consumer must be present before many services can take place. Hence, services are often formed in small and dispersed units, and it can be difficult to take advantage of economies of scale. The rapid developments in computing and communications technologies have changed this in sectors such as banking and insurance, but contact continues to be necessary for many service sectors. Design considerations here include the environment and the systems used. Service facilities, procedures and systems should be designed with the customer in mind, as well as the 'product' and the human resources. Managers need a picture of the total span of the operation, so factors which are crucial to success are not neglected. This clearly means that the functions of marketing, design and operations cannot be separated in services, and this must be taken into account in the design of the operational controls, such as the diagnosing of individual customer expectations. A QFD approach here can be very effective.

Heterogeneity of services occurs in consequence of explicit and implicit service elements relying on individual preferences and perceptions. Differences exist in the outputs of organizations generating the same service, within the same organization, and even the same employee on different occasions. Clearly, unnecessary variation needs to be controlled, but the variation attributed to estimating, and then matching, the consumers' requirements is essential to customer satisfaction and loyalty and must be designed into the systems. This inherent variability does, however, make it difficult to set precise quantifiable standards for all the elements of the service.

In the design of services it is useful to classify them in some way. Several sources from the literature on the subject help us to place services in one of five categories:

- Service factory
- Service shop
- Mass service
- Professional service
- Personal service.

Several service attributes have particular significance for the design of service operations:

1. *Labour intensity* – the ratio of labour costs incurred to the value of assets and equipment used (people versus equipment-based services).
2. *Contact* – the proportion of the total time required to provide the service for which the consumer is present in the system.
3. *Interaction* – the extent to which the consumer actively intervenes in the service process to change the content of the service; this includes customer participation to provide information from which needs can be assessed, and customer feedback from which satisfaction levels can be inferred.
4. *Customization* – which includes *choice* (providing one or more selections from a range of options, which can be single or *fixed* and *adaptation* (the interactions process in which the requirement is decided, designed and delivered to match the need).
5. *Nature of service act* – either tangible, i.e. perceptible to touch and can be owned, or intangible, i.e. insubstantial.
6. *Recipient of service* – either people or things.

Table 6.1 gives a list of some services with their assigned attribute types and Table 6.2 shows how these may be used to group the services under the various classifications.

Parasuraman *et al.* (see Zeithaml, Parasuraman and Berry, 1990) used the relationship between service quality and customer perceptions of product quality; his five dimensions are:

Table 6.1 A classification of selected services

Service	Labour intensity	Contact	Interaction	Customization	Nature of act	Recipient of service
Accountant	High	Low	High	Adapt	Intangible	Things
Architect	High	Low	High	Adapt	Intangible	Things
Bank	Low	Low	Low	Fixed	Intangible	Things
Beautician	High	High	High	Adapt	Tangible	People
Bus/coach service	Low	High	High	Choice	Tangible	People
Cafeteria	Low	High	High	Choice	Tangible	People
Cleaning firm	High	Low	Low	Fixed	Tangible	People
Clinic	Low	High	High	Adapt	Tangible	People
Coaching	High	High	High	Adapt	Intangible	People
College	High	High	Low	Fixed	Intangible	People
Courier firm	High	Low	Low	Adapt	Tangible	Things
Dental practice	High	High	High	Adapt	Tangible	Things
Driving school	High	High	High	Adapt	Intangible	People
Equipment hire	Low	Low	Low	Choice	Tangible	Things
Finance consultant	High	Low	High	Adapt	Intangible	People
Hairdresser	High	High	High	Adapt	Tangible	People
Hotel	High	High	Low	Choice	Tangible	People
Leisure center	Low	High	High	Choice	Tangible	People
Maintenance	Low	Low	Low	Choice	Tangible	Things
Management consultant	High	High	High	Adapt	Intangible	People
Nursery	High	Low	Low	Fixed	Tangible	People
Optician	High	Low	High	Adapt	Tangible	People
Postal service	Low	Low	Low	Adapt	Tangible	Things
Rail service	Low	High	Low	Choice	Tangible	People
Repair firm	Low	Low	Low	Adapt	Tangible	Things
Restaurant	High	High	Low	Choice	Tangible	People
Service station	Low	High	High	Choice	Tangible	People
Solicitors	High	Low	High	Adapt	Intangible	Things
Takeaway	High	Low	Low	Choice	Tangible	People
Veterinary	High	Low	High	Adapt	Tangible	Things

It is apparent that services are part of almost all organizations and not confined to the service sector. What is clear is that the service classifications and different attributes must be considered in any service design process.

The author is grateful to the contribution made by John Dotchin to this section of Chapter 6.

- *Reliability* – ability to perform the service dependably and accurately.
- *Responsiveness* – willingness to help customers and provide prompt service.
- *Assurance* – knowledge and courtesy of employees and their ability to inspire trust and confidence.
- *Empathy* – caring, individualized attention the form provides to its customers.
- *Tangibles* – physical facilities, equipment and appearance.

As a part of their work Parasuraman and his co-researchers developed a generic survey instrument and this is widely recognized as an excellent tool for measuring *service quality*. SERVQUAL scores *service quality* using 22 standardized statements to

Table 6.2 Grouping of similar services

PERSONAL SERVICES	
Driving school	Sports coaching Beautician
	Dental practice Hairdresser
	Optician
SERVICE SHOP	
Clinic	Cafeteria
Leisure center	Service station
PROFESSIONAL SERVICES	
Accountant	Architect
Finance consultant	Management consultant
Solicitor	Veterinary
MASS SERVICES	
Hotel	Restaurant
College	Bus service
Coach service	Rail service
Takeaway	Nursery Courier firm
SERVICE FACTORY	
Cleaning firm	Postal service
Repair firm	Equipment hire
Maintenance	Bank

canvass customer views on the dimensions of *service quality*. Statements from the instrument are shown in Table 6.3.

Responses to these questions using a nine-point Likert scale are used to enable customer satisfaction to be assessed and benchmarked.

It is interesting to reflect on the quality design challenges faced by lawyers in recent times. They are part of a generally honourable profession that is in fundamental transformation. Conventional legal advisers will be much less prominent in the future than in the recent past and there are two major forces that are shaping and characterizing legal services:

- Market pull towards commoditization
- Pervasive development & uptake of information technology.

Just as other industries and sectors are having to adapt to broader change, so too have legal companies which are needing to think more creatively about the way they do business and in particular where they can innovate. To compete, a solid foundation of high-quality, efficient processes will be required:

THERE IS A NEED . . .

- To identify the tasks that the market is increasingly unlikely to tolerate expensive lawyers for that can be delegated to less expert and less expensive people, working with sophisticated processes and systems.
- To identify the new and different client needs emerging.

WHICH MEANS THAT . . .

Table 6.3 SERVQUAL survey statements (Parasuraman *et al.*)

Reliability
1. Providing service as promised
2. Dependability in handling customers' service problems
3. Performing services right the first time
4. Providing services at the promised time
5. Maintaining error free records (e.g. financial)

Responsiveness
6. Keeping customers informed of when services will be performed
7. Prompt service to customers
8. Willingness to help customers
9. Readiness to respond to customers' requests

Assurance
10. Instilling confidence in customers
11. Make customers feel safe in their transactions
12. Being consistently courteous
13. Having the knowledge to answer questions

Empathy
14. Giving customers individual attention
15. Dealing with customers in a caring fashion
16. Having the customers' best interests at heart
17. Understanding the needs of their customers
18. Convenient business hours

Tangibles
19. Modern equipment
20. Visually appealing facilities
21. Having a neat, professional appearance
22. Visually appealing materials associated with the service

- Legal companies will need to be really clear on their strengths, identify their distinctive skills, talents and capabilities that cannot be replaced by advanced systems, or less costly workers supported by IT (standard processes).
- They will also need to be very clear on how to position themselves in the market.

  AND . . .

- They will need to address their weaknesses, drive out excesses, reduce waste, eliminate outdated practice and up-skill or remove outdated lawyers.

In short they will need to ensure they have the best people, processes and technology to support their target operating models.

Figure 6.6 shows the evolution of legal services, as set out in Richard Susskind's excellent book, *The End of Lawyers?* (2008). It is not as simple as choosing where on the spectrum to play, the legal services portfolio is now like a conveyor belt that moves from left to right, though some organizations are polarizing toward the ends.

Figure 6.6
The evolution of legal services (Source: Susskind – *The End of Lawyers*)

FAILURE MODE, EFFECT AND CRITICALITY ANALYSIS (FMECA)

In the design of products, services and processes it is possible to determine possible modes of failure and their effects on the performance of the product or operation of the process or service system. Failure mode and effect analysis (FMEA) is the study of potential failures to determine their effects. If the results of an FMEA are ranked in order of seriousness, then the word CRITICALITY is added to give FMECA. The primary objective of a FMECA is to determine the features of product design, production or operation and distribution that are critical to the various modes of failure, in order to reduce failure. It uses all the available experience and expertise, from marketing, design, technology, purchasing, production/operation, distribution, service, etc., to identify the importance levels or criticality of potential problems and stimulate action to reduce these levels. FMECA should be a major consideration at the design stage of a product or service.

The elements of a complete FMECA are:

- *Failure mode* – the anticipated conditions of operation are used as the background to study the most probably failure mode, location and mechanism of the product or system and its components.
- *Failure effect* – the potential failures are studied to determine their probable effects on the performance of the whole product, process or service, and the effects of the various components on each other.
- *Failure criticality* – the potential failures on the various parts of the product or service system are examined to determine the severity of each failure effect in terms of lowering of performance, safety hazard, total loss of function, etc.

FMECA may be applied to any stage of design, development, production/operation or use, but since its main aim is to prevent failure, it is most suitably applied at the design stage to identify and eliminate causes. With more complex product or service systems, it may be appropriate to consider these as smaller units or sub-systems, each one being the subject of a separate FMECA.

Special FMECA pro-formas and software are available and they set out the steps of the analysis as follows:

1. Identify the product or system components, or process function.
2. List all possible failure modes of each component.
3. Set down the effects that each mode of failure would have on the function of the product or system.
4. List all the possible causes of each failure mode.
5. Assess numerically the failure modes on a scale from 1 to 10. Experience and reliability data should be used, together with judgement, to determine the values, on a scale 1–10, for:

   P the probability of each failure mode occurring (1 = low, 10 = high).
   S the seriousness or criticality of the failure (1 = low, 10 = high).
   D the difficulty of detecting the failure before the product or service is used by the consumer 1 = easy, 10 = very difficult). See Table 6.4.

Design for quality   107

Table 6.4 Probability and seriousness of failure and difficulty of detection

Value	1	2	3	4	5	6	7	8	9	10
P	low chance of occurrence								almost certain to occur	
S	not serious, minor nuisance								total failure, safety hazard	
D	easily detected								unlikely to be detected	

6. Calculate the product of the ratings, C = P x S x D, known as the criticality index or risk priority number (RPN) for each failure mode. This indicates the relative priority of each mode in the failure prevention activities.

7. Indicate briefly the corrective action required and, if possible, which person, department/function or group is responsible and the expected completion date.

When the criticality index has been calculated, the failures may be ranked accordingly. It is usually advisable, therefore, to determine the value of C for each failure mode before completing the last columns. In this way the action required against each item can be judged in the light of the ranked severity and the resources available.

Moments of truth

Moments of truth (MoT) is a concept that has much in common with FMEA. The idea was created by Jan Carlzon, CEO of Scandinavian Airlines (SAS) and was made popular by Albrecht and Zemke. An MoT is the moment in time when a customer comes into contact with the people, systems, procedures, or products of an organization, which leads to the customer making a judgement about the quality of the organization's services or products.

In MoT analysis the points of potential dissatisfaction are identified proactively, beginning with the assembly of process flow chart type diagrams. Every small step taken by a customer in his/her dealings with the organization's people, products or services is recorded. It may be difficult or impossible to identify all the MoTs, but the systematic approach should lead to a minimalization of the number of the number and severity of unexpected failures, and this provides the link with FMEA.

THE LINKS BETWEEN GOOD DESIGN AND MANAGING THE BUSINESS

Research carried out by the European Centre *for* Business Excellence has led to a series of specific aspects that should be addressed to integrate design into the business or organization. These are presented under various business criteria below.

Leadership and management style

- 'Listening' is designed into the organization.
- Management communicates the importance of good design in good partnerships and vice versa.

- A management style is adopted that fosters innovation and creativity, and that motivates employees to work together effectively, whilst avoiding the 'fear of failure'.

Customers, strategy and planning

- The customer is designed into the organization as a focus to shape policy and strategy decisions.
- Designers and customers communicate directly.
- Customers are included in the design process.
- Customers are helped to articulate and participate in the understanding of their own requirements.
- Systems are in place to ensure that the changing needs of the customers inform changes to policy and strategy.
- Design and innovation performance measures are incorporated into policy and strategy reviews.
- The design process responds quickly to customers.

People – their management and satisfaction

- People are encouraged to gain a holistic view of design and product/service creation within the organization.
- There is commitment to design teams and their motivation, particularly in cross-functional teamwork (e.g. Quality Function Deployment teams).
- Training programmes are designed, with respect to design, in terms of people skills training (e.g. interpersonal, management teamwork) and technical training (e.g. resources, software).
- Training helps integrate design activities into the business.
- Training impacts on design (e.g. honing creativity and keeping people up to date with design concepts and activity).
- Design activities are communicated (including new product or service concepts).
- Job satisfaction is harnessed to foster good design.
- The results of employee surveys are fed back into the design process.

Resource management

- Knowledge is managed proactively, including investment in technology.
- Information is shared effectively and efficiently in the organization.
- Past experience and learning is captured from design projects and staff.
- Information resources are available for planning design projects.
- Suppliers contribute to innovation, creativity and design concepts.
- Concurrent engineering and design is integrated through the supply chains.

Process management

- Design is placed at the centre of process planning to integrate different functions within the organization and form partnerships outside the organization.

- 'Process thinking' is used to resolve design problems and foster teamwork within the organization and with external partners.
- An effective and efficient 'end-to-end' Product/Service Creation Process (P/SCP) is developed and deployed with full engagement of all relevant functions, including its improvement.

Impact on society and business performance

- Consideration is given to how the design of a product or service impacts on:
  - the environment
  - the recyclability and disposal of materials
  - packaging and wastage of resources
  - the (local) economy (e.g. reduction of labour requirements).
  - There is understanding of the impact of design on the business results, both financial and non-financial.

This same research showed that strong links exist between good design and proactive flexible deployment of business policies and strategies. These can be used to further improve design by encouraging the sharing of best practice within and across industries, by allowing designers and customers to communicate directly, by instigating new product/service introduction policies, project audits and design/innovation measurement policies and by communicating the strategy to employees. The findings of this work may be summarized by thinking in terms of the 'value chain', as shown in Figure 6.7.

Effective people management skills are essential for good design – these include the ability to listen and communicate, to motivate employees and encourage

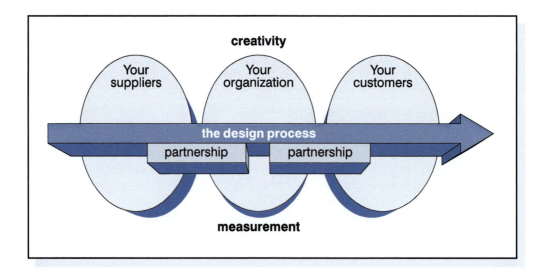

Figure 6.7
The value chain and design process

teamwork, as well as the ability to create an organizational climate which is conducive to creativity and continuous innovation.

The only way to ensure that design actively contributes to business performance is to make sure it happens 'by design', rather than by accident. In short, it needs co-ordinating and managing right across the organization.

BIBLIOGRAPHY

Fox, J., *Quality Through Design*, MGLR, 1993.

Juran, J.J., *Juran on Quality by Design*, Free Press, New York, 1992.

Marsh, S., Moran, J., Nakui, S. and Hoffherr, G.D., *Facilitating and Training in QFD*, ASQ, Milwaukee, 1991.

Susskind, R., *The End of Lawyers – Rethinking the Nature of Legal Services*, Oxford University Press, Oxford, 2008.

Zeithaml, V.A., Parasuraman, A and Berry, L.L., *Delivering Quality Service: Balancing Customer Perceptions and Expectations*, The Free Press (Macmillan), New York, 1990.

CHAPTER HIGHLIGHTS

Innovation, design and improvement

- Design is a multifaceted activity which covers many aspects of an organization.
- All businesses need to update their products, processes and services.
- Innovation entails both invention and design, *and* continuous improvement of existing products, services and processes.
- Leading product/service innovations are market-led, not marketing-led.
- Everything in or from an organization results from design decisions.
- Design in an on-going activity, dynamic not static, a verb not a noun – design is a process.

The design process

- Commitment at the top is required to building in quality throughout the design process. Moreover, the operational processes must be capable of achieving the design.
- State-of-the-art approach to innovation is based on a strategic balance of old and new, top management approach to design and teamwork.
- The 'styling' of products must also be matched by secondary design considerations, such as operating instructions and software support.
- Designing takes in all aspects of identifying the need, developing something to satisfy the need, checking conformance to the need and ensuring the need is satisfied.
- The design process must be carefully managed and can be flow-charted, like any other process, into: planning, practice codes, procedures, activities assignments, identification of organizational and technical interfaces and design input

Design for quality 111

requirements, review investigation and evaluation of new techniques and materials, and use of feedback data from previous designs.

- Total design or 'simultaneous engineering' is similar to quality function deployment and uses multifunction teams to provide an integrated approach to product or service introduction.

Quality function deployment (QFD) – the house of quality

- The 'house of quality' is the framework of the approach to design management known as quality function deployment (QFD). It provides structure to the design and development cycle, which is driven by customer needs rather than innovation in technology.
- QFD is a system for designing a product or service, based on customer demands, and bringing in all members of the supplier organization.
- A QFD team's purpose is to take the needs of the market and translate them into such a form that they can be satisfied within the operating unit.
- The QFD team answers the following question. **WHO** are the customers? **WHAT** do the customers need? **HOW** will the needs be satisfied?
- The answers to the **WHO, WHAT** and **HOW** questions are entered into the QFD matrix or quality table, one of the seven new tools of planning and design.

Specifications and standards

- There is a strong relation between standardization and specifications. If standards are used correctly, the process of drawing up specifications should provide opportunities to learn more about innovations and change standards accordingly.
- The aim of specifications should be to reflect the true requirements of the product/service that are capable of being achieved.

Quality design in the service sector

- In the design of services three distinct elements may be recognized in the service package: physical (facilitating goods), explicit service (sensual benefits) and implicit service (psychological benefits). Moreover, the characteristics of service delivery may be itemised as intangibility, simultaneity and heterogeneity.
- Services may be classified generally as service factory, service shop, mass service, professional service and personal service. The service attributes that are important in designing services include labour intensity, contact interaction, customerization, nature of service act, and the direct recipient of the act. Use of this framework allows services to be grouped under the five classifications.
- Parasuraman's five dimensions of service quality (reliability, responsiveness, assurance, empathy and tangibles) are a very useful framework for assessing service quality weaknesses and for benchmarking service quality.
- A standard set of survey questions have been developed and validated and hence they provide a mature and well-tried performance measurement and benchmarking tool for the services sector.
- Lawyers and other professions are in fundamental transformation with conventional legal advisers much less prominent in the future; there are two major forces that are

shaping and characterizing legal services: market pull towards commoditization and pervasive development & uptake of information technology.

- The evolution of the legal services portfolio is now like a conveyor belt that moves through bespoke, standardized, systemized, packaged and commoditized, although some organizations are polarizing toward the ends

Failure mode, effect and criticality analysis (FMECA)

- FMEA is the study of potential product, service or process failures and their effects. When the results are ranked in order of criticality, the approach is called FMECA. Its aim is to reduce the probability of failure.
- The elements of a complete FMECA are to study failure mode, effect and criticality. It may be applied at any stage of design, development, production/operation or use.
- Moments of truth (MoT) is a similar concept to FMEA. It refers to the moments in time when customers first come into contact with an organization, leading to judgements about quality.

The links between good design and managing the business

- Research has led to a series of specific aspects to address in order to integrate design into an organization.
- The aspects may be summarized under the headings of: leadership and management style; customers, strategy and planning; people – their management and satisfaction; resource management; process management; impact on society and business performance.
- The research shows that strong links exist between good design and proactive flexible deployment of business policies and strategies – design needs co-ordinating and managing right across the organization.

Part II Discussion questions

1. Describe the key stages of integrating total quality into the strategy of an organization of your choice.
2. Explain the difference between the 'Whats' and the 'Hows' of a company in the utilities sector. Identify likely critical success factors for such an organization and list possible key performance indicators for each one.
3. A legal firm is concerned about the changes taking place in its sector and believes that the market is 'polarising' into a high value – high fees end and a 'quick and dirty' – cheaper service associated with lower quality. Prepare a presentation to the senior management of the firm which provides an alternative view and show them how they could look at developing their business accordingly in the future.
4. You are the manager of a busy insurance office. Last year's abnormal winter gales led to an exceptionally high level of insurance claims for house damage caused by strong winds, and you had considerable problems in coping with the greatly increased work load. The result was excessively long delays in both acknowledging and settling customers' claims. Your area manager has asked you to outline a plan for dealing with such a situation should it arise again. The plan should identify what actions you would take to deal with the work, and what, if anything, should be done now to enable you to take those actions should the need arise. What proposals would you make, and why?
5. Discuss the preparations required for the negotiation of a one-year contract with a major material supplier. What are the major factors to consider in partnering with key suppliers?
6. Imagine that you are the chief executive or equivalent in a company which has operations across the globe, and that you plan to introduce best practice supply chain management into the organization.
   a) Prepare a briefing of your senior managers, which should include your assessment of the aims, objectives and benefits to be gained from the approach.
   b) Outline the steps you would take to develop an excellent supply chain and explain how you would attempt to ensure its success.
7. You are a management consultant with particular expertise in the area of product and service design and development. You are at present working on projects for four firms:
   a) a chain of hotels
   b) an internet order goods firm
   c) a furniture manufacturer
   d) a road construction contractor.
   What factors do you consider are important generally in your area of specialization? Compare and contrast how these factors apply to your four current projects.
8. Discuss the application of quality function deployment (QFD) and the 'house of quality' in a fast moving consumer good (fmcg) company which designs and produces personal products.
9. Explain in full how Failure Mode Effect and Criticality Analysis (FMECA) can help a company improve customer satisfaction. Describe in detail the method and consider some of the barriers that may arise.

10. Discuss the application of 'Hoshin Kanri' as a strategic planning methodology. Include a presentation to the senior management of an organization in the service sector showing how the approach could add value and create a platform for performance improvement, paying particular attention to the nature of services and the customer-supplier interfaces.

Part III

Performance

All words, and no performance!

Philip Massinger, 1583–1640,
from 'Parliament of Love', *c.* 1619

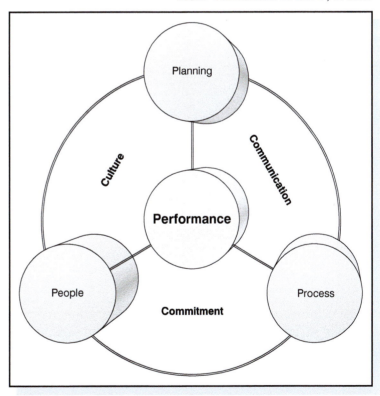

Chapter 7

Performance measurement frameworks

PERFORMANCE MEASUREMENT AND THE IMPROVEMENT CYCLE

Traditionally, performance measures and indicators have been derived from cost-accounting information, often based on outdated and arbitrary principles. These provide little motivation to support attempts to introduce TQM and, in some cases, actually inhibit continuous improvement because they are unable to map process performance. In the organization that is to succeed over the long term, performance must be measured by the improvements seen by the customer.

In the cycle of never ending improvement, measurement plays an important role in:

• Tracking progress against organizational goals.
• Identifying opportunities for improvement.
• Comparing performance against internal standards.
• Comparing performance against external standards.

Measures are used in *process control*, e.g. control charts (see Chapter 13), and in *performance improvement*, e.g. improvement teams (see Chapters 14 and 15), so they should give information about how well processes and people are doing and motivate them to perform better in the future.

The author and his colleagues have seen many examples of so-called performance measurement systems that frustrated improvement efforts. Various problems include systems that:

1. Produce irrelevant or misleading information.
2. Track performance in single, isolated dimensions.
3. Generate financial measures too late, e.g. quarterly, for mid-course corrections or remedial action.

4. Do not take account of the customer perspective, both internal and external.
5. Distort management's understanding of how effective the organization has been in implementing its strategy.
6. Promote behaviour that *undermines* the achievement of the strategic objectives.

Typical potentially harmful summary measures of local performance are purchase price, machine or plant efficiencies, direct labour costs and ratios of direct to indirect labour. In many sectors primary measures used are cost against budget and actual time taken against scheduled time. These can be incompatible with quality and productivity improvement measures because they tend not to provide feedback that will motivate improvement. Measures such as process and throughput times, supply chain performance, inventory reductions and increases in flexibility, which are first and foremost *non-financial*, can do that. Financial summaries provide valuable information, of course, but they should not necessarily be used for control. Effective decision-making requires direct measures for operational feedback and improvement.

One example of a 'measure' with these shortcomings is return on investment (ROI). ROI can be computed only after profits have been totalled for a given period. It was designed therefore as a single-period, long-term measure, but it is often used as a short-term one. Perhaps this is because most executive bonus 'packages' in the West are based on short-term measures. ROI tells us what happened, not what is happening or what will happen, and, for complex and detailed projects, ROI is often inaccurate and irrelevant.

Many managers have a poor or incomplete understanding of their processes and products or services and, looking for an alternative stimulus, become interested in financial indicators. The use of ROI, for example, for evaluating strategic requirements and performance can lead to a discriminatory allocation of resources. In many ways the financial indicators used in many organizations have remained static while the environment in which they operate has changed dramatically.

Traditionally, the measures used have not been linked to the processes where the value-adding activities take place. What has been missing is a performance measurement framework that provides feedback to people in all areas of business operations. Of course, TQM stresses the need to start with the process for fulfilling customer needs.

The critical elements of a good performance measurement framework (PMF) are:

- Leadership and commitment.
- Full employee involvement.
- Good planning.
- Sound implementation strategy.
- Measurement and evaluation.
- Control and improvement.
- Achieving and maintaining standards of excellence.

The Deming cycle of continuous improvement – Plan, Do, Check, Act – clearly requires measurement to drive it, and yet it is a useful design aid for the measurement system itself:

- PLAN: establish performance objective and standards.
- DO: measure actual performance.

- CHECK: compare actual performance with the objectives and standards – determine the gap.
- ACT: take the necessary actions to close the gap and make the necessary improvements.

Before we use performance measurement in the improvement cycle, however, we should attempt to answer four basic questions:

1. Why measure?
2. What to measure?
3. Where to measure?
4. How to measure?

Why measure?

It has been said often that it is not possible to manage what cannot be measured. Whether this is strictly true or not there are clear arguments for measuring. In a quality-driven, never-ending improvement environment, the following are some of the main reasons *why measurement is needed* and why it plays a key role in quality and productivity improvement:

- To ensure customer requirements *have* been met.
- To be able to set sensible *objectives* and comply with them.
- To provide *standards* for establishing comparisons.
- To provide *visibility* and provide a 'balanced scoreboard' for people to *monitor* their own performance levels.
- To highlight *quality problems* and determine which areas require *priority attention*.
- To give an indication of the *costs of poor quality*.
- To justify the *use of resources*.
- To provide *feedback* for driving the improvement effort.

It is also important to know the impact of TQM on improvements in business performance, on sustaining current performance and perhaps on reducing any decline in performance. In the construction environment, for example, there is a need to develop performance measurement frameworks for projects as well as for enterprises. This is also important at the process level, both in design and production or service delivery that are to be targeted for improvement.

What to measure?

A good start-point for deciding what to measure is to look at what are the key goals of senior management; what problems need to be solved; what opportunities are there to be taken advantage of; and what customers perceive to be the key ingredients that influence their satisfaction. In the case studies there are numerous examples of performance measurement in different areas of enterprise and process management. These examples reflect the primary business goals of senior management in each case and they embrace the entire spectrum of issues that are addressed in a company's core values.

In the business of process improvement, process understanding, definition, measurement and management are tied inextricably together. In order to assess and evaluate performance accurately, appropriate measurement must be designed, developed and maintained by people who *own* the processes concerned. They may find it necessary to measure effectiveness, efficiency, quality, impact, timeliness and productivity. In these areas there are many types of measurement, including direct output or input figures, the cost of poor quality, economic data, comments and complaints from customers, information from customer or employee surveys, etc., generally continuous variable measures (such as time or weight) or discrete attribute measures (such as absentees or defectives).

No one can provide a generic list of what should be measured but once it has been decided in any one organization what measures are appropriate, they may be converted into indicators. These include ratios, scales, rankings and financial and time-based indicators. Whichever measures and indicators are used by the process owners, they must reflect the true performance of the process in customer/supplier terms and emphasize continuous improvement. Time-related measures and indicators have great value, especially in the elimination of non-value adding activities – the so-called 'Lean' systems and approaches (see Chapter 15).

Where to measure?

If true measures of the effectiveness of TQM are to be obtained, there are three components that must be examined – the human, technical and business components.

The human component is clearly of major importance and the key tests are that wherever measures are used they must be:

1. *transparent* – understood by all the people being measured
2. *non-controversial* – accepted by the individuals concerned
3. *internally consistent* – compatible with the rewards and recognition systems
4. *objective* – designed to offer minimal opportunity for manipulation
5. *motivational* – trigger a response to improve outcomes.

Technically, the measures should be the ones that truly represent the controllable aspects of the processes at their point of operation, rather than simple output measures that cannot be related to process management. They must also be correct, precise and accurate.

The business component requires that the measures are objective, timely and result-oriented, and above all they must mean something to those working in and around the process, *including the customers.*

How to measure?

Measurement, as any other management system, requires the stages of design, analysis, development, evaluation, implementation and review. The system should be designed to measure *progress*, otherwise it will not engage the improvement cycle. Progress is important in five main areas: effectiveness, efficiency, productivity, quality and impact.

Effectiveness

Effectiveness may be defined as the percentage actual output over the expected output:

$$\text{Effectiveness} = \frac{\text{Actual output}}{\text{Expected output}} \times 100 \text{ per cent}$$

Effectiveness then looks at the *output* side of the process and is about the implementation of the objectives – doing what you said you would do. Effectiveness measures should reflect whether the organization, group or process owner(s) are achieving the desired results, accomplishing the right things. Measures of this may include:

- Quality, e.g. a grade of product, or a level of service.
- Quantity, e.g. tonnes, lots, bedrooms cleaned, accounts opened.
- Timeliness, e.g. speed of response, product lead times, cycle time.
- Cost/price, e.g. unit costs.

Efficiency

Efficiency is concerned with the percentage resource actually used over the resources that were planned to be used:

$$\text{Efficiency} = \frac{\text{Resources actually used}}{\text{Resources planned to be used}} \times 100 \text{ per cent}$$

Clearly, this is a process *input* issue and measures the performance of the process system management. It is, of course, possible to use resources 'efficiently' while being *ineffective*, so performance efficiency improvement must be related to certain output objectives.

All process inputs may be subjected to efficiency measurement, so we may use labour/staff efficiency, equipment efficiency (or utilization), materials efficiency, information efficiency, etc. Inventory data and throughput times are often used in efficiency and productivity ratios.

Productivity

Productivity measures should be designed to relate the process outputs to its inputs:

$$\text{Expected productivity} = \frac{\text{Expected output}}{\text{Resources expected to be consumed}}$$

and this may be quoted as expected or actual productivity:

$$\text{Actual productivity} = \frac{\text{Actual output}}{\text{Resources actually consumed}}$$

There is a vast literature on productivity and its measurement, but simple ratios such as tonnes per man-hour (expected and actual), sales output per telephone operator-day, and many others like this are in use. Productivity measures may be developed for each combination of inputs, e.g. sales/all employee costs.

Quality

This has been defined elsewhere of course (see Chapter 1). The *non-quality* related measures include the simple counts of defect or error rates (perhaps in parts per million, numbers per square meter or per thousand dollars spent), percentage outside specification or Cp/Cpk values, deliveries not on time or, more generally, as the costs of poor quality. When the positive costs of prevention of poor quality are included, these provide a balanced measure of the costs of quality (see next section).

The quality measures should also indicate positively whether we are doing a good job in terms of customer satisfaction, implementing the objectives and whether the designs, systems and solutions to problems are meeting the requirements. These really are 'voice-of-the-customer' measures.

Impact and value added

Impact measures should lead to key performance indicators for the business or organization, including monitoring improvement over time. Value-added management (VAM) requires the identification and elimination of all non-value-adding wastes, including time, often through the use of 'Lean' approaches. Value added is simply the volume of sales (or other measure of 'turnover') minus the total input costs, and provides a good direct measure of the impact of the improvement process on the performance of the business. A related ratio, percentage return on value added (ROVA) is another financial indicator that may be used:

$$\text{ROVA} = \frac{\text{Net profits before tax}}{\text{Value added}} \times 100 \text{ per cent}$$

Other measures or indicators of impact on the business are *growth* in sales, assets, numbers of passengers/students, etc., and *asset-utilization* measures, such as return on investment (ROI) or capital employed (ROCE), earnings per share, etc.

Some of the impact measures may be converted to people productivity ratios, e.g.:

$$\frac{\text{Value added}}{\text{Number of employees (or employee costs)}}$$

Activity-based costing (ABC) is an information system that maintains and processes data on an organization's activities and cost objectives. It is based on the activities performed being identified and the costs being traced to them. ABC uses various 'cost drivers' to trace the cost of activities to the cost of the products or services. The activity and cost-driver concepts are the heart of ABC. Cost drivers reflect the demands placed on activities by products, services or other cost targets. Activities are processes or procedures that cause work and thereby consume resources. This is clearly capable of measuring impact, both on and by the organization.

Following initial enthusiasm, ABC lost ground in the 1990s to alternative metrics, such as Kaplan and Norton's balanced scorecard and economic value add. Some authors have concluded that manually driven ABC was an inefficient use of resources, expensive and difficult to implement for small gains, poor value and that alternative methods should be used.

COSTS OF QUALITY

Manufacturing a quality product, providing a quality service, or doing a quality job – one with a high degree of customer satisfaction – is not enough. The cost of achieving these goals must be carefully managed, so that the long-term effect on the business or organization is a desirable one. These costs are a true measure of the quality effort. A competitive product or service based on a balance between quality and cost factors is the principal goal of responsible management and may be aided by a competent analysis of the costs of quality (COQ).

The analysis of quality related costs is a significant management tool that provides:

- A method of assessing the effectiveness of the management of quality.
- A means of determining problem areas, opportunities, savings and action priorities.

The costs of quality are no different from any other costs. Like the costs of design, sales, production/operations, maintenance and other activities, they can be budgeted, measured and analysed. Having said this, a major difficulty in some sectors is capturing the totality of the costs. Where value-adding processes are fragmented with many parties incurring costs, such as in construction, infrastructure or health care industries, unless costs are to be recovered from another party, it can be difficult to get people interested in recording the costs. Yet a detailed knowledge of these costs is potentially a main driver for improvement and the author and his colleagues have worked recently with different types of organization to develop frameworks that effectively capture the significant costs.

Having specified the quality of design, the operating units have the task of matching it. The necessary activities will incur costs that may be separated into prevention costs, appraisal costs and failure costs, the so-called P-A-F model first presented by Feigenbaum. Failure costs can be further split into those resulting from internal and external failure.

Prevention costs

These are associated with the design, implementation and maintenance of the quality management system. Prevention costs are planned and are incurred before actual operation. Prevention includes:

Product or service requirements
The determination of requirements and the setting of corresponding specifications (which also takes account of process capability) for incoming materials, processes, intermediates, finished products and services.

Quality planning
The creation of quality, reliability, and operational, production, supervision, process control, inspection and other special plans, e.g. pre-production trials, required to achieve the quality objective.

Quality assurance
The creation and maintenance of the quality management systems.

Inspection equipment
The design, development and/or purchase of equipment for use in inspection work.

Training

The development, preparation and maintenance of training programmes for operators, supervisors, staff and managers both to achieve and maintain capability.

Miscellaneous

Clerical, travel, supply, shipping, communications and other general office management activities associated with quality.

Resources devoted to prevention give rise to the *'costs of doing it right the first time'*.

Appraisal costs

These costs are associated with the supplier's and customer's evaluation of purchased materials, processes, intermediates, products and services to assure conformance with the specified requirements. Appraisal includes:

Verification

Checking of incoming material, process set-up, first-offs, running processes, intermediates and final products, including product or service performance appraisal against agreed specifications.

Quality audits

To check that the quality management systems are functioning satisfactorily.

Inspection equipment

The calibration and maintenance of equipment used in all inspection activities.

Supply chain and vendor rating

The assessment and approval of all suppliers, of both products and services.

Appraisal activities result in the *'costs of checking it is right'*.

Internal failure costs

These costs occur when the results of work fail to reach designed quality standards and are detected before transfer to the customer takes place. Internal failure includes the following:

Scrap

Defective products or materials that cannot be repaired, used or sold.

Rework or rectification

The correction of defective material or errors to meet the requirements.

Re-inspection

The re-examination of products or work that have been rectified

Downgrading

A product that is usable but does not meet specifications may be downgraded and sold as 'second quality' at a different price.

Failure analysis

The activity required to establish the causes of internal product or service failure.

External failure costs

These costs occur when products or services fail to reach design quality standards but are not detected until after transfer to the consumer. External failure includes:

Repair and servicing

Either of returned products or those in the field. The repair of defective work in sectors like construction has to be done on site and the costs can be very high proportionately.

Warranty claims

Failed products that are replaced or services re-performed under some form of guarantee or promise. The labour costs involved in 'replacement' can have a greater value than that of the defective product or service!

Complaints

All work and costs associated with handling and servicing of customers' complaints.

Returns and recalls

The handling and investigation of rejected or recalled products or materials including transport costs. In recent times there have examples of huge effect on companies and customers of recalled products – from automotive through to food. The reputational impact of such occurrences can threaten the very existence of the business and even affect the industry as a whole.

Liability

The result of product or service liability litigation and other claims, which may include a change of contract.

Loss of good will

The impact on reputation and image, which impinges directly on future prospects for sales – see comments above on recalls.

External and internal failure produce the 'costs of getting it wrong'.

Order re-entry, conflict, unnecessary travel, transport and telephone calls are just a few examples of the wastage or failure costs often excluded. Every organization should be aware of the costs of getting it wrong, and management needs to obtain some idea how much failure is costing each year.

The classification of cost elements given in this section may be used to interrogate any internal transformation process. Using the internal customer requirements concept as the standard for failure, the cost assessments can be made wherever information, data, materials, service or artefacts are transferred from one person or one department to another. It is the 'internal' costs of lack of quality that lead to the claim that approximately one-third of *all* our efforts are wasted.

The relationship between the quality related costs of prevention, appraisal and failure and increasing quality awareness and improvement in the organization is shown in Figure 7.1. Where the quality awareness is low the total quality related costs

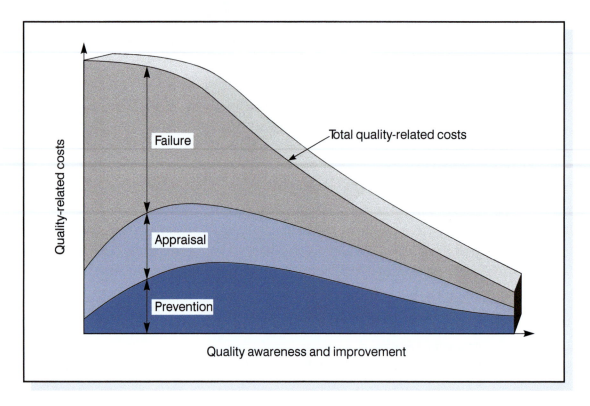

Figure 7.1
Increasing quality awareness and improvement activities (Source: British Standard BS 6143)

are high, the failure costs predominating. As awareness of the cost to the organization of failure gets off the ground, through initial investment in training, an increase in appraisal costs usually results. As the increased appraisal leads to investigations and further awareness, further investment in prevention is made to improve design features, processes and systems. As the preventive action takes effect, the failure *and* appraisal costs fall and the total costs reduce.

The first presentations of the P-A-F model suggested that there may be an optimum operating level at which the combined costs are at the minimum. The author and his colleagues, however, have not yet found one organization in which the total costs have risen following investment in prevention.

Levels of quality costs found

The true and total cost to an organization from a product or service failure can reverberate around that organization and, in some cases, have a devastating long-term effect, both internally and externally. Because the true and total costs of failure are typically either poorly measured, or not measured at all, the effectiveness of quality and operational management is not examined and opportunities to identify priority problem areas and realize significant savings are often missed.

The European Centre *for* Business Excellence (the Research and Education Division of Oakland Consulting) conducted a study into the costs of failure in commercial and public sector organizations, the key findings of which included:

Product organizations – costs of failure

- Nearly 1 in 3 of the organizations that were able to provide figures were spending more than 20 per cent of their turnover on costs of failure, with 10 per cent of the organizations reporting spending over 30 per cent.
- These high costs were in spite of the fact that, on average, only 40 per cent of the failure cost categories were being measured.
- Only 50 per cent of organizations measured the costs of re-inspecting/ re-testing products and dealing with customer complaints and product returns.
- CoQ is more than cost – poor quality impacts customers, damages reputation and distracts management.
- World-leading product manufacturers achieve 5–10 per cent as their CoQ.

Service organizations – costs of failure

- 25 per cent of the service organizations that responded to the survey did not measure any of the failure cost categories at all.
- The total costs of failure for the service organizations that had information available ranged from 2–19 per cent of annual spend.
- The majority of service organizations did not measure the cost of reviewing and auditing their supply chain, IT systems failures, complete service breakdowns or dealing with customer complaints.

The drivers of CoQ are well understood for both manufacturing and service-based organizations and are presented in Figure 7.2. Clearly, both product and service organizations have extensive opportunities to realize significant savings of their turnover or annual spend by preventing quality failures which will translate directly to improved bottom line performance (Figure 7.3). There is growing evidence of an association between lower CoQ and high levels of business maturity.

THE PROCESS MODEL FOR QUALITY COSTING

The P-A-F model for quality costing has a number of drawbacks. In TQM, prevention of problems, defects, errors, waste, etc., is one of the prime functions, but it can be argued that everything a well-managed organization does is directed at preventing quality problems. This makes separation of *prevention costs* very difficult. There are clearly a range of prevention activities in any organization that are integral to ensuring quality but may never be included in the schedule of quality related costs.

It may be impossible and unnecessary to categorize costs into the three categories of P-A-F. For example, a design review may be considered a prevention cost, an appraisal cost or even a failure cost, depending on how and where it is used in the process. Another criticism of the P-A-F model is that it focuses attention on cost reduction and plays down, or in some cases even ignores, the positive contribution made to price and sales volume by improved quality.

Prevention	Appraisal	Internal failure	External failure
Preventive maintenance costs • People hours • Materials costs • Improvements to processing equipment	Inspection costs • Hours spent on QC inspection and test • Hours spent on operator self/peer group inspections	NCR data • Cost of time and materials to repair or rework products by product group • Confirmed complaints	Warranty • Warranty provision • Warranty claims
Training costs • Internal trainer hours • External training costs • Hours spent in training	External inspections • Hours to support customer inspection visits • Factory costs to support visits • Third party examinations	Scrapped product • Materials and labour to point of scrapping including allowance for added value from previous operations	Customer related – direct • Refunds • Product recalls • Excess carriage to replace products
Quality system • Requirements analysis and development of clear specifications • Development and maintenance of clear internal documentation	Checking of incoming raw materials • Hours spent • Testing materials used	Waste above the expected yield from the process • Materials • Time	Customer related – indirect • Loss of reputation and repeat business • Loss of goodwill
Investment in improvement projects including defect investigation and root cause analysis • Hours spent • Costs of materials used	Quality system • Product design reviews • Hours spent on internal audits	Costs of rejected incoming materials – including factory downtime if this is an effect	Technical product support in the field (or via call centre) to answer questions and respond to incorrect user perceptions

Figure 7.2
The drivers of CoQ

The most serious criticism of the original P-A-F model presented by Feigenbaum and used in, for example, British Standard 6143 (1981) 'Guide to the determination and use of quality related costs', is that it implies an acceptable 'optimum' quality level above which there is a trade-off between investment in prevention and failure costs. Clearly, this is not in tune with the never-ending improvement philosophy of TQM. The key focus of TQM is on process improvement, and a cost categorization scheme that does not consider process costs, such as the P-A-F model, has limitations. (BS6143-2 was re-published in 1990 as 'Guide to the economies of quality: prevention, appraisal and failure model'.)

In a total quality related costs system that focuses on processes rather than products or services, the operating costs of generating customer satisfaction will be of prime importance. The so-called 'process cost model', described in the revised

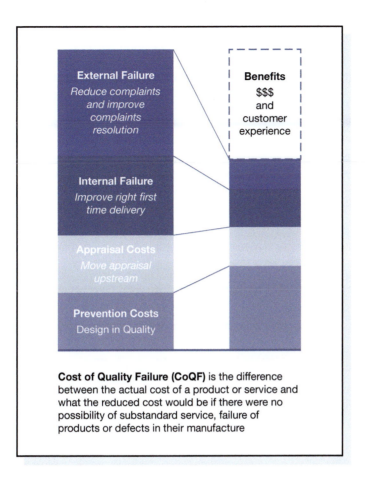

External Failure
Reduce complaints and improve complaints resolution

Benefits
$$$
and
customer
experience

Internal Failure
Improve right first time delivery

Appraisal Costs
Move appraisal upstream

Prevention Costs
Design in Quality

Cost of Quality Failure (CoQF) is the difference between the actual cost of a product or service and what the reduced cost would be if there were no possibility of substandard service, failure of products or defects in their manufacture

Figure 7.3
The costs of quality failure (CoQF)

BS6143-1 (1992), sets out a method for applying quality costing to any process or service. It recognizes the importance of process ownership and measurement, and uses process modelling to simplify classification. The categories of the cost of quality (COQ) have been rationalized into the cost of conformance (COC) and the cost of non-conformance (CONC):

$$COQ = COC + CONC$$

The cost of conformance (COC) is the process cost of providing products or services to the required standards, by a given specified process in the most effective manner, i.e. the cost of the ideal process where every activity is carried out according to the requirements first time, every time. The cost of non-conformance (CONC) is the failure cost associated with the process not being operated to the requirements, or the cost due to variability in the process. Part 2 of BS6143 (1990) still deals with the P-A-F model, but without the 'optimum'/minimum cost theory (see Figure 7.1).

Process cost models can be used for any process within an organization and developed for the process by flowcharting or use of the ICOR (Inputs, Outputs, Controls, Resources) process mapping methodology (see Chapter 10). This will identify the key process steps and the parameters that are monitored in the process.

The process cost elements should then be identified and recorded under the categories of product/service (outputs), and people, systems, plant or equipment, materials, environment, information (inputs). The COC and CONC for each stage of the process will comprise a list of all the parameters monitored.

Steps in process cost modelling

Process cost modelling is a methodology that lends itself to stepwise analysis; while the following example is for the retrieval of medical records, it illustrates the process clearly and could be applied to any routine process in a volume production or service setting within the construction sector. The following are the key stages in building the model.

1. Choose a key process to be analysed, identify and name it, e.g. Retrieval of Medical Records (Acute Admissions).
2. Define the process and its boundaries.
3. Construct an ICOR process diagram, with suppliers and customers:
   a) Identify the outputs and customers (for example see Figure 7.4)
   b) Identify the inputs and suppliers (for example see Figure 7.5)
   c) Identify the controls and resources (for example see Figure 7.6).

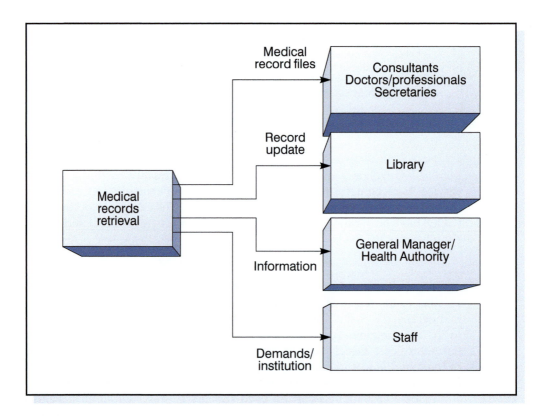

Figure 7.4
Building the model: identify outputs and customers

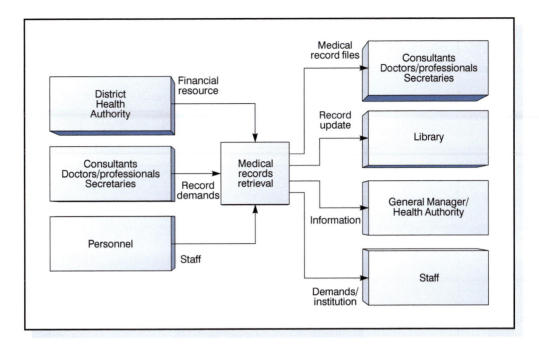

Figure 7.5
Building the model: identify inputs and suppliers

4. Flowchart the process and identify the process owners (for example see Figure 7.7). Note, the process owners will form the improvement team.
5. Allocate the activities as COC or CONC (see Table 7.1).
6. Calculate or estimate the quality costs (COQ) at each stage (COC + CONC). Estimates may be required where the accounting system is unable to generate the necessary information.
7. Construct a process cost report (see Table 7.2). The report summary and results are given in Table 7.3.

There are three further steps carried out by the process owners – the improvement team – which take the process forward into the improvement stage:

8. Prioritize the failure costs and select the process stages for improvement through reduction in costs of non-conformance (CONC). This should indicate any requirements for investment in prevention activities. An excessive cost of conformance (COC) may suggest the need for process re-design.
9. Review the flowchart to identify the scope for reductions in the cost of conformance. Attempts to reduce COC require a thorough process understanding, and a second flowchart of what the new process may help (see Chapter 10).
10. Monitor conformance and non-conformance costs on a regular basis, using the model and review for further improvements.

Performance measurement frameworks 133

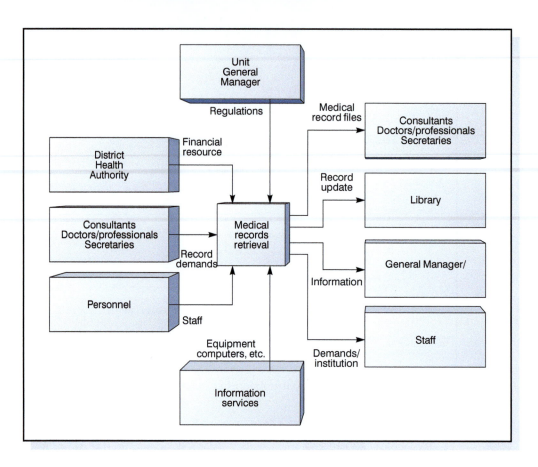

Figure 7.6
Building the model: identify controls and resources

Table 7.1 Building the model: allocate activities as COC or CONC

Key activities	COC	CONC
Search for files	Labour cost incurred finding a record while adhering to standard procedure	Labour cost incurred finding a record while unable to adhere to standard procedure
Make up new files	New patient files	Patients whose original files cannot be located
Rework		Cost of labour and materials for all rework files/records never found as a direct consequence of . . .
Duplication		Cost incurred in duplicating existing files

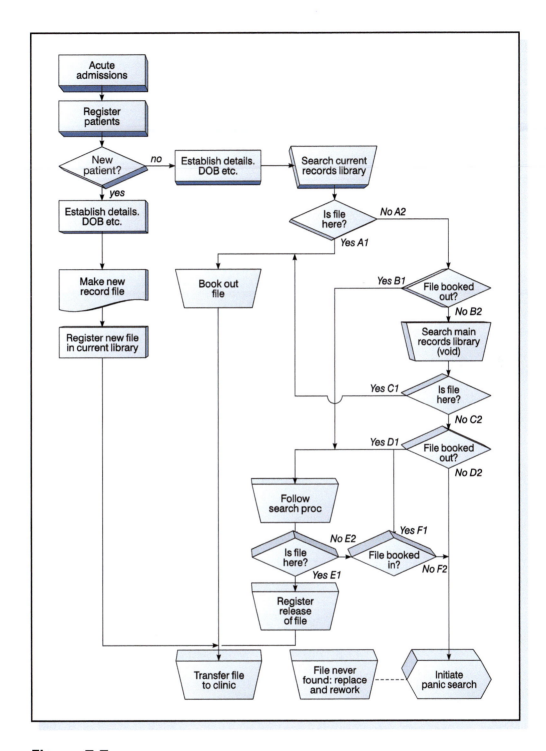

Figure 7.7
Present practice flowchart for acute admissions medical records retrieval

Table 7.2 Building the model: process cost report

Process cost report
Process: medical records retrieval (acute admissions)
Process owner: various

Process COC	Process CONC	Cost details Act	Synth	Definition	Source	
	Labour cost incurred finding records	# ref. Sample		Cost of time required to find missing records	Medical records	£210
	Cost incurred making up replacement files		#	Labour and material costs multiplied by number of files replaced	Medical records	£108
	Rework		#	Labour and material cost of all rework	Medical records	£80
	Duplication		#		Medical records	£24

Time allocation: 4 days (96 hrs)

Table 7.3 Process cost model: report summary

Labour cost
   14 hrs x £12.00/hr = £168
   £168 + overhead and contribution factor 25% = £210

Replacement costs
   No of files unfound 9
   Cost to replace each file £12.00
   Overall cost £108

Rework costs
   2 x Pathology reports to be word processed £80

Duplication costs
   No of files duplicated 2
   Cost per file £12.00
   Overall cost £24

                        TOTAL COST £422
RESULTS
Acute admissions operated 24 hrs/day 365 days/year
This project established a cost of non-conformance of approx. £422
This equates to £422 x 365/4 = £38,507.50
Or two personnel fully employed for 12 months.

The process cost model approach should be seen as more than a simple tool to measure the financial implications of the gap between the actual and potential performance of a process. The emphasis given to the process, improving the understanding, and seeing in detail where the costs occur, should be an integral part of quality improvement. In generating the costs data, the author and his colleagues have found that the involvement of the finance group in the organization is essential.

A PERFORMANCE MEASUREMENT FRAMEWORK (PMF)

A performance measurement framework (PMF) is proposed, based on the strategic planning and process management models outlined in Chapters 4 and 10. The framework has four elements related to: strategy development/goal deployment, process management, individual performance management and review (Figure 7.8). This reflects an amalgamation of the approaches used by a range of organizations in performance measurement.

As we have seen in earlier chapters, the key to strategic planning and goal deployment is the identification of a set of critical success factors (CSFs) and associated key performance indicators (KPIs). These factors should be derived from the organization's vision and mission, and represent a balanced mix of stakeholder issues. Action plans over both the short- and medium-term should be developed, and responsibility clearly assigned for performance. The strategic goals of the organization

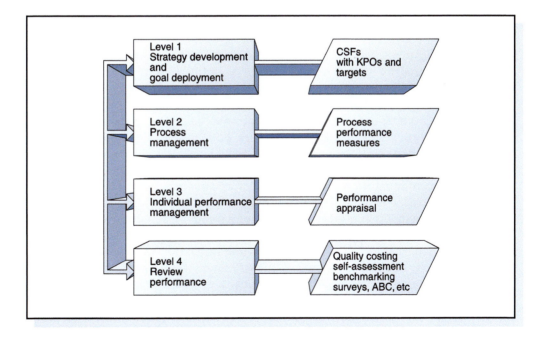

Figure 7.8
Performance measurement framework

should then be clearly communicated to all individuals, and translated into measures of performance at the process/functional/individual level.

The key to successful performance measurement at the process level is the identification and translation of customer requirements and strategic objectives into an integrated set of process performance measures. The documentation and management of processes has been found to be vital in this translation process. Even when a functional organization is retained, it is necessary to treat the measurement of performance between departments as the measurement of customer-supplier performance.

Performance measurement at the individual level usually relies on performance appraisal, i.e. formal planned performance reviews, and performance management, i.e. day-to-day management of individuals. A major drawback with some performance appraisal systems, of course, is the lack of their integration with other aspects of performance measurement.

Performance review techniques are used by many world-class organizations to identify improvement opportunities and to motivate performance improvement. These companies typically use a wide range of such techniques and are innovative in performance measurement in their drive for continuous improvement.

The links between performance measurement at the four levels of the framework are based on the need for measurement to be part of a systematic process of continuous improvement, rather than for 'control'. The framework provides for the development and use of measurement, rather than prescriptive lists of measures that should be used. It is, therefore, applicable in all types of organization.

The elements of the performance measurement are distinct from the budgetary control process, and also from the informal control systems used within organizations. Having said that performance measurement should not be treated as a separate isolated system. Instead measurement is documented as and when it is used at the organizational, process and individual levels. In this way it can facilitate the alignment of the goals of all individuals, teams, departments and processes with the strategic aims of the organization and incorporate the voice of the stakeholders in all planning and management activities.

A number of factors have been found to be critical to the success of performance measurement systems. These factors include the level of top management support for non-financial performance measures, the identification of the vital few measures, the involvement of all individuals in the development of performance measurement, the clear communication of strategic objectives, the inclusion of customers and suppliers in the measurement process and the identification of the key drivers of performance. These factors will need to be taken into account by managers wishing to develop a new performance measurement system, or refine an existing one.

In most world-class organizations there are no separate performance measurement systems. Instead, performance measurement forms part of wider organizational management processes. Although elements of measurement can be identified at many different points within organizations, measurement itself usually forms the 'check' stage of the continuous improvement PDCA cycle. This is important since measurement data that is collected but not acted upon in some way is clearly a waste of resources.

The four elements of the framework in Figure 7.8 are:

Level 1 Strategy development and goal deployment leading to vision/mission/, critical success factors and **key performance outcomes** (KPOs).

Level 2 Process management and **process performance measurement** through **key performance indicators (KPIs)** (including input, in process and output measures, management of internal and external customer-supplier relationships and the use of management control systems).

Level 3 Individual performance management and **performance appraisal**.

Level 4 Performance review (including internal and external benchmarking, self-assessment against quality award criteria and quality costing.)

Level 1 – strategy development and goal deployment

The *first level* of the performance measurement framework is the development of organizational strategy, and the consequent deployment of goals throughout the organization. Steps in the strategy development and goal deployment measurement process are (see also Chapter 4):

1. Develop the vision and a mission statement based on recognising the needs of all organizational stakeholders, customers, employees, shareholders and society. Based on the mission statement, identify those factors critical to the success of the organization achieving its stated mission. The CSFs should represent all the stakeholder groups, customers, employees, shareholders and society.

2. Define performance measures for each CSF – i.e. key performance outcomes (KPOs). There may be one or several KPOs for each CSF. Definition of KPO should include:
   a) Title of KPO
   b) Data used in calculation of KPO
   c) Method of calculation of KPO
   d) Sources of data used in calculation
   e) Proposed measurement frequency
   f) Responsibility for the measurement process.

3. Set targets for each KPO. If KPOs are new, targets should be based on customer requirements, competitor performance or known organizational criteria. If no such data exists, a target should be set based on best guess criteria. If the latter is used, the target should be updated as soon as enough data is collected to be able to do so.

4. Assign responsibility at the organizational level for achievement of desired performance against KPO targets. Responsibility should rest with directors and very senior managers.

5. Develop plans to achieve the target performance. This includes both action plans for one year, and longer-term strategic plans.

6. Deploy mission, CSFs, KPOs, targets, responsibilities and plans to the core business processes. This includes the communication of goals, objectives, plans and the assignment of responsibility to appropriate individuals.

7. Measure performance against organizational KPOs and compare to target performance.

8. Communicate performance and proposed actions throughout the organization.
9. At the end of the planning cycle compare organizational capability to target against all KPOs, and begin again at step 2 above.
10. Reward and recognize superior organizational performance.

Strategy development and goal deployment is clearly the responsibility of senior management within the organization, although there should be as much input to the process as possible by employees to achieve 'buy-in' to the process.

The system outlined above is similar to the policy deployment approach known as Hoshin Kanri, developed in Japan and adapted in the West.

Level 2 – process management and measurement

The second level of the performance measurement framework is process management and measurement, the steps of which are:

1. If not already completed, identify and map processes. This information should include identification by the process owners of:
   a) process customers and suppliers (internal and external)
   b) customer requirements (internal and external)
   c) core and non-core activities
   d) measurement points and feedback loops.
2. Translate organizational goals, action plans and customer requirements into process performance measures (input, in-process and output) – key performance indicators (KPIs). This includes definition of measures, data collection procedures and measurement frequency.
3. Define appropriate performance targets, based on known process capability, competitor performance and customer requirements.
4. Assign responsibility and develop plans for achieving process performance targets.
5. Deploy measures, targets, plans and responsibility to all sub-processes.
6. Operate processes.
7. Measure process performance and compare to target performance.
8. Use performance information to:
   a) implement continuous improvement activities
   b) identify areas for improvement
   c) update action plans
   d) update performance targets
   e) redesign processes, where appropriate
   f) manage the performance of teams and individuals (performance management and appraisal) and external suppliers
   g) provide leading indicators and explain performance against organizational KPIs.
9. At the end of each planning cycle compare process capability to customer requirements against all measures, and begin again at step 2.
10. Reward and recognize superior process performance, including sub-processes, and teams.

The same approach should be deployed to sub-processes and to the activity and task levels.

The above steps should be managed by the process owners, with inputs wherever possible from the owners of sub-processes. The process outlined should be used whether an organization is organized and managed on a process or functional departmental basis. If functionally organized, the key task is to identify the customer-supplier relationships between functions, and for functions to see themselves as part of a customer-supplier chain.

The balanced scorecard

The derivation of KPOs and KPIs may follow the 'balanced scorecard' model, proposed by Kaplan, which divides measures into financial, customer, internal business and innovation and learning perspectives (Figure 7.9).

A balanced scorecard derived from the business excellence model described in Chapters 2 and 8 would include key performance results, customer results (measured via the use of customer satisfaction surveys and other measures, including quality and delivery), people results (employee development and satisfaction) and society results (including community perceptions and environmental performance). In the areas of customers, people and society there needs to be a clear distinction between perception measures and other performance measures.

Financial performance for external reporting purposes may be seen as a result of performance across the other KPOs, the non-financial KPOs and KPIs assumed to be the leading indicators of performance. The only aspect of financial performance that

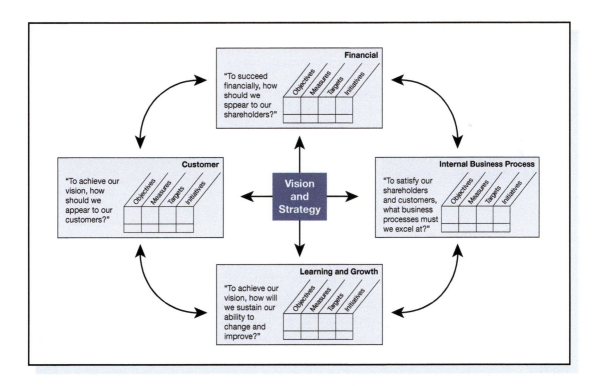

Figure 7.9
The balanced scorecard linking performance measures *(Kaplan and Norton)*

Performance measurement frameworks

is cascaded throughout the organization is the budgetary process, which acts as a constraint rather than a performance improvement measure.

In summary then, organizational KPOs and KPIs should be derived from the balancing of internal capabilities against the requirements of identified stakeholder groups. This has implications for both the choice of KPOs/KPIs and the setting of appropriate targets. There is a need to develop appropriate action plans and clearly define responsibility for meeting targets if they are to be taken seriously.

Performance measures used at the process level differ widely between different organizations. Some organizations measure process performance using a balanced scorecard approach, whilst others monitor performance across different dimensions according to the process. Whichever method is used, measurements should be identified as input (supplier), in-process and output (or results-customers).

It is usually at the process level that the greatest differences can be observed between the measurements used in manufacturing and services organizations. However, all organizations should measure quality, delivery, customer service/satisfaction and cost.

Depending on the process, measurement frequency varies from daily, for example in the measurement of delivery performance, to annual, for example in the measurement of employee satisfaction, which has implications for the PDCA cycle time of the particular process(es). Measurement frequency at the process level may, of course, be affected by the use of information technology. Cross-functional process performance measurement is a vital component in the removal of 'functional silos', and the consequent potential for sub-optimization and failure to take account of customer requirements. The success of performance measurement at the process level is dependent on the degree of management of processes and on the clarity of the deployment of strategic organizational objectives.

Measuring and managing the 'whats' and the 'hows'

Busy senior management teams find it useful to distil as many things as possible down to one piece of paper or one spreadsheet. The use of key performance outcomes (KPOs), with targets, as measures for CSFs, and the use of key performance indicators (KPIs) for processes may be combined into one matrix which is used by the senior management team to 'run the business'.

Figure 7.10 (also shown in Chapter 4) is an example of such a matrix which is used to show all the useful information and data needed.

- the CSFs and their owners – the *'whats'*
- the KPIs and their targets
- the core business processes and their sponsors – the *'hows'*
- the process performance measures – KPIs.

It also shows the impacts of the core processes on the CSFs. This is used in conjunction with a 'business management calendar', which shows when to report/monitor performance, to identify process areas for improvement. This slick process offers senior teams a way of:

- gaining clarity about what is important and how it is measured
- remaining focused on what is important and what the performance is
- knowing where to look if problems occur.

Conduct research	Manage int. systems	Manage financials	Manage our accounts	Develop new business	Develop products	Manage people	Core processes	CSFs: We must have	Measures	Year targets	Target CSF owner
	×	×	×	×		×		Satisfactory financial and non-financial performance	Sales volume. Profit. Costs versus plan. Shareholder return Associate/employee utilisation figures	Turnover £2m. Profit £200k. Return for shareholders. Days/month per person	
×	×		×	×	×	×		A growing base of satisfied customers	Sales/customer Complaints/recommendations Customer satisfaction	>£200k = 1 client. £100k-£200k =5 clients. £50k-£100k= 6 clients <£50k=12 clients	
		×	×			×		A sufficient number of committed and competent people	No. of employed staff/associates Gaps in competency matrix. Appraisal results Perceptions of associates and staff	15 employed staff 10 associates including 6 new by end of year	
×			×			×		Research projects properly completed and published	Proportion completed on time, in budget with customers satisfied. Number of publications per project	3 completed on time, in budget with satisfied customers	
		**	**			**		** = Priority for improvement			
								Process owner			
								Process performance			
								Measures and targets			

Figure 7.10
CSF/core process reporting matrix

Level 3 – individual performance and appraisal management

The third level of the performance measurement framework is the management of individuals. Performance appraisal and management is usually the responsibility of the direct managers of individuals whose performance is to be appraised. At all stages in the process, the individuals concerned must be included to ensure 'buy in'.

Steps in performance and management appraisal are:

1. If not already completed, identify and document job descriptions based on process requirements and personal characteristics. This information should include identification of:
   a) activities to be undertaken in performing the job
   b) requirements of the individual with respect to the identified activities, in terms of competency, skills, experience and training needs
   c) requirements for development of the individual, in terms of personal training and development.
2. Translate process goals and action plans, and personal training and development requirements into personal performance measures.
3. Define appropriate performance targets based on known capability and desired characteristics (or desired characteristics alone if there is no prior knowledge of capability).
4. Develop plans towards achievement of personal performance targets.
5. Document 1 to 4 using appropriate forms, which should include space for the results of performance appraisal.

6. Manage performance. This includes:
   a) planning tasks on a daily/weekly basis
   b) managing performance of the tasks
   c) monitoring performance against task objectives using both quantitative and qualitative information on a daily and/or weekly basis
   d) giving feedback to individuals of their performance in carrying out tasks
   e) giving recognition to individuals for superior performance.
7. Formally appraise performance against range of measures developed, and compare to target performance.
8. Use comparison with target to:
   a) identify areas for improvement
   b) update action plans
   c) update performance targets
   d) redesign jobs, where appropriate. This impacts step 1 of the process.
9. After a suitable period, ideally more than once a year, compare capability to job requirements and begin again at step 2.
10. Reward and recognize superior performance.

The above activities should be undertaken by the individual whose performance is being managed, together with their immediate superior.

The major differences in approaches in the management of individuals lies in the reward of effort, as well as achievement and the consequently different measures used, and in the use of information in continuous improvement required to reward and recognize performance, including teamwork. Unlike management by objectives (MBO), where the focus is on measurement of results, which are often beyond the control of the individual whose performance is appraised, good performance management systems attempt to measure a combination of process/task performance (effort and achievement) and personal development.

The frequency of formal performance appraisal is generally defined by the frequency of the appraisal process, usually with a minimum frequency of six months. Between the formal performance appraisal reviews, most organizations rely on the use of other performance management techniques to manage individuals. Measures of team performance, or of participation in teams should be included in the appraisal systems where possible, to improve team performance. In many organizations, the performance appraisal system is probably the least successfully implemented element of the framework. Appraisal systems are often designed to motivate individuals to achieve process and personal development objectives, but not to perform in teams. One of the limitations of appraisal processes is the frequency of measurement, which could be increased, but few organizations would consider doing so.

Level 4 – performance review

The *fourth level* of the performance measurement framework is the use of Performance Review techniques. Steps in review are as follows:

1. Identify the need for review, which may come from:
   a) poor performance at the organizational or process levels against KPO/KPIs

b) identified superior performance of competitors
c) customer inputs
d) the desire to better direct improvement efforts
e) the desire to concentrate attention on the need for performance improvement.
2. Identify method of performance review to be used. This involves determining whether the review should be carried out internally within the organization, or externally, and the method that should be carried out. Some techniques are mainly internal, e.g. self-assessment, quality costing; whilst others, e.g. benchmarking, involve obtaining information from sources external to the organization. The choice should depend on:
   a) how the need for review was identified (see 1)
   b) the aim of the review e.g. if the aim is to improve performance relative to competitors, external benchmarking may be a better option than internally measuring the cost of quality
   c) the relative costs and expected benefits of each technique.
3. Carry out the review.
4. Feed results into the planning process at the organizational or process level.
5. Determine whether to repeat the exercise. If it is decided to repeat the exercise, the following points should be considered:
   a) frequency of review
   b) at what levels to carry out future reviews e.g. organization-wide or process-by-process
   c) decide whether the review technique should be incorporated into regular performance measurement processes, and if so how this will be managed.

Review methods often require the use of a level of resources greater than that normally associated with performance measurement, often due to the need to develop data collection procedures, train people in their use and the cost of data collection itself. However, review techniques usually give a broader view of performance than most individual measures.

The use of review techniques is most successful when it is based on a clearly identified need, perhaps due to perceived poor performance against existing performance measures or against competitors, and the activity itself is clearly planned and the results used in performance improvement. This is often the difference between the success and failure of quality costing and benchmarking in particular. The use of most of the review techniques has been widely documented, but often without regard to their integration into the wider processes of measurement and management.

Review techniques

Techniques identified for review include:
1. Quality costing, using either presentation-appraisal-failure, or process costing methods, as set out in this chapter.
2. Self-assessment against Baldrige, EFQM excellence Model, or internally developed plant/area assessment criteria (see Chapter 8).
3. CMMI (Capability Maturity Model Integration) at organizational or process levels.

4. Benchmarking, internal or external.
5. Customer satisfaction and loyalty surveys.
6. Activity-based costing (ABC).

THE IMPLEMENTATION OF PERFORMANCE MEASUREMENT SYSTEMS

It has already been established that a good measurement system will start with the customer and measure the right things. The value of any measure clearly needs to be compared with the cost of producing it. There will be appropriate measures for different parts of the organization, but everywhere they must relate process performance to the needs of the process customer. All critical parts of the process must be measured, and it is often better to start with simple measures and improve them.

There must be recognition of the need to distinguish between different measures for different purposes. For example, an operator may measure time, various process parameters and amounts, while at the management level measuring costs and delivery timeliness may be more appropriate.

Participation in the development of measures enhances their understanding and acceptance. Process-owners can assist in defining the required performance measures, provided that senior managers have communicated their vision and mission clearly, determined the critical success factors and identified the critical processes.

If all employees participate, and own the measurement processes, there will be lower resistance to the system, and a positive commitment towards future changes will be engaged. This will derive from the 'volunteered accountability', which will in turn make the individual contribution more visible. Involvement in measurement also strengthens the links in the customer-supplier chains and gives quality improvement teams much clearer objectives. This should lead to greater short-term and long-term productivity gains.

The mnemonic SMART has been associated with designing measures and measurement systems in organizations – they should be Simple, Meaningful, Appropriate, Relevant and Timely. (The same mnemonic is also used to help set Specific, Measureable, Achievable, Realistic and Time based objectives).

There are a number of reasons why measurement systems have been found to fail:

1. They do not define performance operationally.
2. They do not relate performance to the process.
3. The boundaries of the process(es) are not defined.
4. The measures are misunderstood or misused or measure the wrong things.
5. There is no distinction between control and improvement.
6. There is a fear of exposing poor and good performance.
7. It is seen as an extra burden in terms of time and reporting.
8. There is a perception of reduced autonomy.
9. Too many measurements are focused internally and too few are focused externally.
10. There is a fear of the introduction of tighter management controls.

These and other problems are frequently due to poor planning at the implementation stage or a failure to assess current systems of measurement. Before the introduction of a total quality-based performance measurement system, an audit of the existing systems should be carried out. Its purpose is to establish the effectiveness of existing measures, their compatibility with the quality drive, their relationship with the processes concerned and their closeness to the objectives of meeting customer requirements. The audit should also highlight areas where performance has not been measured previously, and indicate the degree of understanding and participation of the employees in the existing systems and the actions that result.

Generic questions that may be asked during the audit include:

- Is there a performance measurement system in use?
- Has it been effectively communicated throughout the organization?
- Is it systematic?
- Is it efficient?
- Is it well understood?
- Is it applied?
- Is it linked to the vision, mission and goals/objectives of the organization?
- Is there a regular review and update?
- Is action taken to improve performance following the measurement?
- Are the people who own the processes engaged in measuring their own performance?
- Have employees been properly trained to conduct the measurement?

Following such an audit, there are twelve basic steps for the introduction of TQM-based performance measurement. Half of these are planning steps and the other half implementation.

Planning

1. Identify the purpose of conducting measurement, i.e. is it for:
   a) reporting, e.g. ROI reported to shareholders
   b) controlling, e.g. using process data on control charts
   c) improving, e.g. monitoring the results of a quality improvement team project.
2. Choose the right balance between individual measures (activity- or task-related) and group measures (process- and sub-process-related) and make sure they reflect process performance.
3. Plan to measure all the key elements of performance, not just one, e.g. time, cost and product quality variables may all be important.
4. Ensure that the measures will reflect the voice of the internal/external customers.
5. Carefully select measures that will be used to establish standards of performance.
6. Allow time for the learning process during the introduction of a new measurement system.

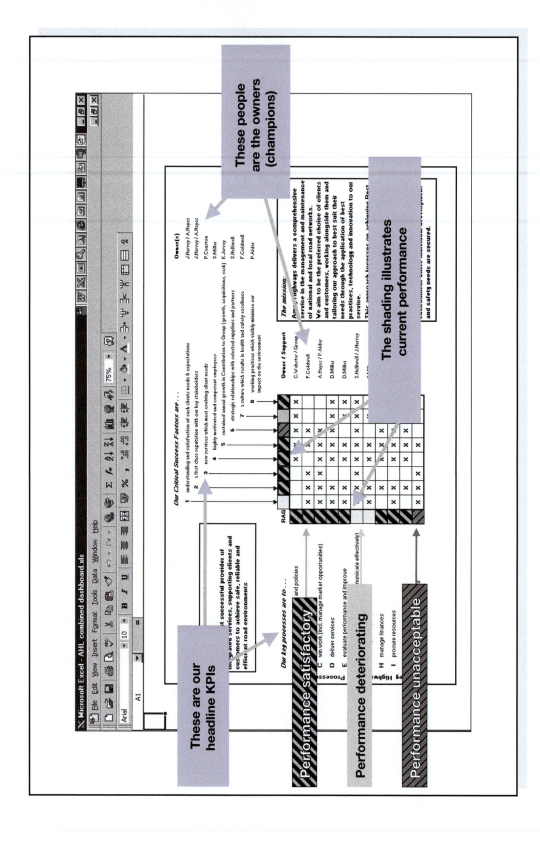

Figure 7.11
Performance dashboard & measurement framework (parts i to iv)

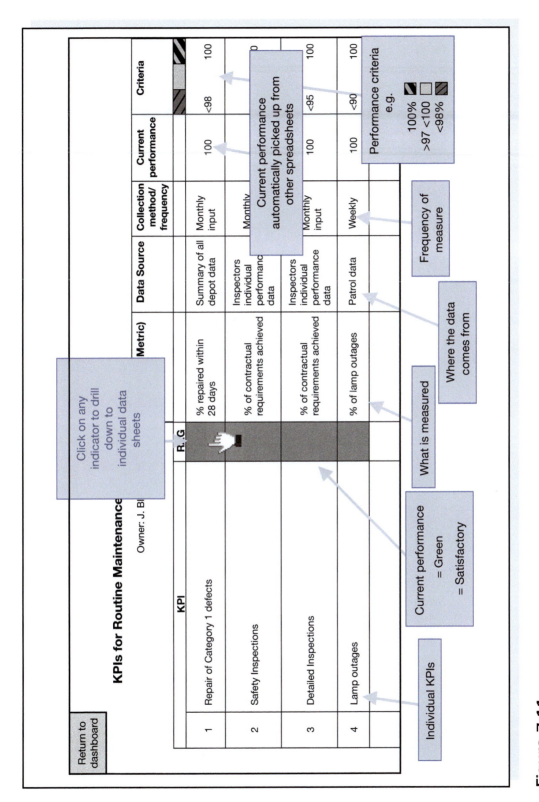

Figure 7.11
continued

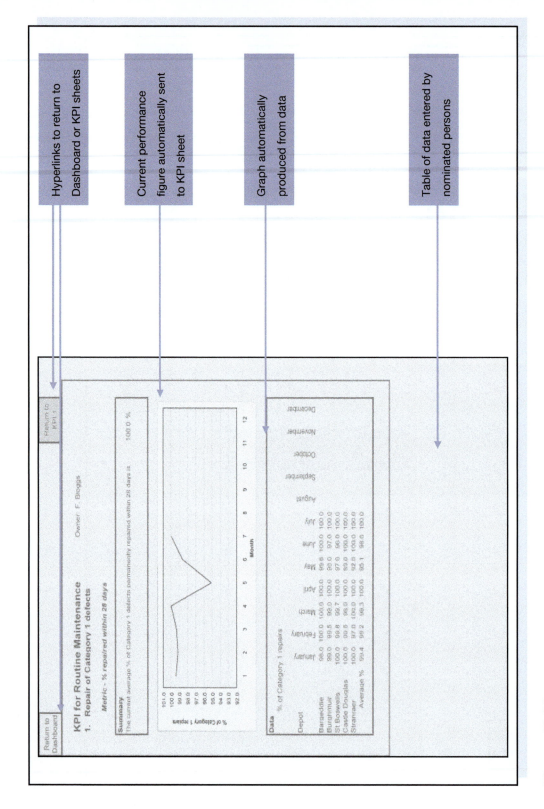

Figure 7.11
continued

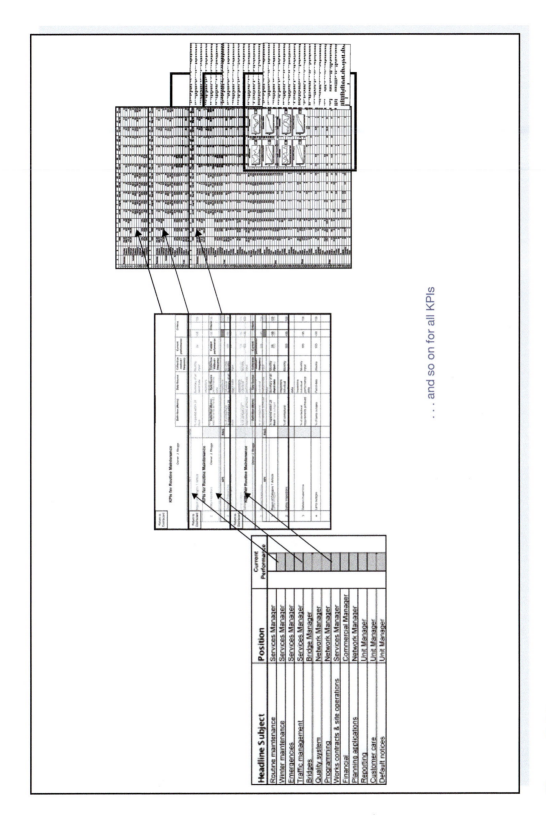

... and so on for all KPIs

Figure 7.11
continued

Implementation

7. Ensure full participation during the introductory period and allow the system to mould through participation.
8. Carry out cost/benefit analysis on the data generation, and ensure measures that have high 'leverage' are selected.
9. Make the effort to spread the measurement system as widely as possible, since effective decision-making will be based on measures from *all* areas of the business operation.
10. Use *surrogate* measures for subjective areas where quantification is difficult, e.g. improvements in morale may be 'measured' by reductions in absenteeism or staff turnover rates.
11. Design the measurement systems to be as flexible as possible, to allow for changes in strategic direction and continual review.
12. Ensure that the measures reflect the quality drive by showing small incremental achievements that match the never-ending improvement approach.

In summary, the measurement system must be designed, planned and implemented to reflect customer requirements, give visibility to the processes and the progress made, communicate the total quality effort and engage the never-ending improvement cycle. So it must itself be periodically reviewed. Figure 7.11 shows a complete practical performance dashboard and measurement framework.

BIBLIOGRAPHY

British Standard BS6143-1: 2002, *Guide to the Economics Of Quality, Process Cost Model*, BSI, London, 2002.
British Standard BS6143-2: 2002, *Guide to the Economics Of Quality, Prevention, Appraisal and Failure Model*, BSI, London, 2002.
Dale, B.G. and Plunkett, J.J., *Quality Costing*, Chapman and Hall, London, 1991.
Dixon, J.R., Nanni, A. and Vollmann, T.E., *The New Performance Challenge – Measuring Operations for World Class Competition*, Business One Irwin, Homewood, 1990.
Hall, R.W., Johnson, J.Y. and Turney, P.B.B., *Measuring Up – Charting Pathways to Manufacturing Excellence*, Business One Irwin, Homewood, 1991.
Kaplan, R.W. (ed.), *Measures for Manufacturing Excellence*, Harvard Business School Press, Boston, MA, 1990.
Kaplan, R.S. and Norton, P., *The Balanced Scorecard*, Harvard Business School Press, Boston, MA, 1996.
Neely, A., *Measuring Business Performance* (2nd edn), Economist Books, London, 2002.
Neely, A., Adams, C. and Kennerly, M., *Performance Prism: The Scorecard for measuring and Managing Business Services*, FT Prentice Hall, 2002.
Porter, L.J. and Rayner, P., 'Quality costing for TQM', *International Journal of Production Economics*, 27, pp. 69–81, 1992.
Wood, D.C. (ed.), *Principles of Quality Costs: Financial Measures for Strategic Implementation of Quality Management* (4th edn), ASQ Press, Milwaukee, 2008.
Zairi, M., *TQM-Based Performance Measurement*, TQM Practitioner Series, Technical Communication (Publishing), Letchworth, 1992.
Zairi, M., *Measuring Performance for Business Results*, Chapman and Hall, London, 1994.

CHAPTER HIGHLIGHTS

Performance measurement and the improvement cycle

- Traditional performance measures based on cost-accounting information provide little to support TQM, because they do not map process performance and improvements seen by the customer.
- Measurement is important in tracking progress, identifying opportunities and comparing performance internally and externally. Measures, typically non-financial, are used in process control and performance improvement.
- Some financial indicators, such as ROI, are often inaccurate, irrelevant and too late to be used as measures for performance improvement.
- The Deming cycle of Plan Do Check Act is a useful design aid for measurement systems, but firstly four basic questions about measurement should be asked i.e. why, what, where and how.
- In answering the question 'how to measure?' progress is important in five main areas: effectiveness, efficiency, productivity, quality and impact.
- Activity-based costing (ABC) is based on the activities performed being identified and costs traced to them. ABC uses cost drivers, which reflect the demands placed on activities.

Costs of quality

- A competitive product or service based on a balance between quality and cost factors is the principal goal of responsible management.
- The analysis of quality related costs may provide a method of assessing the effectiveness of the management of quality and of determining problem areas, opportunities, savings and action priorities.
- Total quality costs may be categorized into prevention, appraisal, internal failure and external failure costs, the P-A-F model.
- Prevention costs are associated with doing it right the first time, appraisal costs with checking it is right and failure costs with getting it wrong.
- When quality awareness in an organization is low, the total quality related costs are high, the failure costs predominating. After an initial rise in costs, mainly through the investment in training and appraisal, increasing investment in prevention causes failure, appraisal and total costs to fall.

The process model for quality costing

- The P-A-F model or quality costing has a number of drawbacks, mainly due to estimating the prevention costs, and its association with an 'optimized' or minimum total cost.
- An alternative – the process costs model – rationalizes cost of quality (COQ) into the costs of conformance (COC) and the cost of non-conformance (CONC). COQ = COC + CONC at each process stage.
- Process cost modelling calls for choice of a process and its definition; construction of a process diagram; identification of outputs and customers, inputs and suppliers,

controls and resources; flowcharting the process and identifying owners; allocating activities as COC or CONC; and calculating the costs. A process cost report with summaries and results is produced.

- The failure costs of CONC should be prioritized for improvements.

A performance measurement framework

- A suitable performance measurement framework (PMF) has four elements related to strategy development, goal deployment, process management, individual performance management and review.
- The key to successful performance measurement at the strategic level is the identification of a set of critical success factors (CSFs) and associated key performance indicators (KPIs).
- The key to success at the process level is the identification and translation of customer requirements and strategic objectives into a process framework, with process performance measures.
- The key to success at the individual level is performance appraisal and planned formal reviews, through integrated performance management
- The key to success in the review stage is the use of appropriate innovative techniques to identify improvement opportunities followed by good implementation.
- A number of factors are critical to the success of performance measurement systems including top management support for non-financial performance measures, the identification of the vital few measures, the involvement of all individuals in the development of performance measurement, the clear communication of strategic objectives, the inclusion of customers and suppliers in the measurement process, and the identification of the key drivers of performance.

The implementation of performance measurement systems

- The value of any measure must be compared with the cost of producing it. All critical parts of the process must be measured, but it is often better to start with the simple measures and improve them.
- Process-owners should take part in defining the performance measures, which must reflect customer requirements.
- Prior to introducing TQM measurement, an audit of existing systems should be carried out to establish their effectiveness, compatibility, relationship and closeness to the customer.
- Following the audit, there are twelve basic steps for implementation, six of which are planning steps. The measurement system, then, must be designed, planned and implemented to reflect customer requirements, give visibility to the processes and progress made, communicate the total quality effort and drive continuous improvement. It must also be periodically reviewed.

Chapter 8

Self-assessment, audits and reviews

FRAMEWORKS FOR SELF-ASSESSMENT

Organizations everywhere are under constant pressure to improve their business performance, measure themselves against world-class standards and focus their efforts on the customer. To help in this process, many are turning to total quality models, such as the European Foundation for Quality Management's (EFQM) Excellence Model (see also Chapter 2), and assessment frameworks, such as CMMI (Capability Maturity Model Integration).

'Total quality' is the goal of many organizations but it has been difficult until relatively recently to find a universally accepted definition of what this actually means. For some people TQ means statistical process control (SPC) or quality management systems, for others teamwork and involvement of the workforce. More recently, in some organizations, it has been replaced by the terms Business Excellence, Six Sigma or Lean. There is even 'Lean Six Sigma'. Clearly there are many different views on what constitutes the 'excellence' organization and, even with an understanding of a framework, there exists the difficulty of calibrating the performance or progress of any organization towards it.

The so-called excellence models recognize that customer satisfaction, business objectives, safety and environmental considerations are mutually dependent and are applicable in any organization. Clearly the application of these ideas involves investment primarily in people and time, time to implement new concepts, time to train, time for people to recognize the benefits and move forward into new or different organizational cultures. But how will organizations know when they are getting close to excellence or whether they are even on the right road, how will they measure their progress and performance?

There have been many recent developments and there will continue to be many more, in the search for a standard or framework, against which organizations may be assessed or measure themselves, and carry out the so-called 'gap analysis'. To many the ability to judge progress against an accepted set of criteria would be most valuable and informative.

Most TQM approaches strongly emphasize measurement, some insist sensibly on the use of cost of quality. The value of a structured discipline using a points system has been well established in quality and safety assurance systems (for example, ISO 9000 and vendor auditing). The extension of this approach to a total quality auditing process has been long established in the Japanese 'Deming Prize,' which is perhaps the most demanding and intrusive auditing process, and there are other excellence models and standards used throughout the world. Perhaps the most famous and widely used framework for self-assessment is the US Baldrige 'Criteria for Performance Excellence'. Many companies have realized the necessity to assess themselves against the Baldrige and Deming criteria, if not to enter for the awards or prizes, then certainly as an excellent basis for self-audit and review, to highlight areas for priority attention and provide internal and external benchmarking. (See Chapter 2 for details of the Deming Prize and Baldrige Award criteria.) CMMI is also in common use, particularly in software development and engineering environments such as aerospace, defence, security and electronics.

The European excellence model for self-assessment

In Europe it has also been recognized that the technique of self-assessment is very useful for any organization wishing to monitor and improve its performance. In 1992 the European Foundation for Quality Management (EFQM) launched a European Quality Award which is now widely used for systematic review and measurement of operations. The EFQM Excellence model recognized that processes are the means by which a company or organization harnesses and releases the talents of its people to produce results performance.

Figure 8.1 displays graphically the principle of the full Excellence model. As described in Chapter 2, customer results, employee results and favourable society results are achieved through leadership driving strategy, people, partnerships and resources, and processes, products and services, which lead ultimately to excellence in key performance results – the enablers deliver the results which in turn drive learning, creativity and innovation. The EFQM have provided a weighting for each criteria which may be used in scoring self-assessments and making awards. The weightings are not rigid and may be modified to suit specific organizational needs.

The EFQM have thus built a model of criteria and a review framework against which an organization may face and measure itself, to examine any 'gaps'. Such a process is known as self-assessment and organizations such as the EFQM, and in the UK the BQF, publish guidelines for self-assessment, including specific ones directed at public sector organizations.

Many managers feel the need for a rational basis on which to measure progress in their organization, especially in those companies 'several years into TQM' which would like the answers to questions such as: 'Where are we now?', 'Where do we need/want to be?' and 'What have we got to do to get there'. These questions need to be answered from internal employees' views, the customers' views and the views of suppliers.

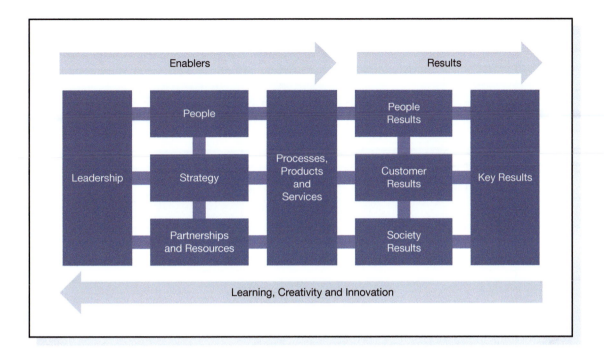

Figure 8.1
The EFQM Excellence Model

Self-assessment promotes business excellence by involving a regular and systematic review of processes and results. It highlights strengths and improvement opportunities and drives continuous improvement.

Enablers

In the Excellence Model, the enabler criteria of: leadership, strategy, people, resources and partnerships and processes, products and services focus on what is needed to be done to achieve results. The structure of the enabler criteria is shown in Figure 8.2. Enablers are assessed on the basis of the combination of two factors (see Figure 8.3, Chart 1, The Enablers):

1. The degree of excellence of the *approach*
2. The degree of *deployment* of the approach.

The detailed criterion parts are as follows:

1 **Leadership**
How leaders develop and facilitate the achievement of the vision and mission, develop values required for long-term success and implement these via appropriate actions and behaviours, and are personally involved in ensuring that the organization's management systems are developed and implemented. Self-assessment should demonstrate how leaders:

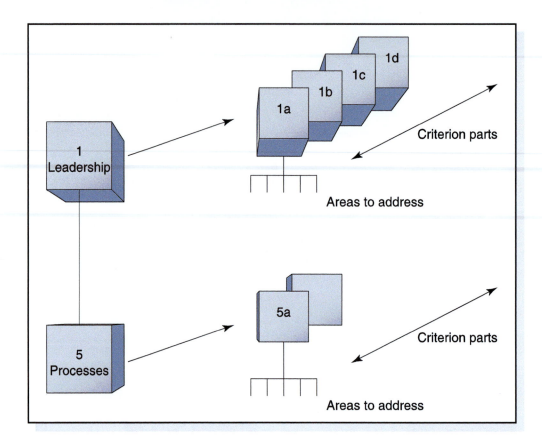

Figure 8.2
Structure of the criteria – enablers

    a) develop the vision, mission and values and are role models of a culture of excellence
    b) are personally involved in ensuring the organization's management systems are developed, implemented and continuously improved
    c) are involved with customers, partners and representatives of society
    d) motivate, support and recognize the organization's people.

2 **Strategy**
How the organization implements its vision and mission via a clear stakeholder focused strategy, supported by relevant policies, plans, objectives, targets and processes. Self-assessment should demonstrate how strategy is:
    a) based on the present and future needs and expectations of stakeholders
    b) based on information from performance measurement, research, learning and creativity related activities
    c) developed, reviewed and updated
    d) deployed through a framework of key processes
    e) communicated and implemented.

Approach	Score	Deployment, assessment and review
Anecdotal or no evidence.	0%	Little effective usage.
Some evidence of soundly based approaches and prevention based processes/systems.	25%	Implemented in about one-quarter of the relevant areas and activities.
Some evidence of integration into normal operations.		Some evidence of assessment and review.
Evidence of soundly based systematic approaches and prevention based processes/systems.	50%	Implemented in about half the relevant areas and activities.
Evidence of integration into normal operations and planning well established.		Evidence of assessment and review.
Clear evidence of soundly based systematic approaches and prevention based processes/systems.	75%	Applied to about three-quarters of the relevant areas and activities.
Clear evidence of integration of approach into normal operations and planning.		Clear evidence of refinement and improved business effectiveness through review cycles.
Comprehensive evidence of soundly based systematic approaches and prevention based processes/systems.	100%	Implemented in all relevant areas and activities.
Approach has become totally integrated into normal working patterns. Could be used as a role model for other organizations.		Comprehensive evidence of refinement and improved business effectiveness through review cycles.

For *Approach, Deployment, Assessment* and *Review* the assessor may choose one of the five levels 0%, 25%, 50%, 75%, or 100% as presented in the chart, or interpolate between these values.

Figure 8.3
Scoring within the self-assessment process: Chart 1, the enablers

3   **People**
How the organization manages, develops and releases the knowledge and full potential of its people at an individual, team-based and organization-wide level, and plans these activities in order to support its policy strategy and the effective operation of its processes. Self-assessment should demonstrate how people:
a)   resources are planned, managed and improved
b)   knowledge and competencies are identified, developed and sustained
c)   are involved and empowered
d)   and the organization have a dialogue
e)   are rewarded, recognized and cared for.

4   **Partnerships and resources**
    How the organization plans and manages its external partnerships and
    internal resources in order to support its strategy and the effective
    operation of its processes. Self-assessment should demonstrate how:
    a) external partnerships are managed
    b) finances are managed
    c) buildings, equipment and materials are managed
    d) technology is managed
    e) information and knowledge are managed.

5   **Processes, products and services**
    How the organization designs, manages and improves its processes,
    products and services in order to support its strategy and fully satisfy, and
    generate increasing value, for its customers and stakeholders. Self-
    assessment should demonstrate how:
    a) processes, products and services are systematically designed and
        managed
    b) processes are improved, as needed, using innovation in order to fully
        satisfy and generate increasing value for customers and other
        stakeholders
    c) products and services are designed and developed based on customer
        needs and expectations
    d) products and services are produced, delivered and serviced
    e) customer relationships are managed and enhanced.

Assessing the enablers criteria

The criteria are concerned with how an organization or business unit achieves its
results. Self-assessment asks the following questions in relation to each criterion part:

- What is currently done in this area?
- How is it done? Is the approach systematic and prevention based?
- How is the approach reviewed and what improvements are undertaken
  following review?
- How widely used are these practices?

Results

The EFQM Excellence Model's result criteria of: customer results, people results,
society results, and key performance results focus on what the organization has
achieved and is achieving in relation to its:

- external customer
- people
- local, national, and international society, as appropriate
- planned performance.

These can be expressed as discrete results, but ideally as trends over a period of
years. The structure of the results criteria is shown in Figure 8.4.

'Performance excellence' is assessed relative to the organization's business
environment and circumstances, based on information which sets out:

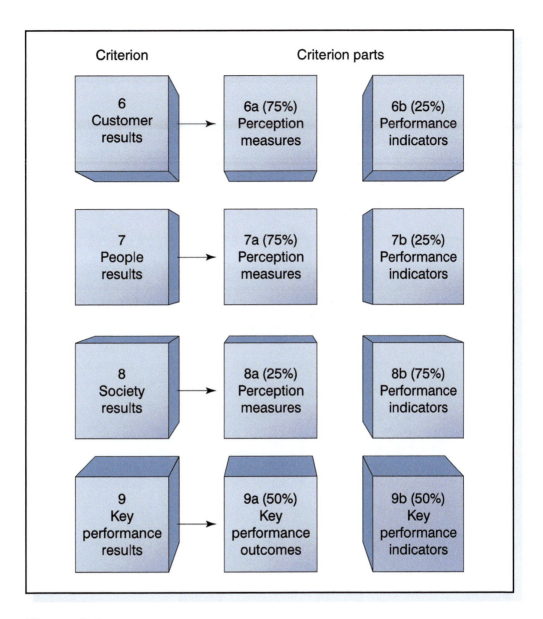

Figure 8.4
Structure of the criteria – results

- the organization's actual performance
- the organization's own targets
  and wherever possible:
- the performance of competitors or similar organizations
- the performance of 'best in class' organizations.

Self-assessment, audits and reviews

Results are assessed on the basis of the combination of two factors (see Figure 8.5, Chart 2, the results):

1   The degree of excellence of the results
2   The scope of the results.

6   **Customer results**
    What the organization is achieving in relation to its external customers.

Results	Score	Scope
No results or anecdotal information.	0%	Results address few relevant areas and activities.
Some results show positive trends and/or satisfactory performance. Some favourable comparisons with own targets/external organizations. Some results are caused by approach.	25%	Results address some relevant areas and activities.
Many results show strongly positive trends and/or sustained good performance over the last 3 years. Favourable comparisons with own targets in many areas. Some favourable comparison with external organizations. Many results are caused by approach.	50%	Results address many relevant areas and activities.
Most results show strong positive trends and/or sustained excellent performance over at least 3 years. Favourable comparisons with own targets in most areas. Favourable comparisons with external organizations in many areas. Most results are caused by approach.	75%	Results address most relevant areas and activities.
Strongly positive trends and/or sustained excellent performance in all areas over at least 5 years. Excellent comparisons with own targets and external organizations in most areas. All results are clearly caused by approach. Positive indication that leading position will be maintained.	100%	Results address all relevant areas and facets of the organization.

For both *Results* and *Scope* the assessor may choose one of the five levels 0%, 25%, 50%, 75%, or 100% as presented in the chart, or interpolate between these values.

Figure 8.5
Scoring within the self-assessment process: Chart 2, the results

Self-assessment should demonstrate the organization's success in satisfying the needs and expectations of its external customers. Areas to consider:

a) Results achieved for the measurement of customer perception of the organization's products, services and customer relationships

b) Internal performance indicators relating to the organization's customers.

7 **People results**

What the organization is achieving in relation to its people. Self-assessment should demonstrate the organization's success in satisfying the needs and expectations of its people. Areas to consider:

a) Results of people's perception of the organization

b) Internal performance indicators relating to people.

8 **Society results**

What the organization is achieving in relation to local, national and international society as appropriate. Self-assessment should demonstrate the organization's success in satisfying the needs and expectations of the community at large. Areas to consider:

a) Society's perception of the organization

b) Internal performance indicators relating to the organization and society.

9 **Key performance results**

What the organization is achieving in relation to its planned performance. Areas to consider:

a) Key performance outcomes, including financial and non-financial

b) Key indicators of the organization's performance which might predict likely key performance outcomes.

Assessing the results criteria

These criteria are concerned with what an organization has achieved and is achieving. Self-assessment addresses the following issues:

- The measures used to indicate performance
- The extent to which the measures cover the range of the organization's activities
- The relative importance of the measures presented
- The organization's actual performance
- The organization's performance against targets

and wherever possible:

- Comparisons of performance with similar organizations
- Comparisons of performance with 'best in class' organizations.

Self-assessment against the Excellence Model may be performed generally using the so-called RADAR system:

Results
Approach
Deployment
Assessment
Review

The RADAR 'screen' with the net level of detail is shown in Figure 8.6.

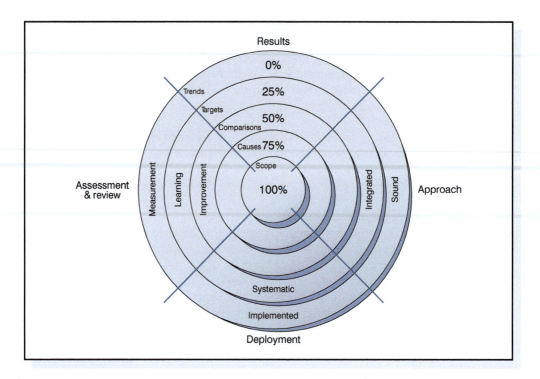

Figure 8.6
The RADAR 'screen'

METHODOLOGIES FOR SELF-ASSESSMENT

A flow diagram of the general steps involved in undertaking self-assessment is shown in Figure 8.7.

There are a number of approaches to carrying out self-assessment including:

- discussion group/workshop methods
- surveys, questionnaires and interviews (peer involvement)
- pro formas
- organizational self-analysis matrices (e.g. see Figure 8.8.)
- an award simulation
- activity or process audits
- hybrid approaches.

Whichever method is used, the emphasis should be on understanding the organization's strengths and areas for improvement, rather than the score. The scoring charts provide a consistent basis for establishing a quantitative measure of performance against the model and gaining consensus promotes discussion and development of the issues facing the organization. It should also gain the involvement, interest and commitment of the senior management, but the scores should not become an end in themselves.

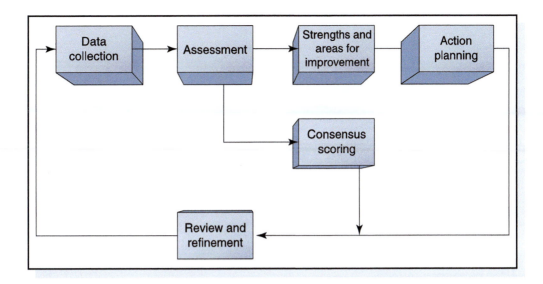

Figure 8.7
The key steps in self-assessment

Using assessment

There is great overlap between the criteria used by the various awards and it may be necessary for an organization to rationalize them. The main components, however, must be the organization's leadership developing and driving its strategy through processes, management systems, people management and results, to deliver customer results and key performance results. Self-assessment can provide an organization with vital information in monitoring its progress towards its goals and 'business excellence'. The external assessments used in the processes of making any awards should be based on these self-assessments that are performed as prerequisites for improvement.

Whatever the main 'motors' are for driving an organization towards its vision or mission, they should be linked to the five stakeholders embraced by the values of any organization, namely:

Customers
Employees
Suppliers
Stakeholders
Community

In any normal business or organization, measurements are continuously being made, often in retrospect, by the leaders of the organization to reflect the value put on the organization by its five stakeholders. Too often, these continuous readings are made by internal biased agents with short-term priorities, not always in the best long-term interests of the organization or its customers, i.e. narrow fire fighting scenarios which can blind the organization's strategic eye. Third party agents, however, can carry out or facilitate periodic assessments from the perspective of one or more of the

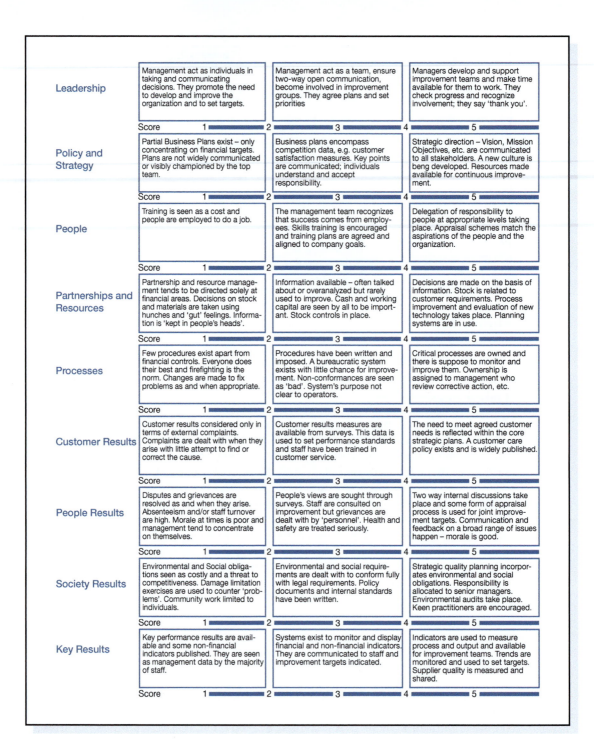

Leadership	Management act as individuals in taking and communicating decisions. They promote the need to develop and improve the organization and to set targets.	Management act as a team, ensure two-way open communication, become involved in improvement groups. They agree plans and set priorities	Managers develop and support improvement teams and make time available for them to work. They check progress and recognize involvement; they say 'thank you'.
	Score 1 ▬▬ 2 ▬▬ 3 ▬▬ 4 ▬▬ 5 ▬▬		
Policy and Strategy	Partial Business Plans exist – only concentrating on financial targets. Plans are not widely communicated or visibly championed by the top team.	Business plans encompass competition data, e.g. customer satisfaction measures. Key points are communicated; individuals understand and accept responsibility.	Strategic direction – Vision, Mission Objectives, etc. are communicated to all stakeholders. A new culture is beng developed. Resources made available for continuous improvement.
	Score 1 ▬▬ 2 ▬▬ 3 ▬▬ 4 ▬▬ 5 ▬▬		
People	Training is seen as a cost and people are employed to do a job.	The management team recognizes that success comes from employees. Skills training is encouraged and training plans are agreed and aligned to company goals.	Delegation of responsibility to people at appropriate levels taking place. Appraisal schemes match the aspirations of the people and the organization.
	Score 1 ▬▬ 2 ▬▬ 3 ▬▬ 4 ▬▬ 5 ▬▬		
Partnerships and Resources	Partnership and resource management tends to be directed solely at financial areas. Decisions on stock and materials are taken using hunches and 'gut' feelings. Information is 'kept in people's heads'.	Information available – often talked about or overanalyzed but rarely used to improve. Cash and working capital are seen by all to be important. Stock controls in place.	Decisions are made on the basis of information. Stock is related to customer requirements. Process improvement and evaluation of new technology takes place. Planning systems are in use.
	Score 1 ▬▬ 2 ▬▬ 3 ▬▬ 4 ▬▬ 5 ▬▬		
Processes	Few procedures exist apart from financial controls. Everyone does their best and firefighting is the norm. Changes are made to fix problems as and when appropriate.	Procedures have been written and imposed. A bureaucratic system exists with little chance for improvement. Non-conformances are seen as 'bad'. System's purpose not clear to operators.	Critical processes are owned and there is suppose to monitor and improve them. Ownership is assigned to management who review corrective action, etc.
	Score 1 ▬▬ 2 ▬▬ 3 ▬▬ 4 ▬▬ 5 ▬▬		
Customer Results	Customer results considered only in terms of external complaints. Complaints are dealt with when they arise with little attempt to find or correct the cause.	Customer results measures are available from surveys. This data is used to set performance standards and staff have been trained in customer service.	The need to meet agreed customer needs is reflected within the core strategic plans. A customer care policy exists and is widely published.
	Score 1 ▬▬ 2 ▬▬ 3 ▬▬ 4 ▬▬ 5 ▬▬		
People Results	Disputes and grievances are resolved as and when they arise. Absenteeism and/or staff turnover are high. Morale at times is poor and management tend to concentrate on themselves.	People's views are sought through surveys. Staff are consulted on improvement but grievances are dealt with by 'personnel'. Health and safety are treated seriously.	Two way internal discussions take place and some form of appraisal process is used for joint improvement targets. Communication and feedback on a broad range of issues happen – morale is good.
	Score 1 ▬▬ 2 ▬▬ 3 ▬▬ 4 ▬▬ 5 ▬▬		
Society Results	Environmental and Social obligations seen as costly and a threat to competitiveness. Damage limitation exercises are used to counter 'problems'. Community work limited to individuals.	Environmental and social requirements are dealt with to conform fully with legal requirements. Policy documents and internal standards have been written.	Strategic quality planning incorporates environmental and social obligations. Responsibility is allocated to senior managers. Environmental audits take place. Keen practitioners are encouraged.
	Score 1 ▬▬ 2 ▬▬ 3 ▬▬ 4 ▬▬ 5 ▬▬		
Key Results	Key performance results are available and some non-financial indicators published. They are seen as management data by the majority of staff.	Systems exist to monitor and display financial and non-financial indicators. They are communicated to staff and improvement targets indicated.	Indicators are used to measure process and output and available for improvement teams. Trends are monitored and used to set targets. Supplier quality is measured and shared.
	Score 1 ▬▬ 2 ▬▬ 3 ▬▬ 4 ▬▬ 5 ▬▬		

Figure 8.8
Organizational self-analysis matrix

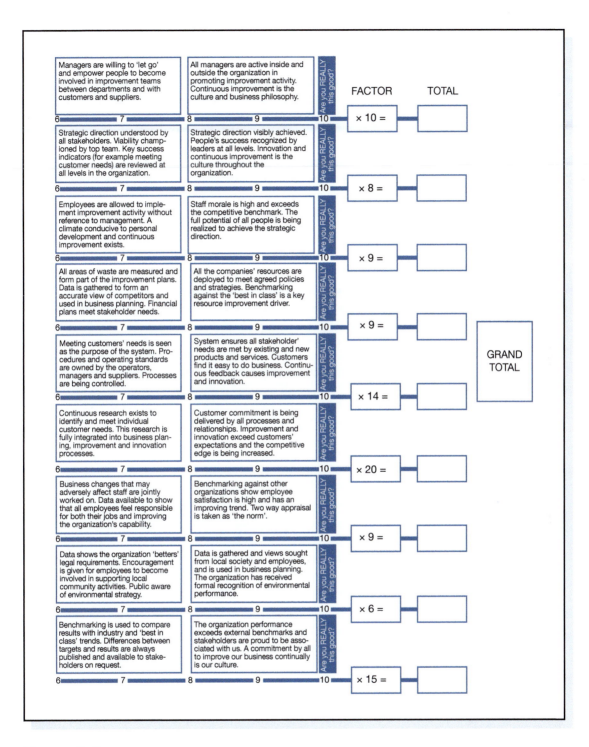

		FACTOR	TOTAL
Managers are willing to 'let go' and empower people to become involved in improvement teams between departments and with customers and suppliers.	All managers are active inside and outside the organization in promoting improvement activity. Continuous improvement is the culture and business philosophy.	× 10 =	
6 — 7 — 8 — 9 — 10	Are you REALLY this good?		
Strategic direction understood by all stakeholders. Viability champ-ioned by top team. Key success indicators (for example meeting customer needs) are reviewed at all levels in the organization.	Strategic direction visibly achieved. People's success recognized by leaders at all levels. Innovation and continuous improvement is the culture throughout the organization.	× 8 =	
6 — 7 — 8 — 9 — 10	Are you REALLY this good?		
Employees are allowed to imple-ment improvement activity without reference to management. A climate conducive to personal development and continuous improvement exists.	Staff morale is high and exceeds the competitive benchmark. The full potential of all people is being realized to achieve the strategic direction.	× 9 =	
6 — 7 — 8 — 9 — 10	Are you REALLY this good?		
All areas of waste are measured and form part of the improvement plans. Data is gathered to form an accurate view of competitors and used in business planning. Financial plans meet stakeholder needs.	All the companies' resources are deployed to meet agreed policies and strategies. Benchmarking against the 'best in class' is a key resource improvement driver.	× 9 =	
6 — 7 — 8 — 9 — 10	Are you REALLY this good?		
Meeting customers' needs is seen as the purpose of the system. Pro-cedures and operating standards are owned by the operators, managers and suppliers. Processes are being controlled.	System ensures all stakeholder' needs are met by existing and new products and services. Customers find it easy to do business. Continu-ous feedback causes improvement and innovation.	× 14 =	
6 — 7 — 8 — 9 — 10	Are you REALLY this good?		
Continuous research exists to identify and meet individual customer needs. This research is fully integrated into business plan-ing, improvement and innovation processes.	Customer commitment is being delivered by all processes and relationships. Improvement and innovation exceed customers' expectations and the competitive edge is being increased.	× 20 =	
6 — 7 — 8 — 9 — 10	Are you REALLY this good?		
Business changes that may adversely affect staff are jointly worked on. Data available to show that all employees feel responsible for both their jobs and improving the organization's capability.	Benchmarking against other organizations show employee satisfaction is high and has an improving trend. Two way appraisal is taken as 'the norm'.	× 9 =	
6 — 7 — 8 — 9 — 10	Are you REALLY this good?		
Data shows the organization 'betters' legal requirements. Encouragement is given for employees to become involved in supporting local community activities. Public aware of environmental strategy.	Data is gathered and views sought from local society and employees, and is used in business planning. The organization has received formal recognition of environmental performance.	× 6 =	
6 — 7 — 8 — 9 — 10	Are you REALLY this good?		
Benchmarking is used to compare results with industry and 'best in class' trends. Differences between targets and results are always published and available to stake-holders on request.	The organization performance exceeds external benchmarks and stakeholders are proud to be asso-ciated with us. A commitment by all to improve our business continually is our culture.	× 15 =	
6 — 7 — 8 — 9 — 10	Are you REALLY this good?		

GRAND TOTAL

Figure 8.8
continued

Self-assessment, audits and reviews

key stakeholders, with particular emphasis on forward priorities and needs. These reviews will allow realignment of the principal driving motors to focus on the critical success factors and continuous improvement, to maintain a balanced and powerful general thrust which moves the whole organization towards its mission.

The relative importance of the five stakeholders may vary in time but all are important. The first three, customers, employees and suppliers, which comprise the core value chain, are the determinant elements. The application of total quality principles in these areas will provide satisfaction as a resultant to the shareholders and the community. Thus, added value will benefit the community and the environment. The ideal is a long way off in most organizations, however, and active attention to the needs of the shareholders and/or community remain a priority for one major reason – they are the 'customers' of most organizational activities and are vital stakeholders.

CAPABILITY MATURITY MODEL INTEGRATION (CMMI) ASSESSMENTS

One of the ways organizations may pursue operational excellence is to consider how mature their operations are. A set of requirements for increasing levels of maturity are defined and assessments are made to determine how mature the assessed entity is. Frameworks exist, such as CMMI for software development (see www.cmmiinstitute.com). Maturity is defined in a framework that sets out the 'criteria' that make up excellence in the area of focus. For each criterion, a set of requirements are arranged in a hierarchy, usually with five levels: basic at level 1 and world class at level 5, with possible intermediate steps in between. Gaining consensus within the organization on criteria/requirements can take time as it is about understanding what is critical to success and then clearly defining what good looks like.

Good definition of criteria/requirements is essential and they should:

* Require quantitative support where possible
* Seek evidence of actual deployment in the entity
* Be mutually exclusive and collectively exhaustive
* Be written in simple language to allow global coverage, and
* Be accompanied by detailed assessor guidelines.

Capability and Maturity Model Integration (CMMI) is an approach which can be used to guide improvements across a project, a division or an entire organization and helps to 'integrate separate organizational functions, appraise the effectiveness of current processes and prioritize improvement activity'. The tools and techniques can be used at the:

* Process Capability level (continuous representation) to satisfy specified product quality, service quality and individual process performance objectives
* Organizational Maturity level (staged representation) to support the application and use of a defined set of capable processes in line with the organization's overall business objectives

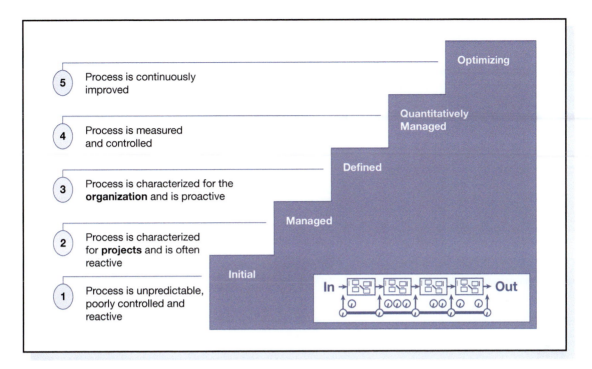

Figure 8.9
CMMI – maturity levels

CMMI uses a scale of 4 levels of capability or 5 levels of maturity to provide a series of well-defined evolutionary plateau for improvement (Figure 8.9). Attaining specific levels will form a firm foundation for further development and embed continuous process improvement – see Figure 8.10 for a process example.

Assessments are usually by either local self-assessment or 'audits' from a 'centre':

- local self-assessment has the advantage of local ownership but with a large training challenge
- audits from a centre have the advantage of global consistency but can meet local resistance.

A synthesis of these two approaches often works using a small central and/or independent core team (objectivity and consistency) working with nominated local assessors (ownership). Every effort should be made to make the assessment a positive learning experience for the entity assessed to ensure motivation to improve maturity.

Maturity assessments using CMMI type frameworks should prompt appropriate action by showing both the current status and what needs to be done next to advance up the maturity curve. The assessments often lead to standardized reports which fulfil three main purposes:

- show the level of maturity of each entity against the set criteria/ requirements

5 – Optimising	Continuous process improvement	Organizational Performance Management (OPM) Causal Analysis and Resolution (CAR)	
4 – Quantitatively Managed	Quantitative management	Organisational Process Performance (OPP) Quantitative Project Management (QPM)	
3 – Defined	Process standardisation	Requirements Development (RD) Technical Solution (TS) Product Integration (PI) Verifcation (VER) Validation (VAL) Organisational Process Focus (OPF) Organisational Process Definition (OPD) Organisational Training (OT) Integrated Project Management (IPM) Risk Management (RSKM) Decision Analysis and Resolution (DAR)	Performance
2 – Managed	Basic process/ project management	Requirements Management (REQM) Project Planning (PP) Project Monitoring and Control (PMC) Supplier Agreement Management (SAM) Measurement and Analysis (MA) Process and Product Quality Assurance (PPQA) Configuration Management (CM)	
1 – Initial			Risk

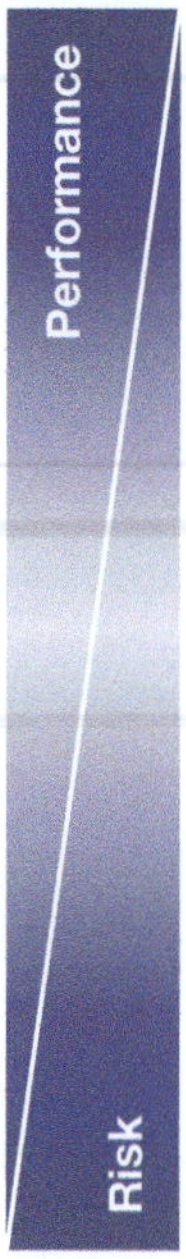

Figure 8.10
CMMI for development – process areas

- indicate where the priorities are for the entity to progress further up the maturity levels
- by showing results from multiple entities, identify where best practice exists so that an entity can learn from others.

Actions should be generated from these reports and mechanisms established to ensure implementation. Periodic re-assessments may need to be scheduled. Clear reporting of all entities' maturity status and progress should be visible to all and reviewed by senior management regularly. The overall goal of any maturity assessments should be to engage people and channel their efforts into activities that will most efficiently improve the operations in their organization.

SECURING PREVENTION BY AUDIT AND REVIEW OF THE MANAGEMENT SYSTEMS

Error or defect prevention is the process of removing or controlling error/defect causes in the management systems. There are two major elements of this:

- Checking the systems
- Error/defect investigation and follow-up.

These have the same objectives – to find record and report *possible* causes of error, and to recommend future preventive or corrective action.

Checking the systems

There are six methods in general use:

1. *Quality Audits and reviews*, which subject each area of an organization's activity to a systematic critical examination. Every component of the total system is included, i.e., policy, attitudes, training, processes, decision features, operating procedures, documentation. Audits and reviews, as in the field of accountancy, aim to disclose the strengths and the main areas of vulnerability or risk – the areas for improvement.
2. *Quality Survey*, a detailed, in-depth examination of a narrower field of activity, i.e. major key areas revealed by system audits, individual sites/plants, procedures or specific problems common to an organization as a whole.
3. *Quality Inspection*, which takes the form of a routine scheduled inspection of a unit or department. The inspection should check standards, employee involvement and working practices, and that work is carried out in accordance with the agreed processes and procedures.
4. *Quality tour*, which is an unscheduled examination of a work area to ensure that, for example, the standards of operation are acceptable, obvious causes of defects or errors are removed, and in general quality standards are maintained.
5. *Quality sampling*, which measures by random sampling, similar to activity sampling, the error/defect potential. Trained observers perform short tours of specific locations by prescribed routes and record the number of potential errors or defects seen. The results may be used to portray trends in the general quality situation.
6. *Quality scrutinizes*, which are the application of a formal, critical examination of the process and technological intentions for new or existing facilities, or to assess the potential for mal-operation or malfunction of equipment and the consequential effects of quality. There are similarities between quality scrutinies and FMECA studies (see Chapter 6)

The design of a prevention programme, combining all these elements, is represented in Figure 8.11.

Error or defect investigations and follow-up

The investigation of errors and defects can provide valuable error prevention information. The general method is based on:

- *Collecting* data and information relating to the error or defect.
- *Checking* the validity of the evidence.
- *Selecting* the evidence without making assumptions or jumping to conclusions.

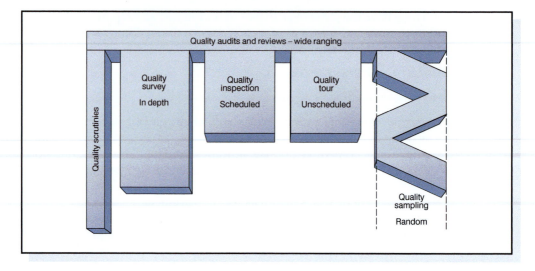

Figure 8.11
A prevention program combining various elements of 'checking' the system

The results of the analysis are then used to:

- *Decide* the most likely cause(s) of the errors or defect.
- *Notify* immediately the person(s) able to take corrective action.
- *Record* the findings and outcomes.
- *Report* them to everyone concerned, to prevent recurrence.

The investigation should not become an inquisition to apportion blame, but focus on the positive preventive aspects. The types of follow-up to errors and their effects are shown in Table 8.1.

It is hoped that errors or defects are not normally investigated so frequently that the required skills are developed by experience, nor are these skills easily learned in a classroom. One suggested way to overcome this problem is the development of a programmed sequence of questions to form the skeleton of an error or defect investigation questionnaire. This can be set out with the following structure:

- *People* – duties, information, supervision, instruction, training, attitudes
- *Processes* – systems, procedures, instructions, monitoring, control methods
- *Technology* – description, condition, controls, maintenance, suitability
- *Environment* – climatic, space, humidity, noise.

INTERNAL AND EXTERNAL MANAGEMENT SYSTEM AUDITS AND REVIEWS

A good management system will not function without adequate audits and reviews. The system reviews, which need to be carried out periodically and systematically, are

Table 8.1 Following up errors

System Type	Aim	General effects
Investigation	To prevent a similar error or defect	*Positive:* identification notification correction
Inquisition	To identify responsibility	*Negative:* blame claims defence

conducted to ensure that the system achieves the required effect, whilst audits are carried out to make sure that actual methods are adhering to the documented procedures. The reviews should use the findings of the audits, for failure to operate according to the plan often signifies difficulties in doing so. A re-examination of the processes and procedures actually being used may lead to system improvements unobtainable by other means.

A schedule for carrying out the *audits* should be drawn up, different activities perhaps requiring different frequencies. All procedures and systems should be audited at least once during a specified cycle, but not necessarily all at the same audit. For example, every three months a selected random sample of the processes could be audited, with the selection designed so that each process is audited at least once per year. There must be, however, a facility to adjust this on the basis of the audit results.

A quality management system *review* should be instituted, perhaps every 12 months, with the aims of:

- ensuring that the system is achieving the desired results
- revealing defects or irregularities in the system
- indicating any necessary improvements and/or corrective actions to eliminate waste or loss
- checking on all levels of management
- uncovering potential danger areas
- verifying that improvements or corrective action procedures are effective.

Clearly, the procedures for carrying out the audits and reviews and the results from them should be documented, and themselves be subject to review. Useful guidance on quality management system audits is given in the international standard ISO 19011, Guidelines for auditing management systems.

The assessment of a quality management system against a particular standard or set of requirements by internal audit and review is known as a *first-party* assessment or approval scheme. If an *external* customer makes the assessment of a supplier against either its own or a national or international standard, a *second-party* scheme is in operation. The assessment by an independent organization, not connected with any contract between customer and supplier, but acceptable to them both, is known as an *independent third-party* assessment scheme. The latter often results in some form of certification or registration by the assessment body.

Self-assessment, audits and reviews

One advantage of the third party schemes is that they obviate the need for customers to make their own detailed checks, potentially saving both suppliers and customers' time and money, and avoiding issues of commercial confidentiality. Just one knowledgeable organization has to be satisfied, rather than a multitude with varying levels of competence. This method can be used to certify suppliers for contracts without further checking, but good customer/supplier relations often include second party extensions to the third party requirements and audits.

Each certification body usually has its own recognized mark, which may be used by registered organizations of assessed capability in their literature, letter headings and marketing activities. There are also publications containing lists of organizations whose quality management systems and/or products and services have been assessed. To be of value, the certification body must itself be recognized and, usually, assessed and registered with a national or international accreditation scheme.

Many organizations have found that the effort of designing and implementing a quality management system, good enough to stand up to external independent third-party assessment, has been extremely rewarding in:

* involving staff and improving morale
* better process control and improvement
* reduced wastage and costs
* reduced customer service costs.

This is also true of those organizations that have obtained third party registrations and supply companies which still insist on their own second party assessment. The reason for this is that most of the standards on quality management systems, whether national, international or company-specific, are now very similar indeed. A system that meets the requirements of the ISO 9001 standard, for example, should meet the requirements of most other standards, with only the slight modifications and small emphases here and there require for specific customers. It is the author's experience, and that of his colleagues, that an assessment carried out by one of the good independent certified assessment bodies is a rigorous and delving process. In certain industries, such as aerospace and pharmaceuticals, there will be specific requirements set out in particular documents, such as the AS 9100 series or 'Good Manufacturing Practice'.

Internal system audits and reviews should be positive and conducted as part of the preventive strategy and not as a matter of expediency resulting from problems. They should not be carried out only prior to external audits, nor should they be left to the external auditor – whether second or third party. An external auditor, discovering discrepancies between actual and documented systems, will be inclined to ask why the internal review methods did not discover and correct them.

Any management team needs to be fully committed to operating an effective quality management system for all the people within the organization, not just the staff in the 'quality department'. The system must be planned to be effective and achieve its objectives in an uncomplicated way. Having established and documented the processes it is necessary to ensure that they are working and that everyone is operating in accordance with them. The system once established is not static, it should be flexible to enable the constant seeking of improvements or streamlining.

Quality auditing standard

The growing use of standards internationally emphasizes the importance of auditing as a management tool for this purpose. There are available several guides to management systems auditing (e.g. ISO 19011) and the guidance provided in these can be applied equally to any one of the three specific and yet different auditing activities:

1. **First party or internal audits,** carried out by an organization on its own systems, either by staff who are independent of the systems being audited, or by an outside agency.
2. **Second party audits,** carried out by one organization (a purchaser or its outside agent) on another with which it either has contracts to purchase goods or services or intends to do so.
3. **Third party audits,** carried out by independent agencies, to provide assurance to existing and prospective customers for the product or service.

Audit objectives and responsibilities, including the roles of auditors and their independence and those of the 'client' or auditee, should be understood. The generic steps involved then are as follows:

- **initiation,** including the audit scope and frequency
- **preparation,** including review of documentation, the programme and working documents
- **execution,** including the opening meeting, examination and evaluation, collecting evidence, observations and closing the meeting with the auditee
- **report,** including its preparation, content and distribution
- **completion,** including report submission and retention.

Attention should be given at the end of the audit to corrective action and follow-up and the improvement process should be continued by the auditee after the publication of the audit report. This may include a call by the client for a verification audit of the implementation of any corrective actions specified.

Any instrument which is developed for assessment, audit or review may be used at several stages in an organization's history:

- before starting an improvement programme to identify 'strengths' and 'areas for improvement', and focus attention (at this stage a parallel cost of quality exercise may be a powerful way to overcome skepticism and get 'buy in')
- as part of a programme launch, especially using a 'survey' instrument
- every one or two years after the launch to steer and benchmark.

The systematic measurement and review of operations is one of the most important management activities of any organization. Assessment, audit and review should lead to clearly discerned strengths and areas for improvement by focusing on the relationship between the planning, people, processes and performance. Within any successful organization these will be regular activities.

BIBLIOGRAPHY

Ahern, D.M., Clouse, A. and Turner, R. *CMMI Distilled: A Practical Introduction to Integrated Process Improvement* (2nd edn), Addison-Wesley, 2004.

Arter, D. R., Cianfrani, C.A and West, J. E., *How to Audit the Process Based BMS* (2nd edn), ASQ, Milwaukee, 2013.

Blazey, M.L., *Insights to Performance Excellence 2013–2014: Understanding the Integrated Management System and the Baldrige Criteria*, NIST, Gaithersburg, 2013.

BQF, *The X-Factor, Winning Performance through Business Excellence*, European Centre for Business Excellence/British Quality Foundation, London, 1999.

BQF, *The Model in Practice* and *The Model in Practice 2*, British Quality Foundation, London, 2000 and 2002 (prepared by the European Centre for Business Excellence).

EFQM, *Assessing for Excellence – A Practical Guide for Self-assessment*, EFQM, Brussels, 2013.

International Standards Organization: ISO 19011: *Guidelines on Quality and/or Environmental Management Systems Auditing*, ISO, 2003.

JUSE (Union of Japanese Scientists and Engineers), *Deming Prize Criteria*, JUSE, Tokyo, 2013.

Keeney, K.A., *The ISO 9001:2000 Auditor's Companion*, Quality Press, Milwaukee, 2002.

NIST (National Institute of Standards and Technology), *The Malcolm Baldrige National Quality Award Criteria*, NIST, Gaithersburg, 2013.

Porter, L.J. and Tanner, S.J., *Assessing Business Excellence* (2nd edn), Butterworth-Heinemann, Oxford, 2003.

Porter, L.J., Oakland, J.S. and Gadd, K.W., *Evaluating the European Quality Award Model for Self-Assessment*, CIMA, London, 1998.

Pronovost, D., *Internal Quality Auditing*, Quality Press, Milwaukee, 2000.

Tricker, R., *ISO 9001 Audit Procedures*, Butterworth-Heinemann, Oxford, 2008.

CHAPTER HIGHLIGHTS

Frameworks for self-sssessment

- Many organizations are turning to total quality models to measure and improve performance. The frameworks include the Japanese Deming Prize, the US Baldrige Award and in Europe the EFQM Excellence Model.
- The nine components of the Excellence Model are: leadership, strategy partnerships, people, resources and processes (ENABLERS), people results, customer results, society results and key results (RESULTS).
- The various award criteria provide rational bases against which to measure progress towards TQM in organizations. Self-assessment against, for example, the EFQM Excellence model should be a regular activity, as it identifies opportunities for improvement in performance through processes and people.

Methodologies for assessment

- Self-assessment against the Excellence Model may be performed using RADAR: results, approach, deployment, assessment and review.
- There are a number of approaches for self-assessment, including groups/workshops, surveys, pro-formas, matrices, award simulations, activity/process audits or hybrid approaches.

Capability Maturity Model Integration (CMMI) assessments

- In CMMI a set of requirements for increasing levels of maturity are defined and assessments are made to determine how mature the assessed entity is.
- Maturity is defined in a framework that sets out the 'criteria' that make up excellence in the area of focus; for each criterion, a set of requirements are arranged in a hierarchy, usually with 5 levels: basic at level 1 and world class at level 5, with possible intermediate steps in between.
- CMMI assessments should prompt appropriate action by showing both the current status and what needs to be done next to advance up the maturity curve; the assessments often lead to standardized reports.

Securing prevention by audit and review of the system

- There are two major elements of error or defect prevention: checking the system, and error/defect investigations and follow-up. Six methods of checking the quality systems are in general use: audits and reviews, surveys, inspections, tours, sampling and scrutinies.
- Investigations proceed by collecting, checking and selecting data, and analysing it by deciding causes, notifying people, recording and reporting findings and outcomes.

Internal and external quality management system audits and reviews

- A good management system will not function without adequate audits and reviews. Audits make sure the actual methods are adhering to documented procedures. Reviews ensure the system achieves the desired effect.
- System assessment by internal audit and review is known as first party, by external customer as second party, and by an independent organization as third party certification. For the latter to be of real value the certification body must itself be recognized.

Chapter 9

Benchmarking and change management

THE WHY AND WHAT OF BENCHMARKING

Product, service and process improvements can take place only in relation to established standards, with the improvements then being incorporated into new standards. *Benchmarking*, one of the most transferable aspects of Rank Xerox's approach to total quality management, and thought to have originated in Japan, measures an organization's operations, products and services against those of its competitors in a ruthless fashion. It is a means by which targets, priorities and operations that will lead to competitive advantage can be established.

There are many drivers for benchmarking including the external ones:

- customers continually demand better quality, lower prices, shorter lead times, etc.
- competitors are constantly trying to get ahead and steal markets
- legislation – changes in our laws place ever greater demands for improvement.
  Internal drivers include:
- targets which require improvements on our 'best ever' performance
- technology – a fundamental change in processes is often required to benefit fully from introducing new technologies
- self-assessment results, which provide opportunities to learn from adapting best practices.

The word 'benchmark' is a reference or measurement standard used for comparison, and benchmarking is the continuous process of identifying, understanding and adapting best practice and processes that will lead to superior performance. Benchmarking is *not*:

- a panacea to cure the organization's problems, but simply a practical tool to drive up process performance

- primarily a cost reduction exercise, although many benchmarking studies will result in improved financial performance
- industrial tourism – study tours have their place, but proper benchmarking goes beyond 'tourism' to really understanding the enablers to outstanding results
- spying – use of a benchmarking code of conduct ensures the work is done with the agreement and openness of all parties
- catching up with the best – the aim is to reach out and extend the current best practice (by the time we have caught up, the benchmark will have moved anyway).

There may be many reasons for carrying out benchmarking. Some of them are set against various objectives in Table 9.1. The links between benchmarking and TQM are clear – establishing objectives based on industry best practice should directly contribute to better meeting of the internal and external customer requirements.

The benefits of benchmarking can be numerous but include:

- creating a better understanding of the current position
- heightening sensitivity to changing customer needs
- encouraging innovation
- developing realistic stretch goals
- establishing realistic action plans.

The American Productivity and Quality Center provides an 'Open Standards Benchmarking' service that was launched in 2004 (see www.aqpc.org/benchmarking). This is powered by a database underwritten by organizations that support the creation of common, open frameworks to measure processes and is based on APQC's widely-adopted Process Classification Framework (see Chapter 10). The database contains more than 1,200 standardized measures spanning people, process, and technology. The same source has identified that the average return from benchmarking is typically *five times* the cost of the study, in terms of reduced costs, increased sales, greater customer retention and enhanced market share.

There are four basic categories of benchmarking:

- **Internal** – the search for best practice of internal operations by comparison, e.g. multi-site comparison of polymerization processes and performance.

Table 9.1 Reasons for benchmarking

Objectives	Without benchmarking	With benchmarking
Becoming Competitive	Internally focused Evolutionary change	Understanding of competitiveness Ideas from proven practices
Industry best practices	Few solutions Frantic catch up activity	Many options Superior performance
Defining customer requirements	Based on history or gut feeling Perception	Market reality Objective evaluation
Establishing effective goals and objectives	Lacking external focus Reactive	Credible, unarguable Proactive
Developing true measures of productivity	Pursuing pet projects Strength and weaknesses not understood Route of least resistance	Solving real problems Understanding outputs Based on industry best practices

- **Functional** – seeking functional best practice outside an industry, e.g. mining company benchmarking preventative maintenance of pneumatic/hydraulic equipment with Disney.
- **Generic** – comparison of outstanding processes irrespective of industry or function, e.g. restaurant chain benchmarking kitchen design with US nuclear submarine fleet to improve restaurant to kitchen space ratios.
- **Competitive** – specific competitor to competitor comparisons for a product, service, or function of interest, e.g. retail outlets comparing price performance and efficiency of internet ordering systems.

THE PURPOSE AND PRACTICE OF BENCHMARKING

The evolution of benchmarking in an organization is likely to progress through four focuses. Initially attention may be concentrated on competitive products or services, including, for example, design, development and operational features. This should develop into a focus on industry best practices and may include, for example, aspects of distribution or service. The real breakthroughs are when organizations focus on all aspects of the in total business performance, across all functions and aspects, and addresses current *and projected* performance gaps. This should lead to the focus on processes and true continuous improvement.

At its simplest, competitive benchmarking, the most common form requires every department/function/process to examine itself against the counterpart in the best competing companies. This includes a scrutiny of all aspects of their activities. Benchmarks which may be important for *customer satisfaction*, for example, might include:

- Product or service quality and consistency.
- Correct and on-time delivery.
- Speed of response or new product development.
- Correct billing.
  For internal *impact* the benchmarks may include:
- Waste, rejects or errors.
- Inventory levels/work in progress.
- Costs of operation.
- Staff turnover.

The task is to work out what has to be done to improve on the competition's performance in each of the chosen areas.

Benchmarking is very important in the 'administration' areas, since it continuously measures services and practices against the equivalent operation in the toughest direct competitors or organizations renowned as leaders in the areas, even if they are in the same organization. An example of quantitative benchmarks in absenteeism is given in Table 9.2.

Technologies and conditions vary between different industries and markets, but the basic concepts of measurement and benchmarking are of general validity. The objective should be to produce products and services that conform to the requirements of the customer in a never-ending improvement environment. The way to accomplish

Table 9.2 Quantitative benchmarking in absenteeism

Organisation's absence level (%)	Productivity opportunity
Under 3	This level matches an aggressive benchmark that has been achieved in 'excellent' organizations.
3–4	This level may be viewed within the organization as a good performance – representing a moderate productivity opportunity improvement.
5–8	This level is tolerated by many organizations but represents a major improvement opportunity.
9–10	This level indicates that a serious absenteeism problem exists.
Over 10	This level of absenteeism is extremely high and requires immediate senior management attention.

this is to use a continuous improvement cycle in all the operating departments – nobody should be exempt. Benchmarking is not a separate science or unique theory of management, but rather another strategic approach to getting the best out of people and processes, to deliver improved performance.

The purpose of benchmarking then is predominantly to:

- **change** the perspectives of executives and managers
- **compare** business practices with those of world class organizations
- **challenge** current practices and processes
- **create** improved goals and practices for the organization.

As a managed process for change, benchmarking uses a disciplined structured approach to identify **what** needs to change, **how** it can be changed, and the **benefits** of the change. It also creates the desire for change in the first place. Any process or practice that can be defined can be benchmarked but the focus should be on those which impact on customer satisfaction and/or business results – financial or non-financial.

For organizations which have not carried out benchmarking before, it may be useful initially to carry out a simple self-assessment of their readiness in terms of:

- how well processes are understood
- how much customers are listened to
- how committed the senior team is.

Table 9.3 provides a simple pro forma for this purpose. The score derived gives a crude guide to the readiness of the organization for benchmarking:

32–48 Ready for benchmarking
16–31 Some further preparation required before the benefits of benchmarking can be fully derived
0–15 Some help is required to establish the foundations and a suitable platform for benchmarking

Table 9.3 Is the organization ready for benchmarking

After studying the statements below tick one box for each to reflect the level to which the statement is true for the organization.	Most	Some	Few	None
Processes have been documented with measures to understand performance.	☐	☐	☐	☐
Employees understand the processes that are related to their own work.	☐	☐	☐	☐
Direct customer interactions, feedback or studies about customers influence decisions about products and services	☐	☐	☐	☐
Problems are solved by teams.	☐	☐	☐	☐
Employees demonstrate by words and deeds that they understand the organization's mission, vision and values.	☐	☐	☐	☐
Senior executives sponsor and actively support quality improvement projects.	☐	☐	☐	☐
The organization demonstrates by words and by deeds that continuous improvement is part of the culture.	☐	☐	☐	☐
Commitment to change is articulated in the organization's strategic plan.	☐	☐	☐	☐
Add the columns:	☐	☐	☐	☐
	x 6 =	x 4 =	x 2 =	Zero
Multiply by the factor	☐	☐	☐	☐
Obtain the grand total?				

The benchmarking process has five main stages which are all focused on trying to measure comparisons and identify areas for action and change (Figure 9.1). The detail is as follows.

PLAN the study

- Select the process(es) for benchmarking.
- Bring together the appropriate team to be involved and establish roles and responsibilities.
- Identify and define benchmarks and measures for data collection.
- Identify best competitors or operators of the process(es), perhaps using customer feedback or industry observers.
- Document the current process(es).

COLLECT data and information

- Decide information and data collection methodology, including desk research.

- Plan
  - Initial research
  - Record current performance

- Collect
  - Identify benchmarking partners
  - Site visits

- Analyse
  - Identify best practices
  - Identify enablers

- Adapt
  - Share best practice learning
  - Adapt and implement best practice

- Review
  - Post-completion review

Figure 9.1
The benchmarking methodology

- Record current performance levels.
- Identify benchmarking partners.
- Conduct a preliminary investigation.
- Prepare for any site visits and interact with target organizations.
- Use chosen data collection methodology.
- Carry out site visits

ANALYSE the data and information

- Normalize the performance data, as appropriate.
- Construct a matrix to compare current performance with benchmarking competitors'/partners' performance.
- Identify outstanding practices.
- Isolate and understand the process enablers, as well as the performance measures.

ADAPT the approaches

- Catalogue the information and create a 'competency profile' of the organization.
- Develop new performance level objectives/targets/standards.
- Vision alternative process(es) incorporating best practice enablers.
- Identify and minimize barriers to change.
- Develop action plans to adapt and implement best practices, make process changes, and achieve goals.
- Implement specific actions and integrate them into the organization.

REVIEW performance and the study

- Monitor the results/improvements.
- Assess outcomes and learnings from the study.
- Review benchmarks.
- Share experiences and best practice learnings from implementation.
- Review relationships with target/partner organizations.
- Identify further opportunities for improving and sustaining performance.

In a typical benchmarking study involving several organizations, the study will commence with the *Plan* phase. Participants will be invited to a 'kick-off' meeting where they will share their aspirations and objectives for the study and establish roles and responsibilities. Participants will analyse their own organization to understand the strengths and areas for improvement. They will then agree appropriate measures for the study.

In the *Collect* phase, participants will collect data on their current performance, based on the agreed measures. The benchmarking partners will be identified, using a suitable screening process, and the key learning points will be shared. The site visits will then be planned and conducted, with appropriate training. Five to seven site visits might take place in each study.

Data collected from the site visits will be *analysed* in the next phase to identify best practices and the enablers which deliver outstanding performance. The reports from this phase will capture the learning and key outcomes from the site visits and present them as the main process enablers, linked to major performance outcomes.

In the *Adapt* phase, the participants will attend a feedback session where the conclusions from the study will be shared, and they will be assisted in adapting them to their own organization. Reports to partners should be issued after this session. (A 'subject expert' is often useful in benchmarking studies, to ensure good learning and adaptation at this stage).

The final phase of the study will be a post-completion *review*. This will give all the participants and partners valuable feedback and establish, above all else, what actions are required to sustain improved performance. Best practice databases may be created to enable further sharing and improvement amongst participants and other members of the organization.

THE ROLE OF BENCHMARKING IN CHANGE

One aspect of benchmarking is to enable organizations to gauge how well they are performing against others who undertake similar tasks and activities. But a more important aspect of best practice benchmarking is gaining an understanding of *how* other organizations achieve superior performance. A good benchmarking study, for example in customer satisfaction and retention, will provide its participants with data and ideas on how excellent organizations undertake their activities and demonstrate best practices that may be adopted, adapted and used.

This new knowledge will result in the benchmarking team being able to judge the gap between leading and less good performance, as well as planning considered actions to bring about changes to bridge that gap. These changes may be things that can be undertaken quickly, with little adaptation and at a minimum of cost and disruption. Such changes, often brought about by the effected operational team, are often called 'quick wins'. This type of change is incremental and carries low levels of risk but usually lower levels of benefit.

Quick wins will often give temporary or partial relief from the problems associated with poor performance and tend to address symptoms not the underlying 'diseases'. They can have a disproportionate favourable physiological impact upon the organizations. Used well, quick wins should provide a platform from which longer lasting changes may be made, having created a feeling of movement and success. All too often however, once quick wins are implemented there is a tendency to move on to other areas, without either fully measuring the impact of the change or getting to the root cause of a performance issue.

Quick wins are clearly an important weapon in effecting change but must be followed up properly to deliver sustainable business improvement through the adoption of best, or at least good practice. The changes needed to do this will usually be of a more fundamental nature and require investment in effort and money to implement. Such changes will need to be carefully planned and systematically implemented as a discrete change project or programme of projects. They carry substantial risk, if not systematically managed, and controlled, but they have the potential for significant improvement in performance. These types of change projects are sometimes referred to as 'step change' or 'breakthrough' projects/programmes (see Figure 9.2).

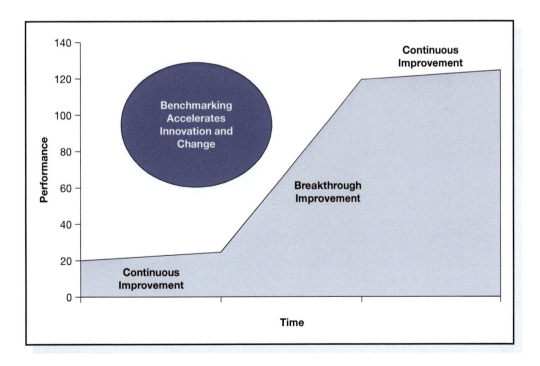

Figure 9.2
Benchmarking, breakthrough and continuous improvement

Benchmarking and change management   185

Whatever type of change is involved, a key ingredient of success is taking the people along. A first class communication strategy is required throughout and beyond any change activity, as well as the linked activity of stakeholder management. The benchmarking efforts need to fit into the change model deployed – such a framework is proposed in Figure 9.3. Many change models exist in diagrammatic form and are often, in both intent and structure, quite similar. Such a model may be considered as a 'footprint' that will lead to the chosen destination, in this case the desired

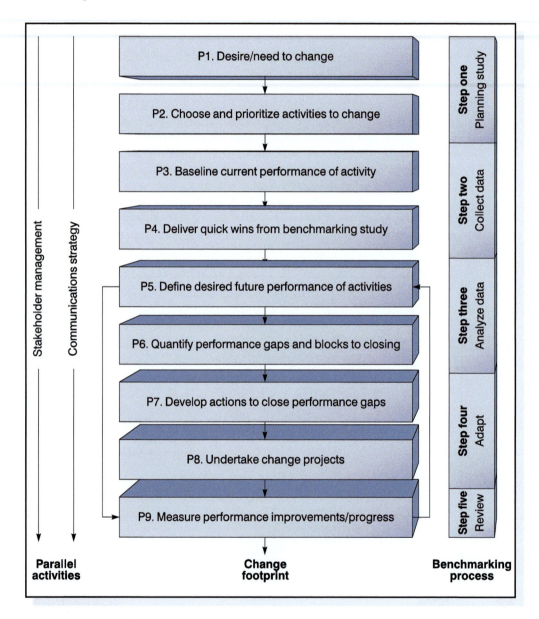

Figure 9.3
The benchmarking change footprint

performance improvements through adoption of best practice. The footprint in Figure 9.3 demonstrates where benchmarking activities link in to the general flow of change activity leading to better results.

The success and benefits derived from any benchmarking and change related activity are directly related to the excellence of the preparation. It is necessary to consider both the 'hard' and 'soft' aspects represented in Figure 9.3 and to systematically plan to meet and overcome any difficulties and challenges identified.

COMMUNICATING, MANAGING STAKEHOLDERS AND LOWERING BARRIERS

The importance of first class communication during benchmarking and resulting change can never be over-emphasized. A vital element of excellent communication is targeting the right audience with the right message in the right way at the right time. A scattergun approach to communication rarely has the intended impact.

In any benchmarking study it is a wise and well-founded investment in time and effort to define and understand the key stakeholders. The increased use of the term 'stakeholder' in business language – used to describe any group or individual that has some, however small, vested interest or influence in the proposed change – is to be welcomed. Stakeholders are frequently referred to in generic groupings and may be either internal or external to an organization or business. The importance of forming, managing and maintaining good working relations with these groups is widely acknowledged and accepted.

The reality is that this activity is frequently not performed well in benchmarking. A disgruntled or ignored stakeholder with high direct organizational power or influence can easily derail the intent and hard work of others. Stakeholders with less direct power or influence can, at best, provide an unwelcome and costly distraction from the main objectives of a benchmarking study. The art of stakeholder management is to proactively head off any major confrontations. This means really understanding the stakeholders' needs and their potential to do both good and ill.

The burden of effective stakeholder management rests with the benchmarking team charged with stimulating change. They may need the ongoing patronage and support of people outside their direct control. In any good benchmarking study early thought will be given to who the stakeholders are and this will be valuable input to developing a robust stakeholder management strategy.

The elements of successful stakeholder management should include:

1. defining and mapping the stakeholder groupings
2. analysing and prioritising these groupings
3. researching the key players in the most important groupings
4. developing a management strategy
5. deploying the strategy by tactical actions
6. reviewing effectiveness of the strategy and improving the future approach.

Objective measurement is also key to targeting change activity wisely. Benchmarking project budgets are often limited and it is good practice to target such discretionary spend at changes and improvements that will deliver the best return for their investment. Systematic measurement will provide a reliable baseline for making such decisions. By relating current performance against desired performance it should be possible to define both the gaps and appreciate the scale of improvements required to achieve the desired change.

Benchmarking studies add an extra dimension by understanding the levels of performance that best practices and leading organizations achieve. This allows realistic and sometimes uncomfortable comparisons with what an organization is currently able to achieve and what is possible. This is especially useful when setting stretch but realistic targets for future performance.

Base lining performance will allow teams to monitor and understand how successful they have been in delivering beneficial change. Used with care, as part of an overall communications strategy, successes on the road to achieving superior performance through change is a powerful motivator and useful influencing tool. Many organizations have clearly defined sets of performance measures, some self-imposed and some statute based – these should be used, if in existence. If the interest is in customer satisfaction and retention, for example, a generic but good starting point might be:

- internal measures (the lead/predictor measures) – production cycle times, unit costs, defect rate found (quality) and complaints resolved
- external measures (the lag/reality measures) – customer satisfaction (perception), customer retention and complaints received, time to market with a new product/service.

The benchmarking activity may provide teams with ideas on how they might change the way goods or services are produced and delivered. They will need to prioritize this opportunity, however, to deliver best value for time and money invested and to ensure the organization does not become paralysed by initiative overload – whilst making improvements the day job has to continue!

The benchmarking data collected will give a clear steer to the areas that require the most urgent attention but decisions will still have to be made. Measurement and benchmarking are tools not substitutes for management and leadership – the data on its own cannot make the decisions.

CHOOSING BENCHMARKING-DRIVEN CHANGE ACTIVITIES WISELY

As we have seen, benchmarking studies should fuel the desire to undertake change activities, but the excitement generated can allow the desire for change to take on a life of its own and irrational and impractical decisions can follow. These can result in full or partial failure to deliver the desired changes and waste of the valuable financial and people resources spent on the benchmarking itself.

Organizations should resist the temptation to start yet another series of improvement initiatives, without any consideration of their impact upon existing initiatives and the 'business as usual' activities, as well as the overall strategy, of course. It is important to target the change wisely and a number of key questions need to be answered including:

- Do we fully understand the scale of the change?
- Do we have the financial resources to support the change?
- Do we have the people resources to undertake the change?
- Do we have the right skills available to undertake the change?
- Do we fully understand the operational impact during the change?
- Can beneficial changes be made without major disruption to the business?
- Will the delivered change support achievement of our business goals?
- What will the new changes do to existing change initiatives?
- Is the organization culturally ready for change?

Table 9.4 shows a simple decision making tool to help consider the opportunities that are presented. The process may be viewed as a series of filters – it is assumed that the organization has defined business goals.

Work to improve quality or business management systems (Q/BMS) benefits from the use of benchmarking. This includes perhaps making the BMS more process based than previously and better use of web-based technology.

Benchmarking studies in various sectors have provided insight on, for example, the potential for new technology to radically change existing broadcasting processes. Benchmarking should be an integral part of each process re-engineering project that is undertaken. The external perspective provided by the benchmarking studies help employees see how things could be different ('thinking outside the box'), and provide valuable input to the steps required to implement new processes (see also Chapters 10 and 11).

Table 9.4 Simple decision tool for choosing change activities

No.	Filter Test	Yes	No
1	Does the benchmarking driven proposed change support the achievement of one or more of the defined business goals?	Allow the opportunity to move forward for consideration	Decline the opportunity or defer taking forward and schedule a review
2.	Does the change require financial and people resources above those agreed for the current budget round?	Prepare a business case within a project definition for consideration by senior management	Pass the opportunity to local operational management to undertake the changes as 'quick win' initiative
3	Will current improvement activity be adversely impacted by the envisaged new changes?	Consider the relative merits and benefits of new and existing change initiates and amalgamate or amend or cancel existing initiates	Allow change project to proceed and add to the controlled list of overall change projects
4.	Is the required additional financial and people resource needed to undertake new change projects available?	Senior management agree and sign off project definition and project begins	Senior management prioritize change activity agreeing necessary slippage or deferment or cancellation of some change projects

The drivers of change are everywhere but properly conducted systematic benchmarking studies can assist in defining clear objectives and help their effective deployment through well-executed change management. Best practice benchmarking and change management clearly are bedfellows. If well understood and integrated they can deliver lasting improvements in performance, which satisfy all stakeholder needs. Benchmarking is an efficient way to promote effective change by learning from the successful experiences of others and putting that learning to good effect.

A FRAMEWORK FOR ORGANIZATIONAL CHANGE

Based on research carried out by the European Centre *for* Business Excellence (www.ecforbe.com) and from other sources, an organizational change framework has been developed by the author and his colleagues. This is a powerful aid for organizations wishing to undertake any change programme, or who are in the processes of delivering change and want to increase their chances of success.

It identifies two main constructs of change management, which can be better understood within the overall framework for change as shown in Figure 9.4. Based on the results of the research, the change framework has two interacting cycles:

- Readiness for change
- Implementing change.

The experiences of many organizations that have launched change programmes, such as Lean, Six Sigma or a combination of the two, or wish to implement change following benchmarking, is that the first part – readiness – is often not at all well understood or developed. This can be caused by a desire to rush into implementation, with huge emphases on training programmes and projects without particular attachment to strategy.

To break into the top circle we need to start with the *Drivers of Change* – it is important to understand what are the key drivers for change inside or outside the organization, in order that the *Need for Change* may be understood and articulated to focus the stakeholders' desire for change. This is where leaders give meaning to the change, without which, as many organizations later discover, initial enthusiasm and energy quickly dissolves. For example, what were the drivers for the introduction of digital technology into BBC World Service – reduced costs, better programme reception, more effective programme making? Good output from the need for change stage will focus the stakeholders' desire for change. Clarity on this is key, as from it derives clear and consistent *Leadership and Direction* to turn the need into expectations – vision, goals, measured objectives and targets, which will set the expectations for change. Robust *Planning* then allows the priorities to emerge and focuses people's minds on the strategic objectives.

The implementation of change is a rich tapestry of potential failure – a minefield for the unsuspecting. Worse than that, most managers tend to find they have entered the minefield at the wrong point. Trying to change behaviours, for example, is a frequent starting point for many 'change programmes' which will include such matters as attitudes and empowerment, without bedding these things in the reality, which is the business.

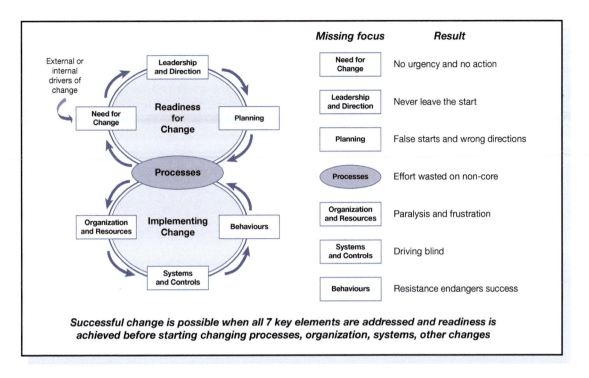

Figure 9.4
The organizational change framework

Following clarity on need, clear and unambiguous leadership, direction and good detailed planning, the first port of call must be the organizational *Processes* in which people live and work. Whether we like it or not, and whether we have worked them out or not, the processes drive the way the *Organization and Resources* work – the structure, roles, competencies and resources deployed. Performance measures and technology then support the organization's *Systems and Controls*. This is where *Behaviour* comes in – all of the above drives behaviour – the way the organization is structured, who my boss is, how I am measured, the processes and systems – good or bad – that I live and work in. When managers talk about attitudes of the people and their assumptions, it might be interesting for them to understand where these come from. Attitudes stem from beliefs and values, both of which are management's responsibility to influence. Most people start work for an organization with positive attitudes and behaviours and it is frequently the systems, the environment and the assumptions and unwritten 'rules' that cause problems and deterioration (see also Chapter 11 and the section on 'Assumption Busting').

The 'figure of 8' closes when we return to the process for it is our behaviour, which makes the processes work or not, resulting in achievements in quality and on time delivery or not and reinforcing the change, or not. Taking another trip round the figure of 8 will verify the change protocols and ensure that anchorage to the strategies is maintained.

It is often difficult for managers to stand back and view their work on change in a holistic fashion. Personal agendas can lead to a push on human resource issues or information technology issues, preventing the holistic view. An output of the 'leadership' box may be scenarios for reacting to the need for change . . . there's usually more than one way to address the need and managers often benefit from a structured approach that drives questions that lead to alternative routes to change.

Figure 9.4 also shows on the right hand side the likely results of missing focus in any of the seven key elements. This can act as a simple diagnostic, starting with the result or effect and working back to the likely area(s) lacking time and attention. The figure of 8 framework can be adapted to any aspect of change management; Figures 9.5 and 9.6 show how it has been used to develop strategies for carbon emission reduction, for example.

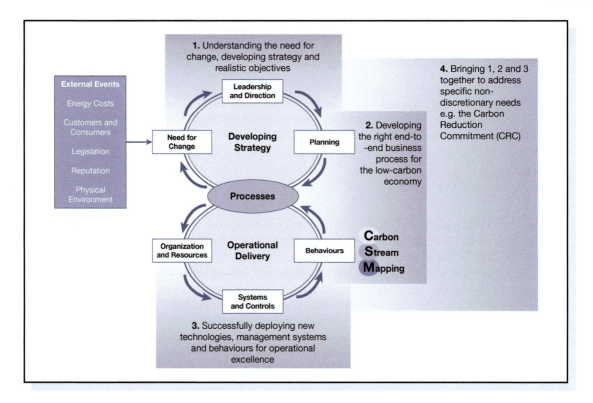

Figure 9.5
Carbon reduction strategy using figure of 8 change framework

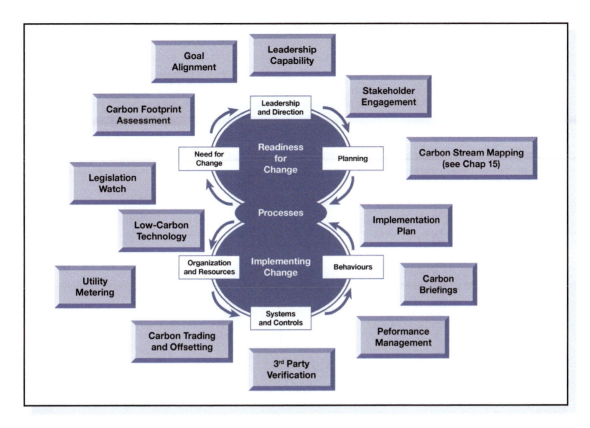

Figure 9.6
Carbon reduction aspects mapped onto figure of 8 framework

ACKNOWLEDGEMENT

The author is grateful for the contribution made in the preparation of this chapter by his colleague Robin Walker.

BIBLIOGRAPHY

Bendell, T., *Benchmarking for Competitive Advantage*, Longman, London, 1993.

Camp, R.C., *Business Process Benchmarking: Finding and Implementing Best Practice*, ASQ Quality Press, Milwaukee, 1995.

Camp, R.C., *Global Cases in Benchmarking: Best Practices from Organisations around the World*, Quality Press, Milwaukee, 1998.

Spendolini, M.J., *The Benchmarking Book*, ASQ, Milwaukee, 1992.

Zairi, M., *Benchmarking for Best Practice*, Butterworth-Heinemann, Oxford, 1996.
Zairi, M., *Effective Management of Benchmarking Projects*, Butterworth-Heinemann, Oxford, 1998.

CHAPTER HIGHLIGHTS

The why and what of benchmarking

- Benchmarking measures an organization's products, services and processes to establish targets, priorities and improvements, leading in turn to competitive advantage and/or cost reductions.
- Benefits of benchmarking can be numerous and include creating a better understanding of the current position, heightening sensitivity to changing customer needs, encouraging innovation, developing stretch goals, and establishing realistic action plans.
- Data from APQC suggests an average benchmarking study takes six months to complete, occupies a quarter of the team members' time and the average return was five times the costs.
- The four basic types of benchmarking are: internal, functional, generic and competitive, although the evolution of benchmarking in an organization is likely to progress through focus on continuous improvement.

The purpose and practice of benchmarking

- The evolution of benchmarking is likely to progress through four focuses: competitive products/services; industry best practices; all aspects of the business; in terms of performance gaps; and ending with focus on processes and true continuous improvement.
- The purpose of benchmarking is predominantly to change perspective, compare business practices, challenge current practices and processes, and to create improved goals and practices, with the focus on customer satisfaction and business results.
- A simple scoring proforma may help an organization to assess whether it is ready for benchmarking, if it has not engaged in it before. Help may be required to establish the right platforms if low scores are obtained.
- The benchmarking process has five main stages: plan, collect, analyse, adapt and review. These are focussed on trying to measure comparisons and identify areas for action and change.

The role of benchmarking in change

- An important aspect of benchmarking is gaining an understanding of how other organizations achieve superior performance. Some of this knowledge will result in 'quick wins', with low risk but relatively low levels of benefit.
- Step changes are of a more fundamental nature, usually require further investment in time and money, will need to be carefully planned and systematically implemented, and typically carry a higher risk.

- A change model or 'footprint' should lead to the chosen destination – improved performance through the adoption of best practice – and show the role of benchmarking.

Communicating, managing stakeholders and lowering barriers

- Communication is vital during change, and a vital element is targeting the right audience, with the right message, in the right way at the right time.
- Defining and understanding the key stakeholders is a wise investment of time. This should be followed by building and managing good relationships. This falls on the benchmarking team.
- Elements of successful stakeholder group management include: defining and mapping; analysing and prioritising; research key players/groups; developing and deploying a strategy; and reviewing effectiveness.
- Objective measurement is key to targeting change wisely and provides a reliable baseline for decisions. Baselining performance allows teams to monitor and understand success in delivering beneficial change.

Choosing benchmarking driven change activities wisely

- Organizations should start benchmarking driven improvement activities only with consideration of their impact on existing initiatives. Questions to be asked include those related to the scale of the change, the financial and people resources (including skills) required, the impact and disruption aspects, the degree of support to the business goals, and the cultural implications.
- Benchmarking may be used to drive revisions in business management systems, facilitate the application of new technologies, and generally to help people to see how processes might be different.
- Properly conducted systematic benchmarking studies can aid the definition of clearer objectives and help their deployment through well executed change management.

A framework for organizational change

- Based on research an organizational change framework has been developed to provide a powerful aid for organizations wishing to undertake any change programme, or who were in the processes of delivering change and want to increase their success.
- The framework, which is in the shape of a figure of 8, identifies two main constructs of change management in the form of two interacting cycles, readiness for change (strategic) and implementing change (operational).
- The figure of 8 framework may be used as a simple diagnostic based on likely outcomes from missing area focus and adapted for use in many areas, including carbon emission reduction strategies.

Part III Discussion questions

1. a) Using the expression: 'if you don't measure you can't improve', explain why measurement is important in service delivery improvement, giving examples.
   b) Using your knowledge of process management, show where measurement should take place in a global manufacturing and distribution company and how should it be conducted.

2. Discuss the important features of a performance measurement system based on a TQM approach. Suggest an implementation strategy for a performance measurement system in a progressive company which is applying TQM principles to its business processes.

3. It is often said that 'you can't control what you can't measure and you can't manage what you can't control'. Measurement is, therefore, considered to be at the heart of managing business processes, activities and tasks. What do you understand by improvement-based performance measurement? Why is it important? Suggest a strategy of introducing TQM-based performance measurement for an organization of your own choice in the public sector.

4. List the main categories of the US Baldrige Performance Excellence Model. How may such criteria be used as the basis for a self-assessment process?

5. Self-assessment using the EFQM Excellence Model criteria enables an organization to systematically review its business processes and results. Briefly describe the criteria and discuss the main aspects of self-assessment.

6. Self-appraisal or assessment against a hybrid 'Excellence Model' can be used by organizations to monitor their progress. Design the criteria for such a hybrid framework and explain the steps that an organization would have to follow to carry out a self-assessment. How could self-assessment against the model be used in a large multi-site organization to drive continuous improvement? What additional requirements would be introduced for an organization that was asked by customers to assess against the CMMI framework?

7. Benchmarking is an important component of many companies' improvement strategies. What do you understand by benchmarking? How does benchmarking link with performance measurement? Suggest a strategy for integrating benchmarking into a TQM approach.

8. a) Some people would argue that benchmarking is not different from competitor analysis and is a practice that organizations have always carried out. Do you agree with this? How would you differentiate benchmarking and what are its key elements?
   b) Suggest and describe in full an approach to change management that would be suitable for implementing the findings from benchmarking studies for a progressive company that has no previous knowledge or experience of doing this.

9 a) What are the major limitations of the 'Prevention-Appraisal-Failure (PAF)' costing model? Why would the process cost model be a better alternative?
   b) Discuss the link between benchmarking and quality costing.

10. A construction company is concerned about its record of completing projects on time. Considerable penalty costs are incurred if the company fails to meet the agreed contractual completion date. How would you investigate this problem and what methodology would you adopt?

Part IV

Processes

I must Create A System, or be enslav'd by another Man's.

*William Blake, 1757–1827,
from 'Jerusalem'*

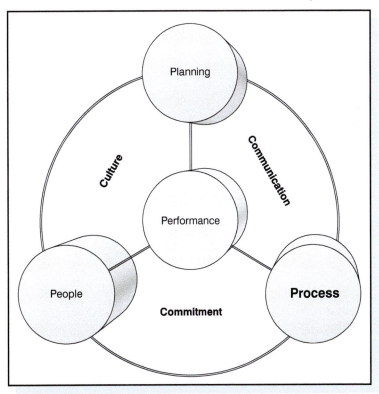

Chapter 10

Process management

THE PROCESS MANAGEMENT VISION

Organizations create value by delivering their products and/or services to customers. Everything they do in that whole chain of events is a process. So to perform well in the eyes of the customers and the stakeholders, all organizations need very good process management – underperformance is primarily caused by poor processes and/or their interaction with people and technology.

In 1999 Fujio Cho, President of the Toyota Motor Company sent this chilling message to the market place: 'We get brilliant results from average people managing brilliant processes – while our competitors get average or worse results from brilliant people managing broken processes.' This comparison between organizations is as true today as it was then, yet we still see many executives and managers in organizations all over the world, large and small, who are struggling to find how to implement the relatively simple but ubiquitously elusive concepts behind this message.

In recent times we have seen organizations adopting a host of different approaches to improving performance: in addition to Total Quality Management (TQM), there has been Statistical Process Control (SPC), Business Process Reengineering (BPR), Lean, Six Sigma (and Lean Six Sigma!) Hoshin Kanri, Taguchi Methods, Business Process Improvement (BPI), etc. Many of these approaches are associated with their own 'technical' jargon; some fall out of favour and others seem to gain ground in particular organizations. What all these methodologies have in common, however, is a focus on processes. At a fundamental level then performance improvement is about changing the way that organizations create and deliver value to their customers through processes, regardless of what labels may be used to describe the approach.

This is recognized, of course, in the EFQM Excellence Model, in which the Processes criterion is the central 'anchor' box linking the other enablers and the results together. The devil is in the detail, of course, and successful exponents of 'process excellence' understand all the dimensions related to:

- Process Strategy – particularly deployment
- Operationalizing Processes – including definition/design/systems

- Process Performance – measurement and improvement
- People and leadership roles – values, beliefs, ownership, responsibilities, accountabilities, authorities and rewards
- Information and knowledge – capturing and leveraging throughout the supply chains.

Where process management is established and working, executives no longer see their organizations as sets of discrete vertical functions with silo-type boundaries. Instead they visualize things from the customer perspective – as a series of inter-connected work and information flows that cut horizontally across the business. Effectively the customers are pictured as 'taking a walk' through some or all of these 'end-to-end' processes and interfacing with the company or service organization, experiencing how it generates demand for products and services, how it fulfils orders, services products, etc (Figure 10.1). All these processes need managing – planning, measuring and improving – sometimes discontinuously.

In monitoring process performance, measurement will inevitably identify necessary improvement actions. In many process managed companies they have shifted the focus of the measurement systems from functional to process goals and even based remuneration and career advancement on process performance.

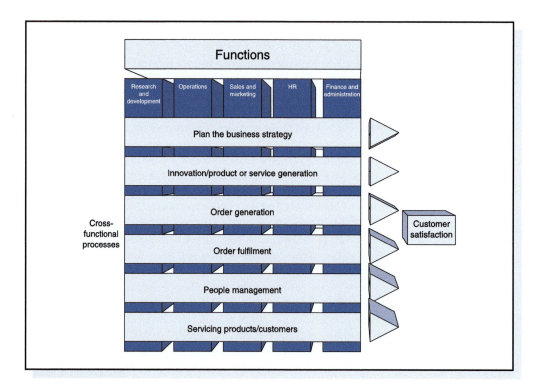

Figure 10.1
Cross-functional approach to managing core business processes

Operationalizing process management

Top management in many organizations now base their approach to business on the effective management of 'key or core business processes'. These are well-defined and developed sequences of steps with clear rationale, which add value by producing required outputs from a variety of inputs. Moreover these management teams have aligned the core processes with their strategy, combining related activities and cutting out ones that do not add value. This has led in some cases to a fundamental change in the way the place is managed and the changes required have caused these organizations to emerge as true 'process enterprises'. There are many such organizations, including those featured in the case studies section of this book.

Companies comprising a number of different business units, such as outsourcing companies, face an early and important strategic decision when introducing process management – should all the business units follow the same process framework and standardization, or should they tailor processes to their own particular and diverse needs? Each organization must consider this question carefully and there can be no one correct approach.

Deployment of a common high-level process framework throughout the organization gives many benefits, including presenting 'one company' to the customers and suppliers, lower costs and increased flexibility, particularly in terms of resource allocation. An example of a high level process framework for a large complex outsourcing company is shown in Figure 10.2.

In continuing research on award-winning companies, the author and his colleagues have identified process management best practices as:

- Identifying the key business processes
  - prioritizing on the basis of the value chain, customer needs and strategic significance, and using process models and definitions
- Managing processes systematically
  - giving process ownership to the most appropriate individual or group and resolving process interface issues through meetings or ownership models
- Reviewing processes and setting improvement targets
  - empowering process owners to set targets and collect data from internal and external customers
- Using innovation and creativity to improve processes
  - adopting self-managed teams, business process improvement and idea schemes
- Changing processes and evaluating the benefits
  - through process improvement or re-engineering teams, project/programme management and involving customers and suppliers.

Too many businesses and organizations generally are still not process oriented, however; they focus instead on tasks, on jobs, on the people who do them and on structures.

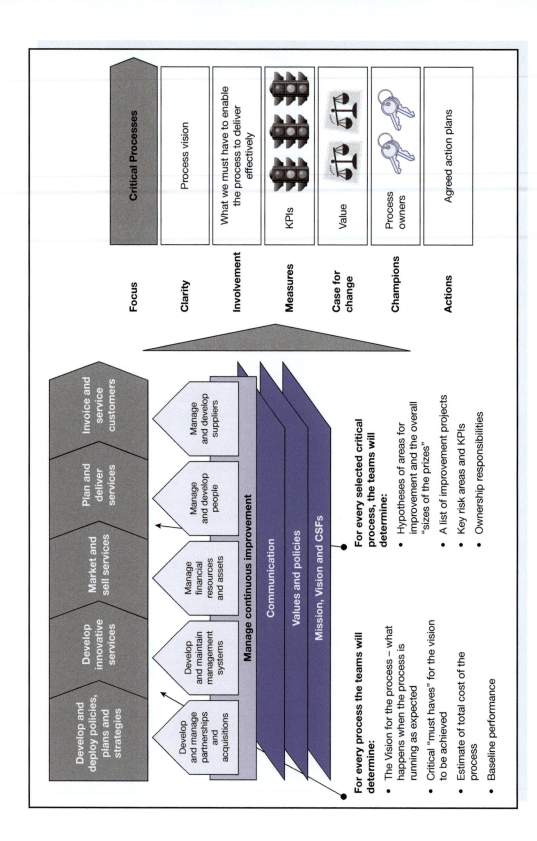

Figure 10.2
Example of a high-level process framework

THE PROCESS CLASSIFICATION FRAMEWORK AND PROCESS MODELLING

In establishing a high level or core process framework, many organizations have found inspiration in the Process Classification Framework developed by the American Productivity and Quality Centre (APQC). With the assistance of several major international corporations, the APQC have created and developed a high-level generic enterprise model, a taxonomy of cross functional business processes that should encourage businesses and other organizations to see their activities from a cross-industry, process viewpoint rather than from a narrow functional viewpoint. The intention is to allow the objective comparison of performance within and among organizations.

The Process Classification Framework supplies a generic view of business processes often found in multiple industries and sectors – manufacturing and service companies, health care, government, education and others. It seeks to represent major processes and sub-processes through its structure (Figure 10.3) and vocabulary

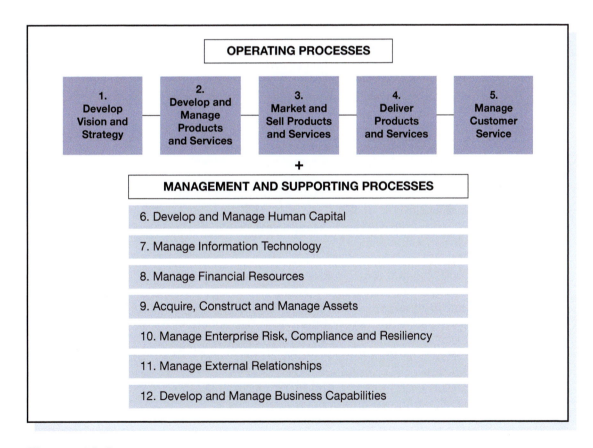

Figure 10.3
APQC process classification framework – overview

(available in detail as a download from APQC: www.apqc.org). The framework does not list all processes within any specific organization; likewise, not every process listed in the framework is present in every organization.

The sub-processes listed under the high level processes shown in Figure 10.3 are as follows:

1. Develop vision and strategy.
   1.1 Define the business concept and long-term vision.
   1.2 Develop business strategy.
   1.3 Manage strategic initiatives.
2. Develop and manage products and services.
   2.1 Manage product and service portfolio.
   2.2 Develop products and services.
3. Market and sell products and services.
   3.1 Understand markets, customers and capabilities.
   3.2 Develop marketing strategy.
   3.3 Develop sales strategy.
   3.4 Develop and manage marketing plans.
   3.5 Develop and manage sales plans.
4. Develop products and services.
   4.1 Plan for and align supply chain resources.
   4.2 Procure materials and services.
   4.3 Produce/manufacture/deliver product.
   4.4 Deliver service to customer.
   4.5 Manage logistics and warehousing.
5. Manage customer services.
   5.1 Develop customer care/customer service strategy.
   5.2 Plan and manage customer service operations.
   5.3 Measure and evaluate customer service operations.
6. Develop and manage human capital.
   6.1 Develop and manage human resources (HR) planning, policies and strategies.
   6.2 Recruit, source and select employees.
   6.3 Develop and counsel employees.
   6.4 Reward and retain employees.
   6.5 Redeploy and retire employees.
   6.6 Manage employees' information.
7. Manage information technology.
   7.1 Manage the business of information technology.
   7.2 Develop and manage IT customer relationships.
   7.3 Develop and implement security, privacy and data protection controls.
   7.4 Manage enterprise information.
   7.5 Develop and maintain information technology.
   7.6 Deploy information technology solutions.
   7.7 Deliver and support information technology solutions.
8. Manage financial resources.
   8.1 Perform planning and management accounting.
   8.2 Perform revenue accounting.

8.3 Perform general accounting and reporting.
8.4 Manage fixed-asset project accounting.
8.5 Process payroll.
8.6 Process accounts payable and expense reimbursements.
8.7 Manage treasury operations.
8.8 Manage internal controls.
8.9 Manage taxes.
8.10   Manage international funds/consolidation.
9. Acquire, construct and manage assets.
9.1 Design and construct/acquire non-productive assets.
9.2 Plan maintenance work.
9.3 Obtain and install assets, equipment and tools.
9.4 Dispose of productive and non-productive assets.
10. Manage enterprise risk, compliance and resiliency.
10.1 Manage enterprise risk.
10.2 Manage business resiliency.
10.3 Manage environmental health and safety.
11. Manage external relationships.
11.1 Build investor relationships.
11.2 Manage government and industry relationships.
11.3 Manage relations with board of directors.
11.4 Manage legal and ethical issues.
11.5 Manage public relations programme.
12. Develop and manage business capabilities.
12.1 Manage business processes.
12.2 Manage portfolio, program and project.
12.3 Manage quality.
12.4 Manage change.
12.5 Develop and manage enterprise-wide knowledge management (KM) capability.
12.6 Measure and benchmark.

The Process Classification Framework can be a useful tool in understanding and mapping business processes. In particular, a number of organizations have used the framework to classify both internal and external information for the purpose of cross-functional and cross-divisional communication. It is a continually evolving document and the APQC will continue to enhance and improve it on a regular basis. To that end, the Center welcomes your comments, suggestions for improvement, and any insights you gain from applying it within your organization. The APQC would like to see the Process Classification Framework receive wide distribution, discussion, and use. Therefore, it grants permission for copying the framework, as long as acknowledgement is made to the American Productivity & Quality Center (www.apqc.org; email: pcf_feedback@apqc.org).

Process modelling

As we saw in Chapter 1 a process is simply something that converts a set of inputs into outputs. The inputs can include materials or information and are supplied into the process externally or internally. The outputs of a process go to a customer – again external or internal. This simple explanation may be represented as the flow of SIPOC: (**Supplier** → **Inputs** → **PROCESS** → **Outputs** → **Customer**). Any substantial process may be broken down into its main steps or sub-processes, all of which have the same SIPOC flow.

Many years ago, the United States Air Force adopted 'Integration DEFinition Function Modelling' (IDEFØ), as part of its Integrated Computer-Aided Manufacturing (ICAM) architecture. IDEFØ is a method designed to model the decisions, actions and activities of an organization or system. IDEFØ was derived from a well-established graphical language, the Structured Analysis and Design Technique (SADT). The US Air Force commissioned the developers of SADT to develop a function modelling method for analysing and communicating the functional perspective of a system.

In December 1993, the Computer Systems Laboratory of the National Institute of Standards and Technology (NIST) released IDEFØ as a standard for Function Modelling in Federal Information Processing Standards (FIPS) Publication 183. This provides a useful structured graphical framework for describing and improving business processes. The associated 'Integration Definition for Information Modelling' (IDEFIX) language allows the development of a logical model of data associated with processes, such as measurement.

These techniques are widely used in business process re-engineering (BPR) and business process improvement (BPI) projects, and to integrate process information. A range of specialist software is also available to support the applications. IDEFØ may be used to model a wide variety of new and existing processes, define the requirements and design an implementation to meet the requirements.

An IDEFØ model consists of a hierarchical series of diagrams, text and glossary, cross-referenced to each other through boxes (process components) and arrows (data and objects). The method is expressive, comprehensive and capable of representing a wide variety of business, service and manufacturing processes. The relatively simple language allows coherent, rigorous and precise process expression, and promotes consistency. Figure 10.4 shows the basis of the approach.

For a full description of the IDEFØ methodology, it is necessary to consult FIPS Publication 183. It should be possible, however, from the simple description given here, to begin process modelling or mapping using the technique.

Processes can be any combination of things, including people, information, software, equipment, systems, products or materials. The IDEFØ model describes what a process does, what controls it, what things it works on, what means it uses to perform its functions and what it produces. The combined graphics and text are comprised of:

- **Boxes** – which provide a description of what happens in the form of an active verb or verb phrase.
- **Arrows** – which convey data or objects related to the processes to be performed (they do not represent flow or sequence as in the traditional process flow model).

Each side of the process box has a standard meaning in terms of box/arrow relationships. Arrows on the left side of the box are **inputs**, which are transformed or consumed by the process to produce **output** arrows on the right side. Arrows entering

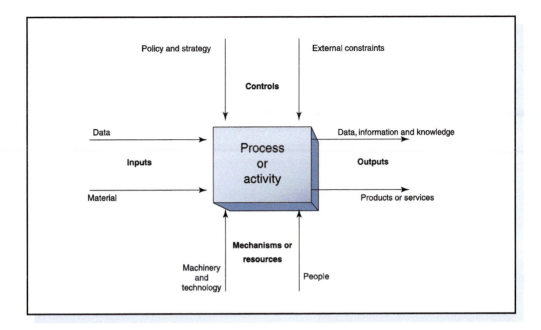

Figure 10.4
IDEFØ model language

the top of the box are **controls** which specify the conditions required for the process to generate the correct outputs. Arrows connected to the bottom of the box represent **'mechanisms'** or **resources**. The abbreviation ICOR (inputs, controls, outputs, resources) is sometimes used.

Using these relationships, process diagrams are broken down or decomposed into more detailed diagrams, the top-level diagram providing a description of the highest level process. This is followed by a series of child diagrams providing details of the sub-processes (see Figure 10.5).

Each process model has a top-level diagram on which the process is represented by a single box with its surrounding arrows (e.g. Figure 10.6). Each sub-process is modelled individually by a box, with parent boxes detailed by child diagrams at the next lower level. An example of the application of IDEF or ICOR modelling in Crime Management and Reporting in a regional police force is given in Figures 10.7 to 10.9.

Text and glossary

An IDEFØ diagram may have associated structured text to give an overview of the process model. This may also be used to highlight features, flows and inter-box connections, and to clarify significant patterns. A glossary may be used to define acronyms, key words and phrases used in the diagrams.

Arrows

Arrows on high-level IDEFØ diagrams represent data or objects as constraints. Only at low levels of detail can arrows represent flow or sequence. These high-level arrows

Figure 10.5
IDEFØ decomposition
structure – sub-
processes

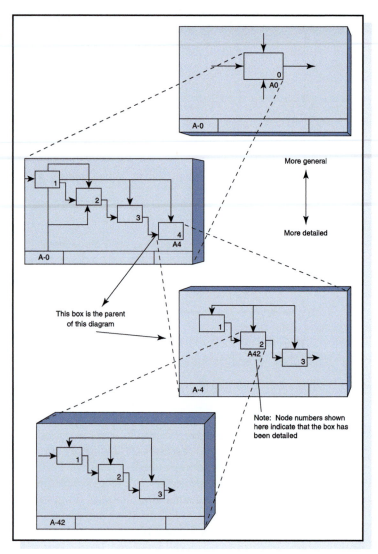

may usefully be thought of as pipelines or conduits with general labels. An arrow may branch, fork or join, indicating that the same kind of data or object may be needed or produced by more than one process or sub-process.

IDEFØ process modelling, improvement and teamwork

The IDEFØ methodology includes procedures for developing and critiquing process models by a group or team of people. The creation of an IDEFØ process model provides a disciplined teamwork procedure for process understanding and improvement. As the group works on the process following the discipline, the diagrams are changed to reflect corrections and improvements. More detail can be added by creating more diagrams, which in turn can be reviewed and altered. The final model represents an agreement on the process for a given purpose and from a given viewpoint, and can be the basis of new process or system improvement projects.

In the police force example (Figures 10.6–10.9), teams involved in constructing these models gained appreciation of the role everyone plays in the overall process. This encouraged ownership of the process and acted as a spur to making improvements.

Initially the overall crime management process was considered at the macro level, identifying its various input, output, control and resource factors through a combination of brainstorming and consultation with managers. This allowed a high level model to be compiled, as shown at Figure 10.8. Although the model reveals little in terms of the process interactions, it is helpful in illustrating the importance of crime management to the police force.

The outputs of the crime management process arguably represent the most important outputs of any police force. They include the levels of recorded crime, as well as the proportion of these offences that have been successfully detected. The publication of this information is of great interest to the public, the force and its political stakeholders, forming perceptions of safety and assessment of organizational performance. Also produced by the process are prosecution files for the courts and crime prevention strategies, often part of community safety partnerships. These outputs, though often less publicised, are measures of activities vital for the medium and long-term reduction of crime.

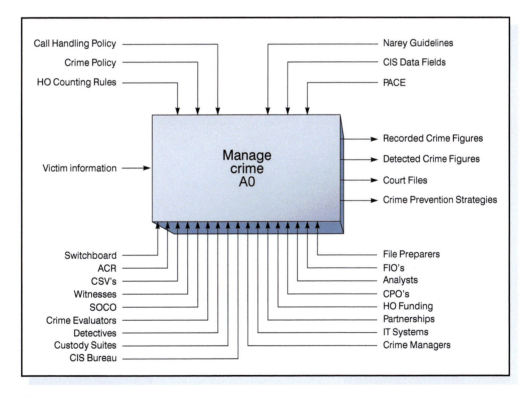

Figure 10.6
A0 crime management

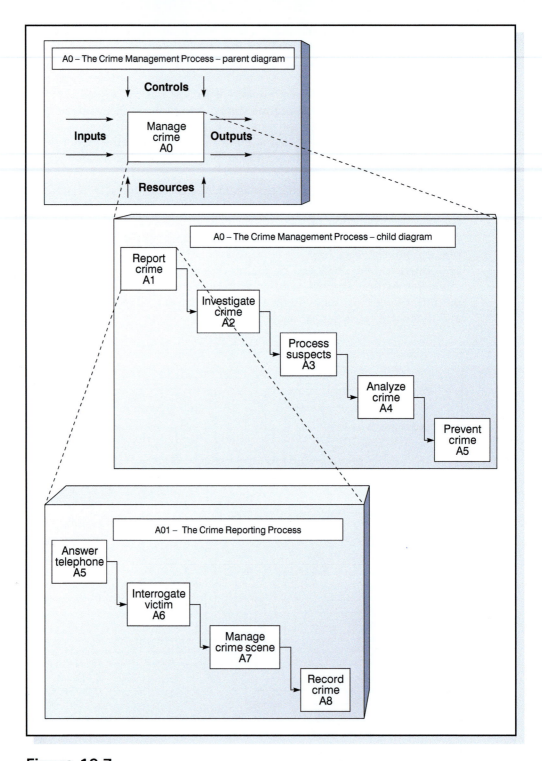

Figure 10.7
IDEFØ decomposition structure – sub-processes – for crime management

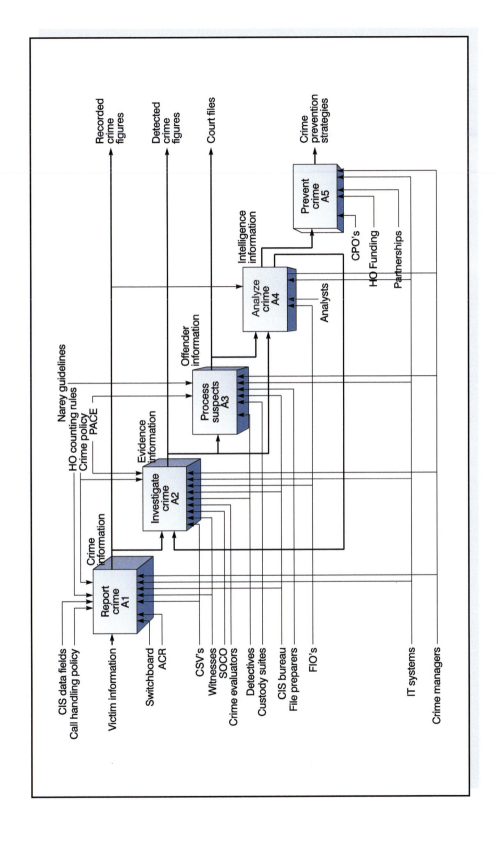

Figure 10.8
A0 crime management child diagram

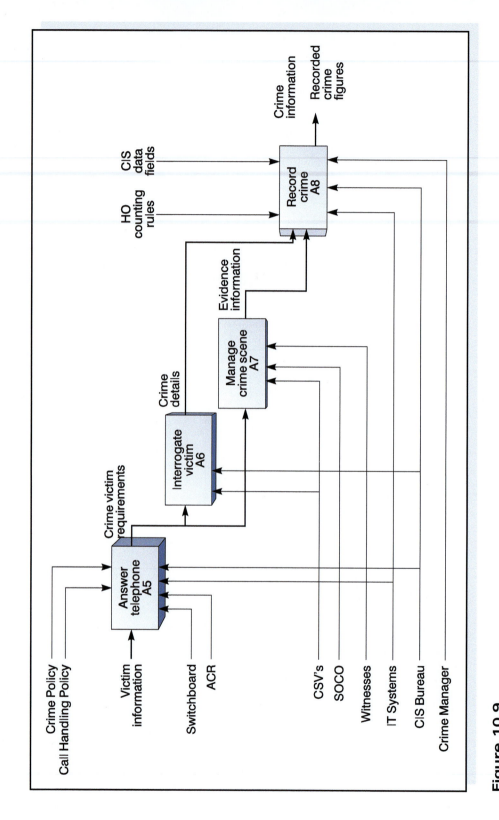

Figure 10.9
A01 – report crime

The range of resources applied by the regional police force to its crime management process covers a large proportion of its functional areas. It includes uniform patrol, Criminal Investigation Department (CID), Intelligence, Communication and Case Preparation. It gives an indication of the complexity, but also the appropriateness of using a process approach to manage improvement across the departmental structures of the force.

Having established the process at its macro level, the process was broken down into the main sub-processes of which it was composed. It was not possible to be prescriptive as to how many sub-processes this entailed. However, they needed to be sufficient to cover the range of the crime management process, and ensure that all the original input, output, control and resource factors could be re-assigned on the child diagram.

Crime reporting was the initial process providing the interface with the victim and included the telephone crime reporting system. Crime investigation was the next process involving the greatest concentration of organizational resources, particularly those of an operational nature. The processing of suspects similarly attracted a heavy concentration of resources, reflecting the sometime onerous and even bureaucratic demands of its main output – court files. The crime analysis process would not have featured in similar process model years ago, but is now an indispensable process in tackling crime, made possible through advances in IT. Finally, the crime prevention process has risen in prominence through community safety partnerships.

The IDEFØ /ICOR methodology allows processes to be broken down to as many levels as are required by the process under investigation. In this study, one further level was required in order to examine the crime reporting process in sufficient detail. This involved taking the crime reporting sub-process A1 from Figure 10.8 and decomposing its own constituent sub-processes.

This final level was composed of four sub-processes: telephone answering where the victim's call was handled between the switchboard, ACR and the CIS Bureau; victim interrogation where information pertaining to the crime was obtained directly through dialogue with the victim, either in person or by phone; crime scene management, which is the obtaining of forensic evidence or identifying potential witnesses at the crime scene; crime recording, where a detailed bundle of crime information is entered onto the CIS database as a recorded crime. The detailed sub-process model is shown in Figure 10.9.

In using such techniques in process management there can be a propensity for maps to assume disproportionate importance. This can result in participants becoming distracted in pursuit of accuracy or even the overall purpose of process improvement being supplanted by the modelling process itself – this is to be avoided at all cost!

IDEFIX

This is used to produce structural graphical information models for processes, which may support the management of data, the integration of information systems, and the building of computer databases. It is described in detail in the FIPS Publication 184 (1993). Its use is facilitated by the introduction of IDEFØ modelling for process understanding and improvement.

A number of commercial software packages, which support IDEFØ and IDEFIX implementation, are available.

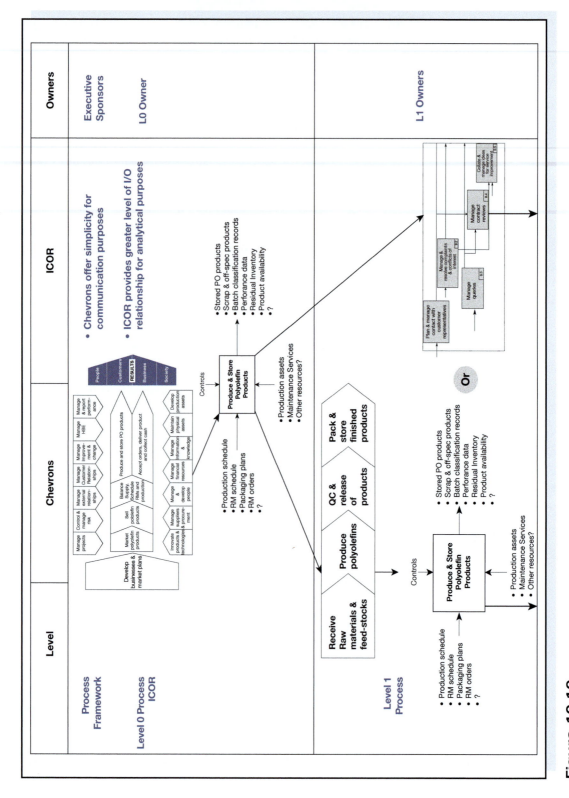

Figure 10.10
Summary of process-mapping approaches

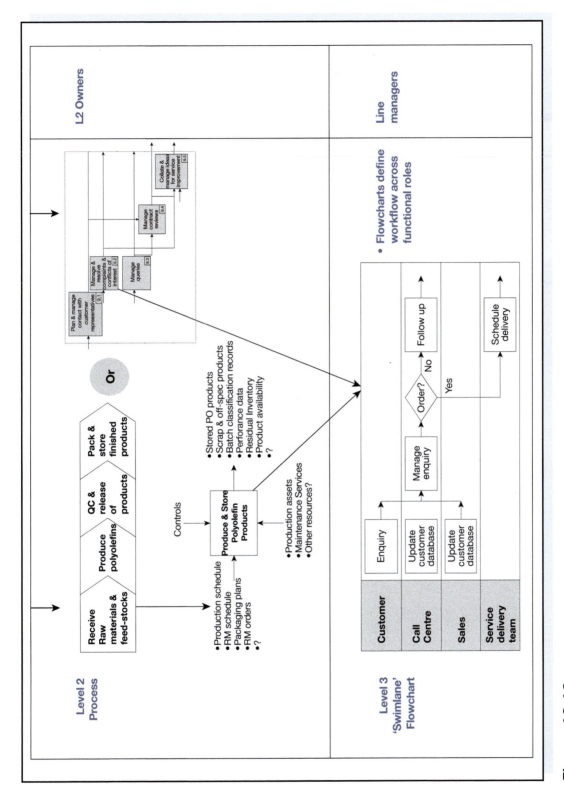

Figure 10.10
Continued

Levels of process detail

The management and ownership of processes takes place at various levels in an organization. For example, the high level process framework for the organization and strategic processes, such as 'define company direction,' are owned at the executive level, with tactical processes such as 'deal with an enquiry,' at lower levels. The detail required at each level can be provided by process chevron diagrams which simply show and list activities and process analysis techniques such as IDEFØ /ICOR maps or flow-charting and swim lane diagrams (see Figure 10.10). As one chief executive commented, 'I need to understand the process, but I don't want to see pages of wiring diagrams!' A key challenge in making best use of process management tools is to pitch the definition at the most appropriate level, bearing in mind the nature of the improvement likely.

PROCESS FLOWCHARTING

Another powerful method of describing a process is flowcharting which owes much to computer programming, where the technique is used to arrange the sequence of steps required for the operation of the programme. It has a much wider application, however, than computing.

Certain standard symbols are used on flowcharts which are shown in Figure 10.11. The starting point of the process is indicated by a circle. Each processing step, indicated by a rectangle, contains a description of the relevant operation, and where the process ends is indicated by an oval. A point where the process branches because of a decision is shown by a diamond. A parallelogram relates to process information but is not a processing step. The arrowed lines are used to connect symbols and to indicate direction of flow. For a complete description of the process, all operation steps (rectangles) and decisions (diamonds) should be connected by pathways to the start circle and end oval. If the flowchart cannot be drawn in this way, the process is not fully understood.

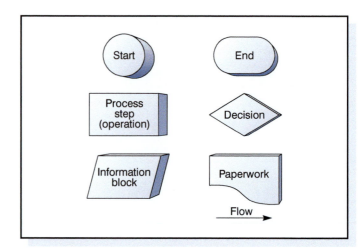

Figure 10.11
Flowcharting symbols

It is a salutary experience for most people to sit down and try to draw the flowchart for a process in which they take part every working day. It is often found that:

- The process flow is not fully understood.
- A single person is unable to complete the flowchart without help from others.

The very act of flowcharting will improve knowledge of the process, and will begin to develop the teamwork necessary to find improvements. In many cases, the convoluted flow and octopus-like appearance of the chart will highlight unnecessary movements of people and materials and lead to common-sense suggestions for waste elimination.

Figures 10.12 and 10.13 provide a before and after example of flow-charting in use to improve a travel booking procedure in a company. The total time taken for the starting or 'as-is' procedure, excluding the correction of any errors and the preparation of overview reports, was 23 minutes per travel request, the flowchart for the process shown in Figure 10.12. An improvement team was set up to analyse the process and make recommendations for improvement, using brainstorming and questioning techniques. They made proposals to change the procedure and the flowchart for the improved or 'to-be' process is shown in Figure 10.13. The proposal reduced the total administrative effort per travel request (or per travel arrangement, because the travel request was eliminated) from 23 minutes to 5 minutes.

The details that appear on a flowchart for an existing process should be obtained from direct observation of the process, not by imagining what is done or what should be done. The latter may be useful, however, in the planning phase, or for outlining the stages in the introduction of a new concept. Such an application is illustrated in Figure 10.14 for the installation of statistical process control (SPC) charting systems (see Chapter 13). Similar charts may be used in the planning the implementation of quality management systems.

It can be surprisingly difficult to draw flowcharts for even the simplest processes, particularly managerial ones, and following the first attempt it is useful to ask whether:

- The facts have been correctly recorded.
- Any over-simplifying assumptions have been made.
- All the factors concerning the process have been recorded.

The author has seen too many process flowcharts that are so incomplete as to be grossly inaccurate. Flowcharts should provide excellent documentation and be useful trouble-shooting tools to determine how each step is related to the others. By reviewing the flowchart, it should be possible to discover inconsistencies and determine potential sources of variation and problems. For this reason, flowcharts are very useful to process improvement teams when examining an existing process to highlight the problem areas. A group of people, with the knowledge about the process, should take the following simple steps:

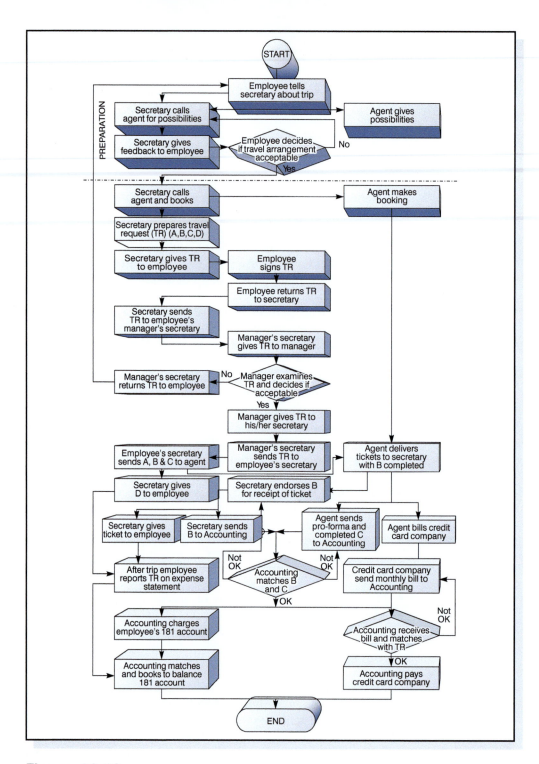

Figure 10.12
Original process for travel procedure

1. Draw a flowchart of existing process.
2. Draw a second chart of the flow the process could or should follow.
3. Compare the two to highlight the changes necessary.

A number of commercial software packages which support process flowcharting are available.

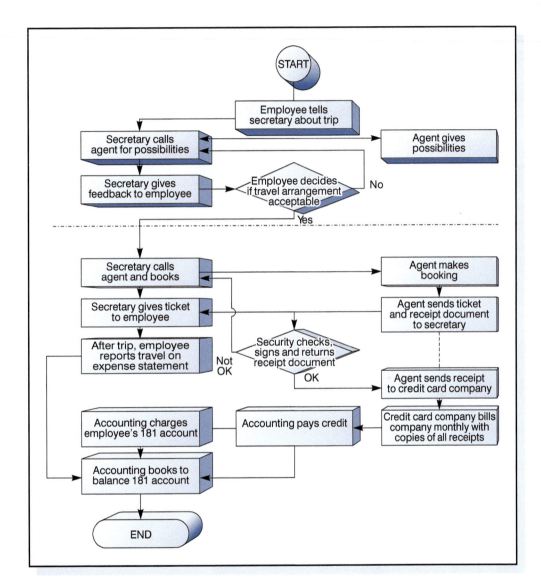

Figure 10.13
Improved travel procedure

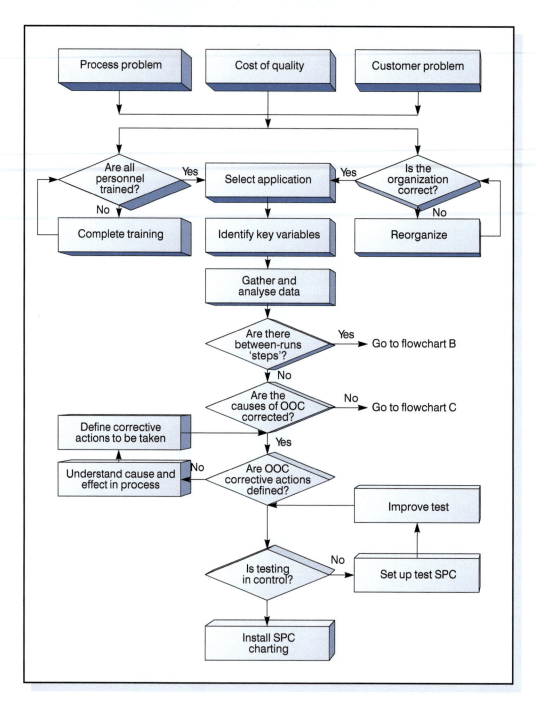

Figure 10.14
Flow chart for SPC implementation

LEADERSHIP, PEOPLE AND IMPLEMENTATION ASPECTS OF PROCESS MANAGEMENT

There are many top executives who have famously used process management to great effect, issuing statements such as:

> *Business processes are designed to be customer driven, cross functional and value based. They create knowledge, eliminate waste and abandon unproductive work yielding world-class productivity and higher perceived service levels for customers.*

and

> *Publishing the process on the wall helps people understand their place in the big picture . . . our continuously improving profits are a consequence of superlative performance that is derived from well-thought-out processes, ongoing measurement of a few carefully selected key indicators, good communications and the full involvement everywhere of all people working in the company.*

These approaches manifest themselves in complete understanding by everyone of the end-to-end process and this was described, for example, by TNT Express in the 'perfect transaction' (Figure 10.15) – see also the TNT Express case study.

Perhaps the most visible difference between a process management enterprise and a more functionally based one is the existence of process owners – management layers with end-to-end responsibilities for individual processes (see Table 10.1). They have real responsibility for and authority over the process design, operation and measurement of performance. This can require changing from being functionally

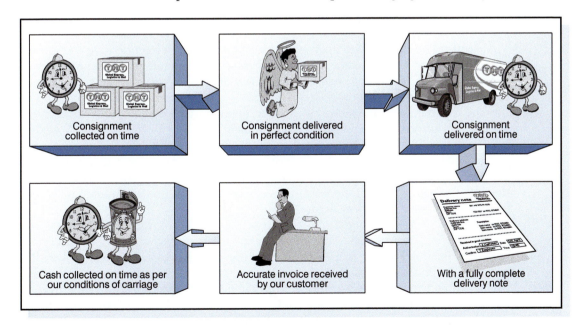

Figure 10.15
TNT Express Delivery Services – the perfect transaction process

Table 10.1 Summary of key process roles

Roles	Descriptions
Executive Sponsor	• Member of the Executive team • Advocate and ambassador for the overall process • Exerts influence inside and outside of the organisation
Process Owner	• Member of the senior management team • End-to-end responsibility and accountability for the performance of the process • Ownership of the overall process
Functional (Line) Manager	• Members of the operational staff • Specific functional (line) responsibilities and accountabilities • Impact within the overall process

driven, which usually means a major cultural challenge for the organization. Process owners cannot simply adopt a command and control style so their ability to influence is at least as important to process management as structure.

Managing the people who work in the processes demands attention to:

- designing, developing and delivering training programmes
- setting performance targets
- regular communication, preferably face-to-face

keeping them informed of changing customer needs

- listening to concerns and ideas
- negotiation and collaboration.

Many organizations have realized that to be cost-effective, competitive and indeed world-class, they must ensure that all processes are understood, measured and in control. Almost everyone needs to be trained in process management and improvement and shown how they are part of a supplier-process-customer chain. The training reinforces that these chains are interdependent and that all processes support the delivery of products or services to customers.

Operators of every process need to be properly trained, have necessary work instructions available and have the appropriate tools, facilities and resources to perform the process to its optimum capability. This applies to all processes throughout the organization, whatever the outputs, including those in finance, human resources and generally shared services areas.

In many process managed organizations this type of approach has changed the way employees are trained and developed, emphasizing the whole process rather than narrowly focussed tasks. It has made fundamental changes to cultures, stressing customers and process-based teamwork rather than functionally driven objectives. Creativity and innovation in process improvement are recognized as core competencies, and the annual performance reviews and personal development plans should be linked to these.

The first thing that top management must recognize is that moving to process management requires much more than re-drawing the organizational chart or structure.

The changes needed are often fundamental, leading to new ways of working and managing, and these will challenge any company or public service organization.

Many organizations today are facing a large number of changes and initiatives, often driven by public–private transitions, customer or government demands, technology, and so on. Before implementing process management, therefore, a senior management team needs to examine closely all its current change initiatives, using simple frameworks, such as the figure of 8 change model shown in Chapters 9 and 18 or the Total Organizational Excellence Model (see Chapter 19) to prune those that are not relevant to a process managed business and combining/rationalizing those that are.

The introduction of process management is often driven or directly connected to a strategic initiative, such as reducing non-value added time through 'Lean' (described in Chapter 15), increasing customer satisfaction, reducing working capital (perhaps tied up in work-in-progress), an enterprise resource planning (ERP) implementation, changes in technology, or introduction of e-business. The application of new enabling technologies is an ideal time to review the design and configuration of key processes. Failure to do so could lead to missed opportunities to extract maximum benefit from the technology.

Implementing process management, like many change initiatives, cannot be a quick fix and it will not happen overnight. Top management need the resolution and commitment for major changes in the way things are structured, carried out and measured. As with all such execution, it needs careful planning and an understanding of what needs to be done first. Things high on the list will be the establishment of a core process framework, aligned to the needs of the business, and the appointment of key process owners. A process based performance measurement framework then needs to be set up to track progress. As in all change initiatives, delivering some tangible measurable benefits early on will help overcome the inevitable resistance. In one pharmaceutical company, for example, the success of the work on the product development and product promotion processes helped significantly the cause of process management and the company extended its approach into the supply-chain management and other processes.

As companies and public service organizations move inexorably towards the wider introduction of e-commerce to do business, this will place a premium on rapid and fault-free execution of business processes. Putting a website in front of an inefficient, ineffectual or even broken process will soon bring it to its knees, together with everyone working in it and around it. This will also bring 'back-office' mistakes to the attention of the marketplace. Some of these processes will, of course, need to be redesigned – from customer order fulfilment to procurement. They will need to change 'shape' as demands, technology and markets change. Without good process management in place this is going to be very difficult for functionally driven organizations.

ACKNOWLEDGEMENT

The author is grateful to Mark Milsom of West Yorkshire Police for permission to use the material related to crime reporting in this chapter.

BIBLIOGRAPHY

Besterfield, D., *Quality Control* (7th edn), Prentice-Hall, Englewood Cliffs, NJ, 2008.

Dimaxcescu, D., *The Seamless Enterprise – Making Cross-functional Management Work,* Harper Business, New York, 1992.

FIPS Publication 184, FIPS Publications – Computer Security Resource Center, csrc.nist.gov/publications, December 1993.

Francis, D., *Unblocking the Organizational Communication,* Gower, Aldershot, 1990.

Harrington, H.J., *Total Improvement Management,* McGraw-Hill, New York, 1995.

Jeston, J and Nelis, J., *Business Process Management – Practical Guidelines to Successful Implementations,* Elsevier, Oxford, 2007.

Rummler, G.A. and Brache, A.P., *Improving Performance: How to Manage the White Space on the Organization Chart* (2nd edn), Jossey-Bass Publishing, San Francisco, CA, 1998.

Senge, P.M., *The Fifth Discipline; The Art and Practice of the Learning Organization,* Random House, 2006.

Senge, P.M., Roberts, C., Ross, R.B., Smith, B.J. and Kleiner A., *The Fifth Discipline Fieldbook – Strategies and Tools for Building a Learning Organization,* Nicholas Brearley, London, 1994.

Warboys, B.C., Kawalek, P., Robertson, I. and Greenwood, R.M., *Business Information Systems: A Process Approach,* McGraw-Hill, New York, 1999.

CHAPTER HIGHLIGHTS

Process management vision

- Everything organizations do to create value for customers of their products or services is a process. Process management is key to improving performance.
- Process managed organizations see things from a customer perspective – as a series of inter-connected work and information flows that cut horizontally across the business functions.
- The key or core business processes are well defined and developed sequences of steps with clear rationale, which add value by producing required outputs from a variety of inputs.
- Deployment of a common high-level process framework throughout the organization gives many benefits, including reduced costs and increased flexibility.
- Process management best practices include: identifying the key business processes, managing processes systematically, reviewing processes and setting improvement targets, using innovation and creativity to improve processes, changing processes and evaluating the benefits.

Process classification framework and process modelling

- The APQC's Process Classification Framework creates a high level generic, cross-functional process view of an enterprise – a taxonomy of business processes.
- The IDEF (Integrated Definition Function Modelling) language provides a useful structured graphical framework for describing and improving business processes. It consists of a hierarchical series of diagrams and text, cross-referenced to each other through boxes. The processes are described in terms of inputs, controls, outputs and resources (ICOR).

Process flowcharting

- Flowcharting is a method of describing a process in pictures, using symbols – rectangles for operation steps, diamonds for decisions, parallelograms for information and circles/ovals for the start/end points. Arrow lines connect the symbols to show the 'flow'.
- Flowcharting improves knowledge of the process and helps to develop the team of people involved.
- Flowcharts document processes and are useful as trouble-shooting tools and in process improvement. An improvement team would flowchart the existing process and the improved or desired process, comparing the two to highlight the changes necessary.

Leadership, people and implementation

- Top management who have used process management to great effect recognize its contribution in creating knowledge and eliminating waste, yet they understand the importance of involving people, measurement and good communications.
- Process owners are key to effective process management. They have responsibility for and authority over process design, operation and measurement of performance.
- Managing the people who work in the processes requires attention to training programmes, performance targets, communicating changing customer needs, negotiation and collaboration.
- Moving to process management requires some challenging fundamental changes, leading to new ways of working and managing. Current initiatives should be carefully examined to ensure good planning and an understanding of what needs to be done first.
- As with all change initiatives, delivering some tangible measurable benefits early on will help overcome the inevitable resistance.
- With the wider introduction of e-commerce systems, there will be greater pressure to run rapid, fault-free business processes. Some of the processes will need to change 'shape' as demands, technologies and markets change.

Chapter 11

Process redesign/engineering

PROCESS REDESIGN, RE-ENGINEERING AND LEAN SYSTEMS

Two of the main movements that have influenced process redesign over recent years are 'lean production' and 'process re-engineering'. The first of these found its roots in a review of the causes of inefficiency in mass production while the latter was born out of the potential provided by information technology for traditional processes to be fundamentally redesigned.

The starting point for the developments that became known as 'Lean Production' was the International Motor Vehicle Programme (IMVP) research at MIT; first reported in *The Machine that Changed the World* (Womack *et al.* 1990). However, as with many management terms, Lean Production is often used loosely. To redefine the term, Jim Womack sent a message to the Lean Enterprise Institute email list entitled 'Deconstructing the Tower of Babel'. He described how in 1987, working with a group of colleagues, they listed the performance attributes of a Toyota-style production system compared with traditional mass production. The Toyota-style production system:

- needed less human effort to design, make, and service products
- required less investment for a given amount of production capacity
- created products with fewer delivered defects and fewer in-process turn-backs
- utilized fewer suppliers with higher skills
- went from concept to launch, order to delivery, and problem to repair in less time with less human effort
- could cost-effectively produce products in lower volume with wider variety to sustain pricing in the market while growing share
- needed less inventory at every step from order to delivery and in the service system
- caused fewer employee injuries, etc.

The group very quickly ascertained that this system needed less of everything to create a given amount of value, so they called it 'lean,' hence the term was born. In the intervening period, the term has become loosely applied to a great variety of improvement activity and so to set the record straight, Jim Womack wrote (Womack *et al.* 1990):

here's what lean means to me:

- it always begins with the customer;
- the customer wants value: the right goods or service at the right time, place and price with perfect quality.
- value in any activity – goods, services or some combination – is always the end result of a process (design, manufacture, and service for external customers, and business processes for internal customers);
- every process consists of a series of steps that need be taken properly in the proper sequence at the proper time;
- to maximize customer value, these steps must be taken with zero waste;
- to achieve zero waste, every step in a value-creating process must be valuable, capable, available, adequate and flexible, and the steps must flow smoothly and quickly from one to the next at the pull of the downstream customer;
- a truly lean process is a perfect process; perfectly satisfying the customer's desire for value with zero waste;
- none of us have ever seen a perfect process nor will most of us ever see one; but lean thinkers still believe in perfection, the never-ending journey toward the truly lean process.

Note that identifying the steps in the process, getting them to flow, letting the customer pull, etc. are not the objectives of lean practitioners; these are simply necessary steps to reach the goal of perfect value with zero waste.

Essentially what Womack defined was simply *a focus on the customer, on creating value and on eliminating waste* – the ideal of any production process. The wastes he referred to were those defined by Engineer Ohno of Toyota:

- overproduction
- waiting
- excess conveyance
- extra processing
- excessive inventory
- unnecessary motion
- defects requiring rework or scrap.

The elimination of which will:

- reduce the proportion of non-value adding activities
- reduce lead time
- reduce variation
- simplify processes
- increase flexibility
- increase transparency.

Process redesign/engineering

227

Over recent years, 'Lean' has been broadened out to include all the causes of waste and the strategies for their elimination. The agenda has an increasing amount in common with TQM, the main difference being that the Lean focus tends to be on production/delivery processes whereas the TQM focus is on organizations as a whole.

See Chapter 15 – Lean for a full treatment of this important topic.

RE-ENGINEERING THE ORGANIZATION?

When it has been recognized that a major business process requires radical re-assessment, business process re-engineering or redesign (BPR) methods are appropriate. In their book *Re-Engineering the Corporation* (1993), Hammer and Champy talked about re-inventing the nature of work, 'starting again – re-inventing our corporations from top to bottom.' BPR was launched on a wave of organizations needing to completely re-think how and why they do what they do in order to cope with the ever-changing world, particularly the development of technology based solutions.

The reality is, of course, that many processes in many organizations are very good and do not need re-engineering, redesigning or re-inventing, not for a while anyway. These processes should be subjected to a regime of continuous improvement (Chapter 13) at least until we have dealt with the very poorly performing processes that clearly do need radical review.

Some businesses and industries more than others have been through some pretty hefty changes – technological, political, financial and/or cultural. Customers of these organizations may be changing and demanding certain new relationships. Companies are finding leaner competitors encroaching into their market place, increased competition from other countries where costs are lower and start-up competitors which do not share the same high bureaucracy and formal structures.

Enabling an organization, whether in the public or private sector, to be capable of meeting these changes is not a case of working harder but working differently. There have been many publicized BPR success stories and, equally, there have been some abject failures. In some cases, radical changes to major business processes have brought corresponding radical improvements in productivity. However, knowing how to reap such benefit, or indeed knowing if and how to apply BPR, has proved difficult for some organizations.

Many companies adopted TQM initiatives in the 1980s hoping to win back business lost to Japanese competition. When Ford benchmarked Mazda's accounts payable department, however, they discovered a business process being run by five people, compared to Ford's 500. Even with the difference in scale of the two companies, this still demonstrated the relative inefficiency of Ford's accounts payable process. At Xerox, taking a customer's perspective of the company identified the need to develop systems rather than stand-alone products, which highlighted Xerox's own inefficient office systems.

Both Ford and Xerox realized that incremental improvement alone was not enough. They had developed high infrastructure costs and bureaucracies that made them relatively unresponsive to customer service. Focussing on internal customer-supplier interfaces improved quality, but preserved the current process structure and they could not hope to achieve in a few years what had taken the Japanese 30 years.

To achieve the necessary improvements required a radical rethink and redesign of these processes.

What was being applied by organizations such as Ford and Xerox was *discontinuous improvement*. In order to respond to the competitive threats of Canon and Honda, Xerox and Ford needed TQM to catch up, but to get ahead they felt they required radical breakthroughs in performance. Central to these breakthrough improvements was information technology.

Information technology as a driver for BPR

BPR is often based on new possibilities for breakthrough performance provided by the emergence of new enabling technologies. The most important of these, the one that is the nominal ingredient in many BPR recipes, is IT. Explosive advances in IT have enabled the dissemination, analysis and use of information from and to customers and suppliers and within enterprises, in new ways and in time frames that impact processes, organization designs and strategic competencies. Computer networks, open systems, client-server architecture, groupware and electronic data interchange have opened up the possibilities for the integrated automation of business processes. Neural networks, enterprise analyser approaches, computer-assisted software engineering and mobile object-oriented programming now facilitate systems design around many processes in most organizations.

The pace of change has, of course, been enormous and IT systems unavailable just a few years ago have enabled sweeping changes in business process improvement, particularly in business systems. Just as statistical process control (SPC) enabled manufacturing processes to be improved by controlling variation and improving efficiency, so IT is enabling all types of processes to be fundamentally restructured.

IT in itself, however, does not offer all the answers. Many companies putting in major new computer systems have achieved only the automation of existing processes. Frequently, different functions within the same organization have systems that are incompatible with each other. Locked into traditional functional structures, some managers have spent large amounts on IT systems that have not been integrated cross-functionally. Yet it is in this cross-functional area that the big improvement gains through IT are to be made. Once a process view is taken to designing and installing an IT system, it becomes possible to automate cross-functional, cross-divisional, even cross-company processes.

WHAT IS BPR AND WHAT DOES IT DO?

There are almost as many definitions of BPR as there are of TQM! However most of them boil down to the same substance – the fundamental rethink and radical redesign of a business process, its structure and associated management systems, to deliver major or step improvements in performance (which may be in process, customer or business performance terms).

Of course, BPR and TQM programmes are complementary under the umbrella of 'strategic process management.' The continuous and step change improvements must live side by side – when does continuous change become a step change anyway? There has been over the years much debate, including some involving the author, about this issue. Whether it gets resolved is not usually the concern of the organization facing today's uncertainties with the realization that 'business as usual' will not do and some major changes in the ways things are done are required.

Put into a strategic context, BPR is a means of aligning work processes with customer requirements in a dynamic, flexible way, in order to achieve long-term corporate objectives. This requires the involvement of customers and suppliers and thinking about future requirements. Indeed the secrets to redesigning a process successfully lie in thinking about how to reshape it for the future.

BPR then challenges managers to rethink their traditional methods of doing work and to commit to customer-focussed processes. Many outstanding organizations have achieved and/or maintained their leadership through process re-engineering, especially where they found processes which were not customer focussed. Companies using these techniques have reported significant bottom-line results, including better customer relations, reductions in time to market, increased productivity, fewer defect/errors and increased profitability. BPR uses recognized methods for improving business results and questions the effectiveness of the traditional organizational structures. Defining, measuring, analysing and re-engineering work processes to improve customer satisfaction can pay off in many different ways.

For example, Motorola had set stretch goals of ten-fold improvement in defects and two-fold improvement in cycle time within five years. The time period was subsequently revised to three years and the now famous Six Sigma goal of 3.4 defects per million became a slogan for the company and probably one of the real drivers (see also Chapter 14). These stretch goals represent a focus on discontinuous improvement and there are many examples of other companies that have made dramatic improvements following major organizational and process redesign as part of TQM initiatives, including approaches such as the 'clean sheet' design of a 'green field' plant around work cells and self-managed teams.

Most organizations have 'vertical' functions: experts of similar backgrounds grouped together in a pool of knowledge and skills capable of completing any task in that discipline. This focus, however, fosters a vertical view and limits the organization's ability to operate effectively. Barriers to customer satisfaction evolve, resulting in unnecessary work, restricted sharing of resources, limited synergy between functions, delayed development time and no clear understanding of how one department's activities affect the total process of attaining customer satisfaction. Managers remain tied to managing singular functions with rewards and incentives for their narrow missions, inhibiting a shared external customer perspective.

BPR breaks down these internal barriers and encourages the organization to work in cross-functional teams with a shared horizontal view of the business. As we have seen in earlier chapters this requires shifting the work focus from managing functions to managing processes. Process owners, accountable for the success of major cross-functional processes, are charged with ensuring that employees understand how their individual work processes affect customer satisfaction. The interdependence between one group's work and the next becomes quickly apparent when everyone understands who the customer is and the value they add to the entire process of satisfying that customer.

IT provided the means to achieve the breakthrough in process performance in some organizations. The inspiration, however, came from understanding both the current and potential processes. This required a more holistic view than that taken in traditional quality programmes, involving wholesale redesigns of the processes concerned.

Ford estimated a 20 per cent reduction in head count if it automated the existing processes in accounts payable. Taking an overall process perspective, Ford achieved a 75 per cent reduction in one department. Xerox took an organizational view and concentrated on the cross-functional processes to be re-engineered, radically changing the relationship between supplier and external customer.

Clearly, the larger the scope of the process, the greater and farther reaching is the consequences of the redesign. At a macro level, turning raw materials into a produce used by a delighted customer is a process made up of subsets of smaller processes. The aim of the overall process is to add value to the raw materials. Taking a holistic view of the process makes it possible to identify non-value-adding elements and remove them. It enables people to question why things are done, and to determine what should be done.

Some of the re-engineering literature advises starting with a blank sheet of paper and redesigning the process anew. The problems inherent in this approach are:

- the danger of designing another inefficient system
- not appreciating the scope of the problem.

Therefore, the author and his colleagues recommend a thorough understanding of current processes before embarking on a re-engineering project.

Current processes can be understood and documented by process mapping and flowcharting. As processes are documented, their inter-relationships become clear

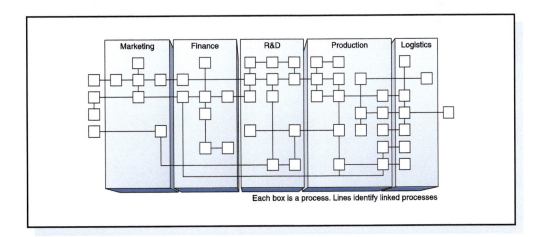

Figure 11.1
Simplified process map

and a map of the organization emerges. Figure 11.1 shows a much simplified process map. As the aim of BPR is to make discontinuous, major improvements, this invariably means organizational change, the extent of which depends on the scope of the process re-engineered.

Taking the organization depicted in Figure 11.1 as an example, if the decision is made to redesign the processes in finance, the effect may be that in Figure 11.2a: eight individual processes have become three. There has been no organizational effect on

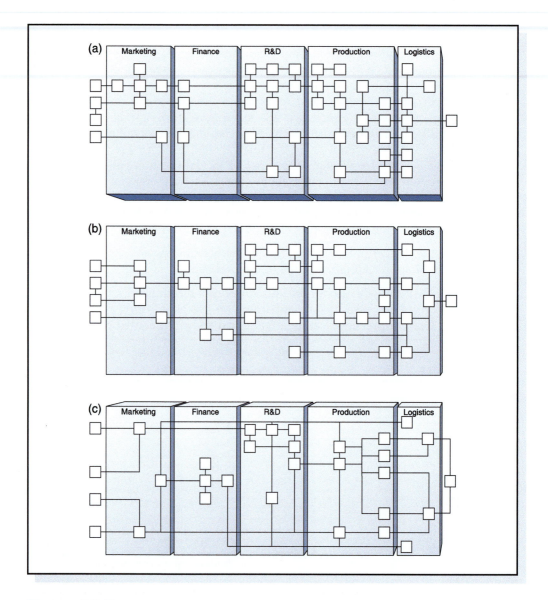

Figure 11.2
(a) Process redesign in finance, (b) cross-functional process design,
(c) organizational process redesign

the processes in the other functions, but finance has been completely restructured. In Figure 11.2b, a chain of processes, crossing all the functions has been re-engineered. The effect has been the loss of redundant processes and possibly many heads but much of the organization has been unaffected. Figure 11.2c shows the organization after a thorough re-engineering of all its processes. Some elements may remain the same, but the effect is organization-wide.

Whatever the scope of the redesign, head count is not the only change. When work processes are altered, the way people work alters. Figures 11.1 and 11.2 show an organization's functional departments with process running through them. These are the handful of core processes that make up what an organization does (see Figure 11.3) and in many organizations these would benefit from re-engineering to improve added value output and efficiency.

Focus on results

BPR is not intended to preserve the status quo, but to fundamentally and radically change what is done; it is *dynamic*. Therefore, it is essential for a BPR effort to focus on required customers which will determine the scope of the BPR exercise. A simple requirement may be a 30 per cent reduction in costs or a reduction in delivery time of two days. These would imply projects with relatively narrow scope, which are essentially inwardly focussed and probably involve only one department; for example, the finance department in Figure 11.2a.

When Wal-Mart focussed on satisfying customer needs as an outcome it started a redesign that not only totally changed the way it replenished inventory, but also made this the centrepiece of its competitive strategy. The system put in place was radical, and required tremendous vision. In a few years, Wal-Mart grew from being a small niche retailer to the largest and most profitable retailer in the world.

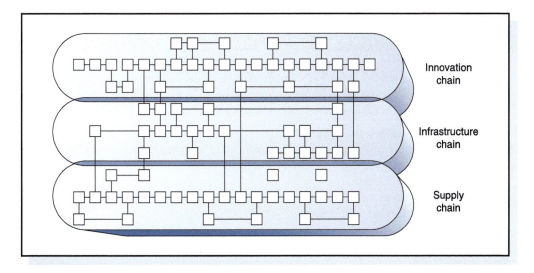

Figure 11.3
Process organization

Focussing on results rather than just activities can make the difference between success and failure in change projects. The measures used, however, are crucial. At every level of redesign and re-engineering, a focus on results gives direction and measurability; whether it be cost reduction, head count reduction, increase in efficiency, customer focus, identification of core processes and non-value-adding components, or strategic alignment of business processes. Benchmarking is a powerful tool for BPR and is the trigger for many BPR projects, as in Ford's accounts payable process. As shown in Chapter 9, the value of benchmarking does not lie in what can be copied, but in its ability to identify goals. If used well, benchmarking can shape strategy and identify potential competitive advantage.

The redesign process

Central to BPR is an objective overview of the processes to be redesigned. Whereas information needs to be obtained from the people directly involved in those processes it is never initiated by them. Even at its lowest level, BPR has a top-down approach and most BPR efforts, therefore, take the form of a project. There are numerous methodologies proposed, but all share common elements. Typically, the project takes the form of seven phases, shown in Figure 11.4.

1. Discover and define

This involves first identifying a problem or unacceptable outcome, followed by determining the desired outcome. This usually requires an assessment of the business need and will certainly include determining the processes involved, including the scope, identifying process customers and their requirements, and establishing effectiveness measurements.

2. Establish redesign team

Any organization, even a small company, is a complex system. There are customers, suppliers, employees, functions, processes, resources, partnerships, finances, etc. and many large organizations are incomprehensible – no one person can easily get a clear picture of all the separate components. Critical to the success of the redesign is the make-up of a redesign team.

The team should comprise as a minimum the following:

- senior manager as sponsor
- steering committee of senior managers to oversee overall re-engineering strategy
- process owner
- team leader
- redesign team members.

It is generally recommended that the redesign team have between five and ten people; represent the scope of the process (that is, if the process to be re-engineered is a cross-functional, so is the team); only work on one redesign at a time; and is made up of both insiders and outsiders. Insiders are people currently working within the process concerned who help gain credibility with co-workers. Outsiders are people

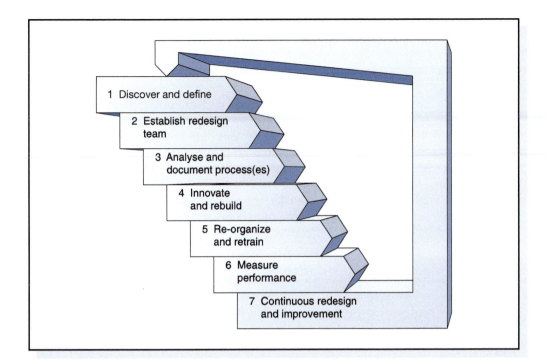

Figure 11.4
The seven phases of BPR

from outside the organization who bring objectivity and can ask the searching questions necessary for the creative aspects of the redesign. Many companies use consultants for this purpose.

3. Analyse and document process(es)

Making visible the invisible, documenting the process through mapping and flow-charting is the first crucial step that helps an organization see the way work really is done and not the way one thinks or believes it is done. Seeing the process as it is provides a baseline from which to measure, analyse, test and improve.

Collecting supporting process data, including benchmarking information and IT possibilities, allows people to weigh the value each task adds to the total process, to rank and select areas for the greatest improvement, and to spot unnecessary work and points of unclear responsibility. Clarifying the root causes of problems, particularly those that cross department lines, safeguards against quick-fix remedies and assures proper corrective action, including the establishment of the right control systems.

4. Innovate and rebuild

In this phase the team rethink and redesign the new process, using the same process-mapping techniques, in an iterative approach involving all the stakeholders, including senior management. A powerful method for challenging existing practices and generating breakthrough ideas is 'assumption busting' – see later section.

5. Re-organize and re-train

This phase includes piloting the changes and validating their effectiveness. The new process structure and operation/system will probably lead to some re-organization, which may be necessary for reinforcement of the process strategy and to achieve the new levels of performance. Training and/or re-training for the new technology and roles play a vital part in successful implementation. People need to be equipped to assess, re-engineer and support – with the appropriate technology – the key processes that contribute to customer satisfaction and corporate objectives. Therefore, BPR efforts can involve substantial investment in training but they also require considerable top management support and commitment.

6. Measure performance

It is necessary to develop appropriate metrics for measuring the performance of the new process, sub-processes, activities and tasks. These must be meaningful in terms of the inputs and outputs of the process, and in terms of the customers of and suppliers to the process (see Chapter 7).

7. Continuous redesign and improvement

The project approach to BPR suggests a one-off approach. When the project is over, the team is disbanded and business returns to normal, albeit a radically different normal. It is generally recommended that an organization does not attempt to re-engineer more than one major process at a time, because of the disruption and stress caused. Therefore, in major re-engineering efforts of more than one process, as one team is disbanded, another is formed to redesign yet another process. Considering that Ford took five years to redesign its accounts payable process, BPR on a large scale is clearly a long-term commitment.

In a rapidly changing, ever more competitive business environment, it is becoming more likely that companies will re-engineer one process after another. Once a process has been redesigned, continuous improvement of the new process by the team of people working in the process should become the norm.

Assumption busting

Within BPR is a powerful method for challenging existing practices and generating breakthrough ideas for improvement. 'Assumption busting', as it was named by Hammer and Champy, aims to identify the *rules* that govern the way we do business and then uncover the real underling *assumptions* behind the adoption of these rules. Business processes are governed by a number of rules that determine the way the process is designed, how it interfaces with other activities within the organization and how it is operated. These rules can exist in the form of explicit policies and guidelines or, what is more often the case, in the mind of the people who operate the process. These unwritten rules are the product of assumptions about the process environment that have been developed over a number of years and often emerge from uncertainties surrounding trading relationships, capabilities, resources, authorities, etc. Once these underlying assumptions are uncovered they can be challenged for relevance and, in many cases, can be found to be false. This opens up new opportunities for process redesign and, as a consequence, the creation of new value and improved performance.

For example Resources Ltd., a supplier of TV and radio studio and outside broadcast resource services, faced the requirement to improve business performance. The business was losing money and faced stiff competition from independent providers. They needed to improve the efficiency of their processes whilst retaining their core capability that created competitive advantage. A team was commissioned to review the core value adding processes, setting challenging targets for improvements in performance in order to stimulate breakthrough thinking. The team decided to take a more radical approach by using assumption busting and prepared a six-week programme of work. Within that timeframe they used an established eight-step method cycle (shown in Figure 11.5) to redesign the core end-to-end service delivery processes for two major business units – Studios and Outside Broadcasts. The work involved identifying the key areas of cost consumption, challenging the rules and assumptions that governed the existing process and generating a set of improvement opportunities. When they had evaluated their findings, the team presented ideas to deliver an improvement in excess of 15 per cent in process efficiency.

One of the process rules concerned the use of a highly technically qualified member of staff for the planning and delivery of all of the programmes supported. The core underlying assumption was that all of the programmes were complex in nature. When this assumption was challenged the team in Resources Ltd realized that, as not all programmes were so complex in nature, less technically qualified members of staff could be utilized at lower cost to the business.

Application of the technique
In practice the author and his colleagues have found this technique to be of greatest value when applied by a cross-functional group of process operators and supervisors who are given a specific problem to fix. In using the technique, care must be exercised in the use of terms such as 'rule' and 'assumption'. They often cause initial confusion and there can be real difficulty in uncovering the core underlying assumptions. Rules should be clearly stated and tested for validity before proceeding down what eventually could become a blind alley. Furthermore, a rigorous approach to the identification of the core assumption is vital to uncovering the real opportunities for improvement. An assumption by definition 'is a statement/belief that is accepted or supposed to be true without roof or demonstration'. In some cases, rules are created from specific knowledge about the business and its environs and not based on assumptions. Assumptions spring from our beliefs about the environment and not our specific knowledge.

Other applications of assumption busting
Assumption busting is of particular benefit when applied by partners within a supply chain. The trading relationships and practices that exist in a modern supply chain, such as a supermarket and its multiple tiers of suppliers, are the product of a number of assumptions made by the supply chain partners about what is possible. Once teams from each of the partnering businesses work collaboratively to uncover the rules and

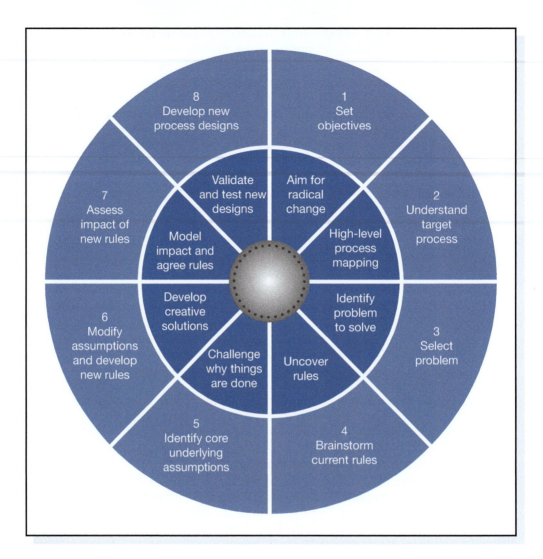

Figure 11.5
The assumption busting cycle

assumptions that govern their trading relationships the door is unlocked to new methods and economies.

The method can also be immensely powerful when companies are introducing new technology. Breakthrough technologies can lead to breakthrough performance as they make possible what is considered impossible today. Hence, a number of current rules and assumptions are there to be challenged as processes are redesigned to take advantage of the new technology. We are often just as constrained by our lack of imagination regarding the possibilities of tomorrow as we are by our knowledge of what is possible today. One example of this was in BBC World Service where the introduction of digital technology to replace analogue was accompanied by assumption busting led process redesign to take advantage of the new technological capability.

While assumption busting has been primarily applied to the generation of new process designs, it exists in its own right as a method for developing more 'lateral' solutions to problems. In the early 1970s, Dr Edward de Bono introduced the concept of lateral thinking as an alternative method of generating ideas to that of the more traditional logical or 'vertical' thinking. Dr de Bono argued that our thinking is constrained by patterns that form in our minds over time and channel our future thoughts. Assumption busting helps people break out of this 'channelled thinking' to develop creative ideas. Managers could benefit from applying assumption busting to a number of problems or opportunities in their businesses – assumptions constrain us everywhere, not just within our business processes.

Whether it is in response to specific customer requirements new technology, or in the quest for competitive advantage, assumption busting provides a simple but effective method for breaking into new areas of adding value. World-class performance will not be achieved by effort alone; creativity and innovation are cornerstones of future success. Innovative ways of delivering new value will be rewarded. Assumption busting provides a powerful method for generating new ideas from looking at today and tomorrow in a different way.

BPR – THE PEOPLE AND THE LEADERS

For an organization to focus on its core processes almost certainly requires an understanding of its core competencies. Moreover, core process redesign can channel an organization's competencies into an outcome that gives it strategic competitive advantage and the key element is visioning that outcome. Visioning the outcome may not be enough, however, since many companies' 'vision' desires results without simultaneously 'visioning' the systems that are required to generate them. Without a clear vision of the processes, systems, methods and approaches that will allow achievement of the desired results, dramatic improvement is frequently not obtained as the organization fails to align around a common tactical strategy. Such an 'operational' vision is lacking in many organizations.

The fallout from BPR has profound impacts on the employees in any enterprise at every level – from executives to operators. In order for BPR to be successful, therefore, significant changes in organization design and enterprise culture are also often required. Unless the leaders of the enterprise are committed to undertake these changes, the BPR initiative will flounder. The point is, of course, that organization design and culture change are much more difficult than modifying processes to take advantage of new IT.

While enabling IT is often necessary and is clearly going to play a role in many BPR exercises, it is by no means sufficient, nor is it the most difficult hurdle on the path to success. Thanks to new technologies we can radically change the processes an organization operates and, hopefully, achieve dramatic improvements in performance. However, in any BPR project there will be considerable risk attached to building the information system that will support the new, redesigned processes. Information systems should be but rarely are described so that they are easy for people to understand.

While BPR may be a distinct, short-term activity for a specific business function, the record indicates that BPR activities are most successful when they occur within the framework of a long-term thrust for excellence. Within a TQM culture, a BPR effort is more likely to find the process focused supportive workforce, organization design, and mind-set changes needed for its success.

Process improvement is sometimes positioned as a bottom-up activity. In some contrast, TQM involves setting longer-term goals at the top and modifying the business as necessary to achieve the goals. Often, the modifications to the business required to achieve the goals are extensive and groundbreaking. The history of successful TQM thrusts in award-winning companies in Europe and the United States is replete with new organization designs, with flattened structures and with empowered employees in the service of end customers. In many successful organizations, BPR has been an integral part of the culture – a process-driven change dedicated to the ideals and concepts of TQM. That change must create something that did not exist before, namely a 'learning organization,' capable of adapting to a changing competitive environment. When processes, or even the whole business, needs to be re-engineered, the radical change may not, probably will not, be readily accepted.

ACKNOWLEDGEMENT

The author is grateful to the contribution made by his colleague Mike Turner to the preparation of this chapter.

BIBLIOGRAPHY

Braganza, A. and Myers, A., *Business Process Redesign – A View from the Inside*, International Thomson Business Press, London, 1997.

Hammer, M. and Champy, J., *Re-engineering the Corporation*, Nicholas Brearley, London, 1993.

Hammer, M. and Stanton, S.A., *The Re-engineering Revolution – The Handbook*, BCA, Glasgow, 1995.

Harmon, P., *Business Process Change*, Morgan Kaufmann, 2003.

Jacobson, I., *The Objective Advantage – Business Process Re-engineering with Object Technology*, J. Wiley, Chichester, 1993.

Johansson, H.J., McHugh, P., Pendlebury, A.J. and Wheeler, W.A., *Business Process Re-engineering*, John Wiley, London, 1993.

Womack, J.P., Jones, D.T. and Roos, D., *The Machine that Changed the World*, New York: Rawson MacMillan, 1990.

CHAPTER HIGHLIGHTS

Process redesign, re-engineering and lean systems

- The main movements influencing process redesign over recent years are 'lean production' and 'process re-engineering'. The first has its roots in reviews of causes of inefficiency in mass production; the second was born out of the potential provided by information technology for traditional processes to be fundamentally redesigned.
- Lean focuses on the customer, the process, the elimination of waste and the maximization of value.
- The primary forms of waste are overproduction, waiting, excess conveyance, extra processing, excessive inventory, unnecessary motion and defects requiring rework or scrap.

Re-engineering the organization?

- When a major business process requires radical re-assessment, perhaps through the introduction of new technology, discontinuous methods of business process re-engineering or redesign (BPR) are appropriate.
- The opportunity for radical change in processes may involve collaboration across the supply chain and this might be best achieved through partnerships among organizations.
- Drivers for process change include information technology (IT), political, financial, cultural and competitive aspects. These often require a change of thinking about the ways processes are and could be operated.
- IT often creates opportunities for breakthrough performance but BPR is needed to deliver it. Successful practitioners of BPR have made striking improvements in customer satisfaction and productivity in short periods of time.
- Inter-organizational integration of IT is one of the greatest opportunities and challenges facing the sector – it has the potential to unlock significant value.

What is BPR and what does it do?

- There are many definitions of BPR but the basic elements involve a fundamental re-think and radical redesign of a business process, its structure and associated management systems to deliver step improvements in performance.
- BPR and TQM are complementary under the umbrella of process management – the continuous and discontinuous improvements living side by side. Both require the involvement of customers and suppliers and their future requirements.
- BPR challenges managers to re-think their traditional methods of doing work and to commit to customer-focussed processes. This breaks down organizational barriers and encourages cross-functional teams.

Processes for redesign/focus on results

- Much larger savings and head count reductions are possible through properly applied BPR than simply automating existing processes. The larger the scope of the process, the greater and farther reaching the consequences of the redesign.

Process redesign/engineering   **241**

- A thorough understanding of the current process is needed before embarking on a re-engineering project. Documentation of processes through mapping and flowcharting allows inter-relationships to be clarified.
- Focussing on results rather than activities can make the difference between success and failure in BPR and other change projects, but the measures used are critical. Benchmarking is a powerful tool for BPR and often the trigger for many projects.

The redesign process/assumption busting

- BPR has a top-down approach and needs an objective overview of the process to be redesigned to drive the project.
- Typically a BPR project will have seven phases: discover – identifying the problem or unacceptable outcome; establish redesign team; analyse and document processes; innovate and rebuild; re-organize and re-train; measure performance; continuous redesign and improvement.
- Assumption busting is a useful eight-step BPR method which aims to identify and challenge the 'rules' and assumptions that govern and underlay the way business is done. A team is formed to: identify the core value to be delivered to customers and stakeholders; map the process at high level; select problems to resolve and collect performance data; brainstorm and test the rules; rigorously review each rule to uncover underlying assumptions; identify modified assumptions and process rules; identify impact and construct new set of process principles; develop revised process and test validity.

BPR – the people and the leaders

- For an organization to focus on its core processes requires an understanding of its core competencies, and the channelling of these into outcomes that deliver strategic competitive advantage.
- BPR has profound impacts on employees from the top to the bottom of an organization. In order to be successful, significant changes in organization design and enterprise culture are also often required. This requires commitment from the leaders to undertake these changes.
- TQM ideals and concepts provide a perfect platform for BPR projects and the creation of a 'learning organization' capable of adapting to a radically changing environment.

Chapter 12

Quality management systems

WHY A QUALITY MANAGEMENT SYSTEM?

In recent years the value of formal quality management systems has been widely questioned and it is small wonder. In some countries government policy has been to simply mandate certification against the ISO 9000 family of standards on Quality Management Systems for all government suppliers. Consequently, some company managements saw the need to obtain certification as simply part of getting onto government tender lists and, hence, primarily as a marketing problem. In many cases, where senior management gave the issue scant attention, marketing and/or other managers hired a consultant to develop a compliant system.

In the early 1990s, quality consultants generally understood little or nothing about the specific industries in which they were operating. However, they knew how to put together a top-down ISO 9000 based quality system and that satisfied the needs of senior managers faced with an urgent need to obtain certification, so there was little or no workplace involvement in early quality system development in many sectors. Frequently system manuals were very thick and did not reflect the business goals or management needs of the enterprise. Because of this inappropriate start, to this day, there are organizations worldwide that have not grasped the strategic significance for their business of quality management systems and their underlying philosophy.

In earlier chapters we have seen how the keystone of quality management is the concept of customer and supplier working together for their mutual advantage. For any particular organization this can become 'total' quality management if the supplier/customer interfaces extend beyond the immediate customers, back inside the organization and beyond into the supply chain. In order to achieve this, a company must organize itself in such a way that the human, administrative and technical factors affecting quality will be under control. This leads to the requirement for the development and implementation of a quality management system that enables the objectives set out in the quality policy to be accomplished. Clearly, for maximum effectiveness and to meet individual customer requirements, the management system in use must be appropriate to the type of activity and product or service being offered.

It may be useful to reflect on why such a device is necessary to achieve control of processes. The author still remembers being at a table in a restaurant with eight people who all ordered the 'chef's special individual soufflé'. All eight soufflés arrived together at the table, magnificent in their appearance and consistency, each one exhibiting an almost identical size and shape – a truly remarkable demonstration of culinary skill. How had this been achieved? The chef had *managed* such consistency by making sure that, for each soufflé, he used the same ingredients (materials), the same equipment (plant), the same method (procedure) in exactly the same way every time. The process was under control. This is the aim of a good quality management system, to provide the 'operator' of the process with consistency and satisfaction in terms of methods, materials, equipment, etc. (Figure 12.1). Two feedback loops are also required: the 'voice' of the customer (marketing activities) and the 'voice' of the process (measurement activities).

The chef's soufflés – they were not British Standard, NIST Standard, Australian Standard, or ISO Standard soufflés – they were the 'chef's special soufflés'. It is not conceivable that the chef sat down with a blank piece of paper to invent a soufflé recipe. Why re-invent wheels? He probably used a standard formula and changed it slightly to make it his own. This is exactly the way in which successful organizations use the international standards on quality management systems that are available. The 'wheel' has been invented but it must be built in a way that meets the specific organizational and product or service requirements. The international family of standards ISO 9000 'Quality Management Systems' specifies systems which can be implemented in an organization to ensure that all the product/service performance requirements and needs of the customer are fully met.

Let us return to the chef in the restaurant and propose that his success leads to a desire to open eight restaurants in which are served his special soufflés. Clearly he cannot rush from each one of these establishments to another every evening making soufflés. The only course open to him to ensure consistency of output, in all eight restaurants, is for him to write down in some detail the system he uses, and then

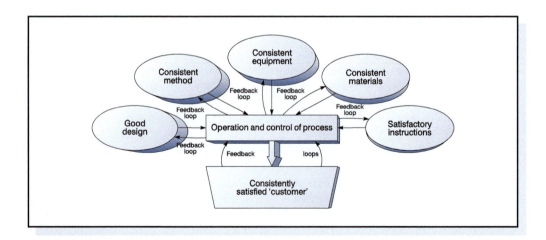

Figure 12.1
The systematic approach to process management

make sure that it is used on all sites, every time a soufflé is produced. Moreover, he must periodically visit the different sites to ensure that:

1. The people involved are operating according to the designed system (a system audit).
2. The soufflé system still meets the requirements (a system review).

If in his system audits and reviews he discovers that an even better product or less waste can be achieved by changing the method or one of the materials, then he may wish to effect a change. To maintain consistency, he must ensure that the appropriate changes are made to the management system *and* that everyone concerned is issued with the revision is re-trained and begins to operate accordingly.

A good quality management system will ensure that two important requirements are met:

- *the customer's requirements* – for confidence in the ability of the organization to deliver the desired product or service consistently.
- *the organization's requirements* – both internally and externally including regulatory, and at an optimum cost, with efficient utilization of the resources available – material, human, technological, and information.

The requirements can be truly met only if objective evidence is provided, in the form of information and data, which supports the system activities, from the ultimate supplier through to the ultimate customer.

A *quality management system* may be defined, then, as an assembly of components, such as the management, responsibilities, processes and resources for implementing total quality management. These components interact and are affected by being in the system, so the isolation and study of each one in detail will not necessarily lead to an understanding of the system as a whole. Often the interactions between the components – such as materials and processes, people and responsibilities – are just as important as the components themselves, and problems can arise from these interactions as much as from the components. Clearly, if one of the components is removed from the system, the whole thing will change.

The adoption of a quality management system is, of course, a strategic decision and its design should be influenced by the organization's objectives, structure and size, the products or services offered and its processes.

QUALITY MANAGEMENT SYSTEM DESIGN AND ISO 9000

The quality management system should apply to and interact with all processes in the organization. It begins with the identification of the customer requirements and ends with their satisfaction, at every transaction interface. The activities may be classified in several ways – generally as processing, communicating and controlling, but more usefully and specifically as shown in the quality management process model described in ISO 9001, Figure 12.2. This reflects graphically the integration of four major areas:

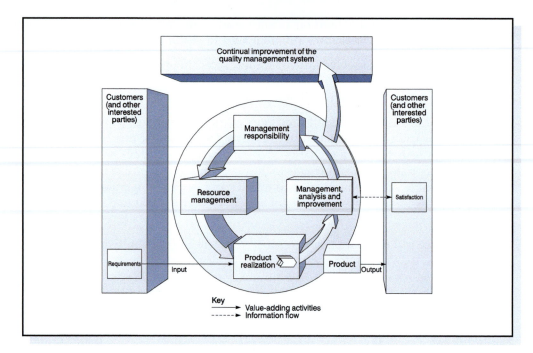

Figure 12.2
Model of a process-based quality management system

- Management responsibility
- Resource management
- Product realization
- Measurement, analysis and improvement.

The management system requirements under these headings are specified in the ISO 9000 family of standards published by the International Organization for Standardization and available through national standards bodies. ISO 9000 deals with the fundamentals of quality management systems, including the eight management principles on which the family of standards is based. These are given in outline below, together with chapter references in this book where the topics are expanded:

- Customer focus – see Chapter 1
- Leadership – see Chapter 3
- Involvement of People – see Chapters 15 to 17
- Process approach – see Chapters 10 and 11
- System approach to management – see Chapters 4 and 7
- Continual improvement – see Chapters 13 and 14
- Factual approach to decision making – see Chapters 7 and 13
- Mutually beneficial supplier relationships – see Chapter 5.

ISO 9001 deals with the requirements that organizations wishing to meet the standard have to fulfil. Third party certification bodies provide independent confirmation that organizations meet the requirements of ISO 9001. Over a million organizations worldwide are independently certified, making ISO 9001 one of the most widely used management tools in the world today.

A number of studies have identified significant financial benefits for organizations certified to ISO 9001, with a recent survey from the British Assessment Bureau which showed that nearly 50 per cent of their certified clients had won new business as the result of having certification. Other studies have shown that certified organizations achieved superior return on assets, compared to otherwise similar organizations without certification, superior operational performance in the US motor car industry, improvements in operating performance and superior financial performance, superior stock market performance. It has even been suggested that shareholders are richly rewarded for the investment in an ISO 9001 system.

ISO 9000 was first published in 1987 and was based on the BS 5750 series of standards from BSI that were proposed to ISO in 1979. Its history can be traced before that to the publication of the United States Department of Defense MIL-Q-9858 standard in 1959, which was revised into the NATO AQAP series of standards in 1969, which in turn were revised into the BS 5179 series of guidance standards published in 1974, and finally revised into the BS 5750 series of requirements standards in 1979 before being submitted to ISO. BSI has been certifying organizations for their quality management systems since 1978 and claims to certify organizations at nearly 70,000 sites globally.

In the 2000 version, ISO 9001:2000 replaced all three former standards of the 1994 issue: ISO 9001, ISO 9002 and ISO 9003. Design and development procedures were required only if a company engages in the creation of new products. The 2000 version made a radical change in thinking by placing the concept of process management at the core, this being the monitoring and optimization of a company's tasks and activities, instead of just inspection of final products. The 2000 version also demanded involvement by senior executives in order to integrate quality into the business system and avoid delegation of quality functions to junior administrators. Another goal was to improve effectiveness via process performance metrics, numerical measurement of the effectiveness of tasks and activities. Expectations of continual process improvement and tracking customer satisfaction were made explicit.

In the 2008 version, ISO 9001:2008 basically re-narrated ISO 9001:2000, introducing only clarifications to the existing requirements of the 2000 version and some changes intended to improve consistency with the environmental standard ISO 14001. There were no new requirements so that a quality management system being upgraded to the 2008 version just needed to be checked against the clarifications introduced. The standard will continue to be revised periodically to ensure successful implementation of quality management systems provides the appropriate risk management required in particular organizations.

ISO 9001 is supplemented directly by two other standards of the family:

- ISO 9000:2005 'Quality management systems. Fundamentals and vocabulary'
- ISO 9004:2009 'Managing for the sustained success of an organization. A quality management approach'.

Other standards, such as the ISO 19011 and the ISO 10000 series, may also be used for specific parts of the quality system.

ISO 9001:2008 Quality Management Systems – Requirements is a document of approximately 30 pages which is available from the national standards organization in each country. Whilst it is supplemented by the two other standards mentioned above, ISO 9000:2005 and ISO 9004:2009, which contain detailed information on how to sustain and improve quality management systems, only ISO 9001 can be directly audited against for third party assessment purposes. Outline 'Requirements' for ISO 9001 are as follows:

- Section 1: *Scope*
- Section 2: *Normative reference*
- Section 3: *Terms and definitions* (specific to ISO 9001, not specified in ISO 9000)
- Section 4: *Quality management system*
- Section 5: *Management responsibility*
- Section 6: *Resource management*
- Section 7: *Product realization*
- Section 8: *Measurement, analysis and improvement.*

Fundamentally, ISO 9001 requires that

- The quality policy is a formal statement from management, closely linked to the business and marketing plan and to customer needs.
- The quality policy is understood and followed at all levels and by all employees, each employee working towards measurable objectives.
- The business makes decisions about the quality system based on recorded data.
- The quality system is regularly audited and evaluated for conformance and effectiveness.
- Records show how and where materials and products were processed to allow products and problems to be traced to the source.
- The business determines customer requirements.
- The business has created systems for communicating with customers about product information, inquiries, contracts, orders, feedback and complaints.
- When developing new products, the business plans the stages of development, with appropriate testing at each stage; it tests and documents whether the product meets design requirements, regulatory requirements and user needs.
- The business regularly reviews performance through internal audits and meetings. The business determines whether the quality system is working and what improvements can be made; it has a documented procedure for internal audits.
- The business deals with past problems and potential problems; it keeps records of these activities and the resulting decisions, and monitors their effectiveness.

- The business has documented procedures for dealing with actual and potential non-conformances (problems involving suppliers, customers or internal problems).
- The business:
  1. makes sure no one uses a poor quality product
  2. determines what to do with a poor quality product
  3. deals with the root cause of problems
  4. keeps records to use as a tool to improve the system.

Industry-specific interpretations of ISO 9001

The ISO 9001 standard is generalized and abstract and its parts must be carefully interpreted to make sense within a particular organization. Constructing buildings is not like making cars or offering legal services, yet the ISO 9001 guidelines can be applied to each of these because they are business management guidelines. Diverse organizations – police departments, professional sports teams and city councils – have all successfully implemented ISO 9001 systems.

Over time, various industry sectors have sought to standardize interpretations of the guidelines within their own marketplace. This is partly to ensure that their versions of ISO 9000 have their specific requirements, but also to try and ensure that more appropriately trained and experienced auditors are sent to assess them:

- The **TickIT** guidelines are an interpretation of ISO 9000 produced by the UK Board of Trade to suit the processes of the information technology industry, especially software development.
- **AS9100** is the Aerospace Basic Quality System Standard, an interpretation developed by major aerospace manufacturers, including AlliedSignal, Allison Engine, Boeing, General Electric Aircraft Engines, Lockheed-Martin, McDonnell Douglas, Northrop Grumman, Pratt & Whitney, Rockwell-Collins, Sikorsky Aircraft and Sundstrand.
- **PS 9000 * QS 9000** is an interpretation agreed upon by major automotive manufacturers (GM, Ford, Chrysler). It includes techniques such as FMEA and APQP. QS 9000 is now replaced by ISO/TS 16949.
- **ISO/TS 16949** is an interpretation agreed upon by major automotive manufacturers (American and European manufacturers). The emphasis on a process approach is stronger than in ISO 9001 so ISO/TS 16949 contains the full text of ISO 9001 plus automotive industry-specific requirements.
- **TL 9000** is the Telecom Quality Management and Measurement System Standard, an interpretation developed by the telecom consortium, QuEST Forum in 1998 to meet the supply chain quality requirements of the worldwide telecommunications industry. Unlike ISO 9001 or other sector-specific standards, TL 9000 includes standardized product and process measurements that must be reported into a central repository, which allow organizations to benchmark their performance in key process areas against peer organizations. TL 9000 contains the full text of ISO 9001.
- **ISO 13485** is the medical industry's equivalent of ISO 9001. Whereas the standards it replaces were interpretations of how to apply ISO 9001 and ISO 9002 to medical devices, ISO 13485 stands alone. Because ISO 13485 is

relevant to medical devices manufacturers (unlike ISO 9001, which is applicable to any industry) and because of the differences between the two standards relating to continual improvement, compliance with ISO 13485 does not necessarily mean compliance with ISO 9001 (and vice versa).

- **ISO/IEC 90003** provides guidelines for the application of ISO 9001 to computer software.
- **ISO/TS 29001** provides quality management system requirements for the design, development, production, installation and service of products for the petroleum, petrochemical and natural gas industries.
Other related standards and documents include:
- **ISO 10006** – Quality management –Guidelines to quality management in projects
- **ISO 14001** – Environmental management standards
- **ISO 19011** – Guidelines for quality management systems auditing and environmental management systems auditing
- **ISO/TS 16949** – Quality management system requirements for automotive-related products suppliers
- **ISO/IEC 27001** – Information security management
- **ISO 39001** – Road traffic safety management
- **ISO 50001** – Energy audit
- **AS 9100** – Aerospace industry implementation of ISO 9000/1.

It is interesting to bring together the concept of Deming's Cycle of continuous improvement – PLAN DO CHECK ACT – and quality management systems. A simplification of what a good management system is trying to do is given in Figure 12.3, which follows the improvement cycle.

In many organizations established methods of working already exist around identified processes, and all that is required is the *documenting of what is currently done.*

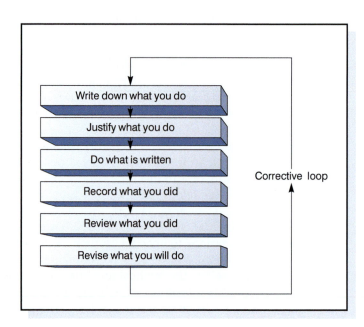

Figure 12.3
The quality management system and never-ending improvement

In some instances companies may not have procedures to satisfy the requirements of a good standard, and they may have to begin to devise them. Alternatively, it may be found that two people, supposedly performing the same task, are working in different ways, and there is a need to standardize procedures. Some organizations use the effective slogan 'If it isn't written down, it doesn't exist'. This can be a useful discipline, provided it doesn't lead to bureaucracy.

Justify that the *system* as it is designed *meets the requirements of a good international standard,* such as ISO 9001. There are other excellent standards that are used, and these provide similar checklists of things to consider in the establishment of the quality system.

One person alone cannot document a quality management system; the task is the job of all personnel who have responsibility for any part of it. This means that a quality system, by definition, has to be built from the operational level up and cannot be imposed by external consultants from without the organization. The quality system must be a *practical working one* – that way it ensures that consistency of operation is maintained and it may be used as a training aid.

In the operation of any process, a useful guide is:

- No process without data collection (Measurement)
- No data collection without Analysis
- No analysis without Decisions
- No decisions without actions (Improvement) – which can include doing nothing.

This excellent discipline is built into any good quality management system, primarily through the audit and review mechanism. The requirement to *audit or 'check'* that the system is functioning according to plan, and to *review* possible system improvements, utilising audit results, should ensure that the *improvement* cycle is engaged through the *corrective action* procedures. The overriding requirement is that the systems must reflect the established practices of the organization, improved where necessary to bring them into line with current and future requirements.

QUALITY MANAGEMENT SYSTEM REQUIREMENTS

The quality management system that needs to be documented and implemented will be determined by the nature of the process carried out to ensure that the product or service conforms to customer requirement. Certain fundamental principles are applicable, however, throughout industry, commerce, and the services. These fall into generally well defined categories which are detailed in ISO 9001.

1. Leadership and management responsibility

Customer needs/requirements (see Chapter 1)
The organization must focus on customer needs and specify them as defined requirements for the organization. The aim of this is to achieve customer confidence in the products and/or services provided. It is also necessary to ensure that the defined requirements are understood and fully met.

Quality policy (see Chapter 3)

The organization should define and publish its quality policy, which forms one element of the corporate policy. Full commitment is required from the most senior management to ensure that the policy is communicated, understood, implemented and maintained at all levels in the organization. For every project, the quality plans must fully reflect the company quality policy and the project leadership must be responsible for implementing company quality policy and goals within the supply chain at the project level. The company quality policy should be authorized by top management and signed by the Chief Executive, or equivalent, who must also ensure that it:

- is suitable for the needs/requirements of the customers and the purpose of the organization
- includes commitment to meeting requirements and continual improvement for all levels of the organization
- provides a framework for establishing and reviewing quality objectives
- is regularly reviewed for its suitability and objectiveness.

Quality objectives and planning

Organizations should establish written quality objectives and define the responsibilities of each function and level in the organization.

One manager (the 'management representative') reporting to top management, with the necessary authority, resources, support and ability, should be given the responsibility to co-ordinate, implement and maintain the quality management system, resolve any problems and ensure prompt and effective corrective action. This includes responsibility for ensuring proper handling of the system and reporting on needs for improvement. Those who control sales, service operations, warehousing, delivery and reworking of non-conforming product or service processes should also be identified.

At the project level, responsibility for conformance with the quality policy and its implementation lies fully with line management and should not be separated out as a special responsibility. Doing so would only lead to a conflict within the project organization.

Management review

Management reviews of the system must be carried out, by top management at defined intervals, with records to indicate the actions decided upon. The effectiveness of these actions should be considered during subsequent reviews. Reviews typically include data on the internal quality audits, customer feedback, product conformance analysis, process performance and the status of preventive, corrective and improvement actions.

Quality manual

The organization should prepare a 'quality manual' that is appropriate. It should include but not necessarily be limited to:

a) the quality policy
b) definition of the quality management system – scope, exclusions, etc.
c) description of the interaction between the processes of the quality management system

d)   documented procedures required by the quality management system, or reference to them.

In the quality manual for a large organization it may be convenient to indicate simply the existence and contents of other manuals, those containing the details of procedures and practices in operation in specific areas of the system.

Before an organization can agree to supply to a specification, it must ensure that:

a)   The processes and equipment (including any that are subcontracted) are capable of meeting the requirements.
b)   The operators have the necessary skills and training.
c)   The operating procedures are documented and not simply passed on verbally.
d)   The plant and equipment instrumentation is capable (e.g. measuring the process variables with the appropriate accuracy and precision).
e)   The quality-control procedures, and any inspection, check or test methods available, provide results to the required accuracy and precision, and are documented.
f)   Any subjective phrases in the specification, such as 'finely ground', 'low moisture content', 'in good time', are understood, and procedures to establish the exact customer requirement exist.

Control of documents

The organization needs to establish procedures for controlling the new and revised documents required for the operation of the quality management system. Documents of external origin must also be controlled. These procedures should be designed to ensure that:

a)   documents are approved
b)   documents are periodically reviewed, and revised as necessary
c)   the current versions of relevant documents are available at all locations where activities essential to the effective functioning of the processes are performed
d)   obsolete documents are promptly removed from all points of issue and use, or otherwise controlled to prevent unplanned use
e)   any obsolete documents retained for legal or knowledge-preservation purposes are suitably identified.

Documentation needs to be legible, revision controlled, readily identifiable and maintained in an orderly manner. Of course, the documentation may be in any form or any type of media.

Control of quality records

Quality records are needed to demonstrate conformance to requirements and effective operation of the quality management system. Quality records from suppliers also need to be controlled. This aspect includes record identification, collection, indexing, access, filing, storage and disposition. In addition, the retention times of quality records needs to be established.

2. Resource management

The organization should determine and provide the necessary resources to establish and improve the quality management system, including all processes and projects. The general infrastructure needed to achieve conformity to the product or service requirements needs to be provided and maintained, including buildings, equipment, any supporting services and the work environment.

Human resources and capability

The organization needs to select and assign people who are competent, on the basis of applicable education, training, skills and experience, to those activities which impact the conformity of product and/or service. On construction sites, for example, where much of the work is undertaken by subcontractors, this includes workers and supervisors across the supply chain.

The organization also needs to:

a) determine the training needed to achieve conformity of product and/or service
b) provide the necessary training to address these needs
c) evaluate the effectiveness of the training on a continual basis.

Individuals clearly need to be educated and trained to qualify them for the activities they perform. Competence, including qualification levels achieved, needs to be demonstrated and documented. It may be beneficial to conduct joint training of supervisors and managers on a project to ensure that they are working within a consistent framework of values and expectations.

Information is ever increasingly a vital resource and any organization needs to define and maintain the current information and the infrastructure necessary to achieve conformity of products and/or services. The management of information, including access and protection of information to ensure integrity and availability, need also to be considered. Once again, the seamless use of real-time information across the supply chain is essential to the efficient operation of the delivery process as a whole.

3. Product or service realization

As we have seen in Figure 12.2, any organization needs to determine the processes required to convert customer requirements into customer satisfaction, by providing the required product and/or service. In determining such processes the organization needs to consider the outputs from the quality planning process.

The sequence and interaction of these processes need to be determined, planned and controlled to ensure they operate effectively, and there is a need to assign responsibilities for the operation and monitoring of the product/service generating processes.

These processes clearly need to be operated under controlled conditions and produce outputs which are consistent with the organization's quality policy and objective and it is necessary to:

a) determine how each process influences the ability to meet product and/or service requirements

b) establish methods and practices relevant to process activities, to the extent necessary to achieve consistent operation of the process
c) verify processes can be operated to achieve product and/or service conformity
d) determine and implement the criteria and methods to control processes related to the achievement of product and/or service conformity
e) determine and implement arrangements for measurement, monitoring and follow-up actions, to ensure processes operate effectively and the resultant product/service meets the requirements
f) ensure availability of process documentation and records which provide operating criteria and information, to support the effective operation and monitoring of the processes. (This documentation needs to be in a format to suit the operating practices, including written quality plans)
g) provide the necessary resources for the effective operation of the processes.

Customer-related processes

One of the first processes to be established is the one for identifying customer requirements: both internal and external customers and immediate and end-user customers. This needs to consider the:

a) extent to which customers have specified the product/service requirements
b) requirements not specified by the customer but necessary for fitness for purpose
c) obligations related to the product/service, including regulatory and legal requirements
d) other customer requirements e.g. for availability, delivery and support of product and/or service.

The identified customer requirements need also to be reviewed before a commitment to supply a product/service is given to the customer (e.g. submission of a tender, acceptance of a contract or order). This should determine that:

a) identified customer requirements are clearly defined for the product and/or service
b) the order requirements are confirmed before acceptance, particularly where the customer provides no written statement of requirements
c) the contract or order requirements differing from those in any tender or quotation are resolved.

This should also apply to amended customer contracts or orders. Moreover, each commitment to supply a product/service, including amendment to a contract or order, needs to be reviewed to ensure the organization will have the ability to meet the requirements.

Any successful organization needs to implement effective communication and liaison with customers, particularly regarding:

a) product and/or service information
b) enquiry and order handling, including amendments
c) customer complaints and other reports relating to non-conformities
d) recall processes, where applicable
e) customer responses relating to conformity of product/service.

Quality management systems

Where an organization is supervising or using customer property, care needs to be exercised to ensure verification, storage and maintenance. Any customer product or property that is lost, damaged or otherwise found to be unsuitable for use should, of course, be recorded and reported to the customer. Customer property may, of course, include intellectual property e.g. information provided in confidence.

Design and development

The organization needs to plan and control design and development of products and/or services, including:

a) stages of the design and development process
b) required review, verification and validation activities
c) responsibilities for design and development activities.

Interfaces between different groups involved in design and development need to be managed to ensure effective communication and clarity of responsibilities, and any plans and associated documentation should be:

a) made available to personnel that need them to perform their work
b) reviewed and updated as design and development evolves.

The requirements to be met by the product/service need to be defined and recorded, including identified customer or market requirements, applicable regulatory and legal requirements, requirements derived from previous similar designs and any other requirements essential for design and development. Incomplete, ambiguous or conflicting requirements must be resolved.

The outputs of the design and development process need to be recorded in a format that allows verification against the input requirements. So, the design and development output should:

a) meet the design and development input requirements
b) contain or make reference to design and development acceptance criteria
c) determine characteristics of the design essential to safe and proper use, and application of the product or service
d) design and development output documents should also be reviewed and approved before release.

Validation needs to be performed to confirm that resultant product/service is capable of meeting the needs of the customers or users under the planned conditions. Wherever possible, validation should be defined, planned and completed prior to the delivery or implementation of the product or service. Partial validation of the design or development output may be necessary at various stages to provide confidence in their correctness, using such methods as:

a) reviews involving other interested parties
b) modelling and simulation studies
c) pilot production, construction or delivery trials of key aspects of the product and/or service.

Design and development changes or modifications need to be determined as early as possible, recorded, reviewed and approved, before implementation. At this stage, the effect of changes on compatibility requirements and the usability of the product or service throughout its planned life need to be considered.

Because of the size and complexity of constructed products in-process evaluation may not provide sufficient feedback regarding design quality. Evaluations of design effectiveness are often best conducted by interviewing end-users and owners, and undertaking an expert evaluation some years after a project has been completed. Only in this way can we ensure that the cycle of learning about design quality is continuous and that the lessons learned are fed back into the initiation and development process. The responsibility for initiating this kind of improvement cycle lies both with the designers (who need the information to enable them to improve their services) and the developers (who need to ensure that future projects are informed by lessons from current practice).

Procurement/purchasing

Procurement or purchasing processes need to be controlled to ensure purchased products/services conform to the organization's requirements. The type and extent of methods to do this are dependent on the effect of the purchased product/service upon the final product/service. Clearly suppliers need to be evaluated and selected on their ability to supply the product or service in accordance with the organization's requirements. Supplier evaluations, supplier audit records and evidence of previously demonstrated ability should be considered when selecting suppliers and when determining the type and extent of supervision applicable to the purchased materials/services.

The procurement/purchasing documentation should contain information clearly describing the product/service ordered, including:

a) requirements for approval or qualification of product and/or service, procedures, processes, equipment and personnel
b) any management system requirements.

Review and approval of purchasing documents, for adequacy of the specification of requirements prior to release, is also necessary.

Any purchased products/services need some form of verification. Where this is to be carried out at the supplier's premises, the organization needs to specify the arrangements and methods for product/service release in the purchasing documentation.

Production and service delivery processes

The organization needs to control production and service delivery processes through:

a) information describing the product/service characteristics
b) clearly understandable work standards or instructions
c) suitable production, installation and service provision equipment
d) suitable working environments
e) suitable inspection, measuring and test equipment, capable of the necessary accuracy and precision
f) the implementation of suitable monitoring, inspection or testing activities
g) provision for identifying the status of the product/service, with respect to required measurement and verification activities
h) suitable methods for release and delivery of products and/or services.

Quality management systems 257

Where applicable, the organization needs to identify the product/service by suitable means throughout all processes. Where traceability is a requirement for the organization, there is a need to control the identification of product/service. There is also a need to ensure that, during internal processing and final delivery of the product/service, the identification, packaging, storage, preservation and handling do not adversely affect conformity with the requirements. This applies equally to parts or components of a product and elements of a service.

Where the resulting output cannot be easily or economically verified by monitoring, inspection or testing, including where processing deficiencies may become apparent only after the product is in use or the service has been delivered, the organization needs to validate the production and service delivery processes to demonstrate their effectiveness and acceptability.

The arrangements for validation might include:

a) processes being qualified prior to use
b) qualification of equipment or personnel
c) use of specific procedures or records.

Evidence of validated processes, equipment and personnel needs to be recorded and maintained, of course.

Post delivery services

Where there is a requirement for the organization to provide support services, after delivery of the product or service, this needs to be planned and in with the customer requirement.

Monitoring and measuring devices

There is a need to control, calibrate, maintain, handle and store the applicable measuring, inspection and test equipment to specified requirements. In construction this applies in the main to survey measuring equipment, quality assessment measuring equipment and precision manufacturing equipment. Measuring, inspection and test equipment should be used in a way which ensures that any measurement uncertainty, including accuracy and precision, is known and is consistent with the required measurement capability. Any test equipment software should meet the applicable requirements for the design and development of the product.

The organization certainly needs to:

a) Calibrate and adjust measuring, inspection and test equipment at specified intervals or prior to use, against equipment traceable to international or national standards. Where no standards exist, the basis used for calibration needs to be recorded.
b) Identify measuring, inspection and test equipment with a suitable indicator or approved identification record to show its calibration status.
c) Record the process for calibration of measuring, inspection and test equipment.
d) Ensure the environmental conditions are suitable for any calibrations, measurements, inspections and tests.
e) Safeguard measuring, inspection and test equipment from adjustments which would invalidate the calibration.

f) Verify validity of previous inspection and test results when equipment is found to be out of calibration.

g) Establish the action to be initiated when calibration verification results are unsatisfactory.

4 Measurement, analysis and improvement

Any organization needs to define and implement measurement, analysis and improvement processes to demonstrate that the products, services and processes conform to the specified requirements. The type, location and timing of these measurements needs to be determined and the results recorded based on their importance. The results of data analysis and improvement activities should be an input to the management review process, of course.

Measurement and monitoring

There is a need to determine and establish processes for measurement of the quality management system performance. Customer satisfaction must be a primary measure of system output and the internal audits should be used as a primary tool for evaluating ongoing system compliance.

The organization needs to establish a process for obtaining and monitoring information and data on both immediate and end-user customer satisfaction for all essential processes. The methods and measures for obtaining customer satisfaction information and data and the nature and frequency of reviews need to be defined to demonstrate the level of customer confidence in the delivery of conforming product and/or service supplied by the organization. Suitable measures for establishing internal improvement need to be implemented and the effectiveness of the measures periodically evaluated.

The organization must establish a process for performing internal audits of the quality management system and related processes. The purpose of the internal audit is to determine whether:

• the quality management system established by the organization conforms to the requirements of the International Standard, and
• the quality management system has been effectively implemented and maintained.

The internal audit process should be based on the status and importance of the activities, areas or items to be audited, and the results of previous audits.

The internal audit process should include:

• planning and scheduling the specific activities, areas or items to be audited
• assigning trained personnel independent of those performing the work being audited
• assuring that a consistent basis for conducting audits is defined.

The results of internal audits should be recorded including:

• activities, areas and processes audited
• non-conformities or deficiencies found
• status of commitments made as the result of previous audits, such as corrective actions or product audits
• recommendations for improvement.

Quality management systems

The last point is a critical one for construction projects. The authors have observed that on many construction projects even though errors are detected and rectified on a regular basis they are repeated again and again. Although the project organization may have set processes in place for error identification, they are not avoided. This suggests that the focus is on detection rather than avoidance.

The results of the internal audits should be communicated to the area audited and the management personnel responsible need to take timely corrective action on the non-conformities recorded.

Suitable methods for the measurement of processes necessary to meet customer requirements need to be applied, including monitoring the output of the processes that control conformity of the product or service provided to customers. The measurement results then need to be used to determine opportunities for improvements.

The organization needs to also apply suitable methods for the measurement of the product or service to verify that the requirements have been met. Evidence from any inspection and testing activities and the acceptance criteria used need to be recorded. If there is an authority responsible for release of the product and/or service, this should also be recorded.

Products or service should not be dispatched until all the specified activities have been satisfactorily completed and the related documentation is available and authorized. The only exception to this is when the product or service is released under positive recall procedures.

Control of non-conforming products

Product and services which do not conform to requirements need to be controlled to prevent unplanned use, application or installation, and the organization needs to identify, record and review the nature and extent of the problem encountered, and determine the action to be taken. This needs to include how non-conforming service will be:

a)   corrected or adjusted to conform to requirements, or
b)   accepted under concession, with or without correction, or
c)   re-assigned for an alternative, valid application, or
d)   rejected as unsuitable.

The responsibility and authority for the review and resolving of non-conformities needs to be defined, of course.

When required by the contract, the proposed use or repair of non-conforming product or a modified service needs to be reported for concession to the customer. The description of any corrections or adjustments, accepted non-conformities, product repairs or service modifications also need to be recorded. Where it is necessary to repair or re-work a product or modify a service, verification requirements need to be determined and implemented.

Analysis of data

Analysis of data needs to be established as a means of determining where system improvements can be made. Data needs to be collected from relevant sources, including internal audits, corrective and preventive action, non-conforming product service, customer complaints and customer satisfaction results.

The organization should then analyse the data to provide information on:

a) the effectiveness of the quality management system
b) process operation trends
c) customer satisfaction
d) conformance to customer requirements of the product/service
e) suppliers.

There is also a need to determine the statistical techniques to be used for analysing data, including verifying process operations and product service characteristics. Of course the statistical techniques selected should be suitable and their use controlled and monitored.

Improvement

The organization needs to establish a process for eliminating the causes of non-conformity and preventing recurrence. Non-conformity reports, customer complaints and other suitable quality management system records are useful as inputs to the corrective action process. Responsibilities for corrective action need to be established together with the procedures for the corrective action process, which should include:

a) identification of non-conformities of the products, services, processes, quality management system and customer complaints
b) investigation of causes of non-conformities and recording results of investigations
c) determination of corrective actions needed to eliminate causes of non-conformities
d) implementation of corrective action
e) follow-up to ensure corrective action taken is effective and recorded.

Corrective actions also need to be implemented for products or services already delivered, but subsequently discovered to be non-conforming and customers need to be notified where possible.

The organization needs to establish a process for eliminating the causes of potential non-conformities to prevent their occurrence. Quality management system records and results from the analysis of data should be used as inputs for this and responsibilities for preventive action established. The process should include:

a) identification of potential product, service and process non-conformities
b) investigation of the causes of potential non-conformities of products/services, process and the quality management system, and recording the results
c) determination of preventive action needed to eliminate causes of potential non-conformities
d) implementation of preventive action needed
e) follow-up to ensure preventive action taken is effective, recorded and submitted for management review.

Processes for the continual improvement of the quality management system need to be established including methods and measures suitable for the products/services.

Compliance certification

The International Standards Organization, ISO does not certify organizations itself. Numerous certification bodies exist, which audit organizations and, upon success, issue ISO 9001 compliance certificates. Although commonly referred to as 'ISO 9000' certification, the actual standard to which an organization's quality management system can be certified is ISO 9001 (under further revision at time of publication). Many countries have formed accreditation bodies to authorize ('accredit') the certification bodies. Both the accreditation bodies and the certification bodies charge fees for their services. The various accreditation bodies have mutual agreements with each other to ensure that certificates issued are accepted worldwide. Certification bodies themselves operate under another quality standard, ISO/IEC 17021, while accreditation bodies operate under ISO/IEC 17011.

An organization applying for ISO 9001 certification is audited based on an extensive sample of its sites, functions, products, services and processes. The auditor presents a list of problems (defined as 'nonconformities', 'observations' or 'opportunities for improvement') to management. If there are no major nonconformities, the certification body will issue a certificate. Where major nonconformities are identified, the organization will present an improvement plan to the certification body, e.g. corrective action reports showing how the problems will be resolved. Once the certification body is satisfied that the organization has carried out sufficient corrective action, it will issue a certificate. The certificate is limited by a certain scope, e.g. production of golf balls, and will display the addresses to which the certificate refers.

An ISO 9001 certificate is not a once-and-for-all award, but must be renewed at regular intervals recommended by the certification body, usually once every three years. There are no grades of competence within ISO 9001: either a company is certified (meaning that it is committed to the method and model of quality management described in the standard) or it is not. In this respect, ISO 9001 certification contrasts with measurement-based quality systems.

OTHER MANAGEMENT SYSTEMS AND MODELS

Organizations of all kinds are increasingly concerned to achieve and demonstrate sound environmental performance. Many have undertaken environmental audits and review to assess this. To be effective these need to be conducted within a structured management system, which in turn is integrated with the overall management activities dealing with all aspects of desired environmental performance.

Such a system should establish processes for setting environmental policy and objectives, and achieving compliance to them. It should be designed to place emphasis on the prevention of adverse environmental effects, rather than on detection after occurrence. It should also identify and assess the environmental effects arising from the organization's existing or proposed activities, products or services and from incidents, accidents and potential emergency situations. The system must identify the relevant regulatory requirements, the priorities and pertinent environmental objective and targets. It needs also to facilitate planning, control, monitoring, auditing and review activities to ensure that the policy is complied with, that it remains relevant and is capable of evolution to suit changing circumstances.

The international standard ISO 14001 contains a specification for environmental management systems for ensuring and demonstrating compliance with stated policies and objectives. The standard is designed to enable any organization to establish an effective management system, as a foundation for both sound environmental performance and participation and environmental auditing schemes.

ISO 14001 shares common management system principles with the ISO 9001 standard and organizations may elect to use an existing management system, developed in conformity with that standard, as a basis for environmental management. The ISO 14001 standard defines environmental policy, objectives, targets, effect, management, systems, manuals, evaluation, audits and reviews. It mirrors the ISO 9001 standard requirements in many of its own requirements, and it includes a guide to these in an informative annex. ISO 9001 also gives a useful annex of 'correspondence between ISO 9001 and ISO 14001' under each of the main headings of both standards. This shows that the quality management system standard has been aligned with requirements of the environmental management system standard in order to enhance the compatibility of the two for the benefit of the users. As such ISO 9001 enables an organization to align and integrate its own quality management systems with other related management system requirements. In this way, it should be possible for any organization to adapt its existing management system(s) in order to establish one that complies with ISO 9001.

ISO 9000 also makes comments on the relationship between quality management systems and excellence models. It claims that the approaches set down in the ISO 9000 family of standards and in the various excellence models (see Chapter 2) are based on common principles and that the only differences are in the scope of application. Excellence models and ISO 9000:

i)   help organizations identify strengths and weaknesses
ii) aid the evaluation of organizations
iii) establish a basis for continuous improvement
iv) allow and support external recognition.

It is also recognized that excellence models add value to the QMS approach by providing criteria that allow comparative evaluation of organizational performance with other organizations.

Management systems are needed in all areas of activity, whether large or small businesses, manufacturing, service or public sector. The advantages of systems in manufacturing are obvious, but they are just as applicable in areas such as marketing, sales, personnel, finance, research and development, as well as in the service industries and public sectors. No matter where it is implemented a good management system will improve process control, reduce wastage, lower costs, increase market share (or funding), facilitate training, involve staff and raise morale.

BIBLIOGRAPHY

Arter, D.R., Cianfrani, C.A and West, J.E., *How to Audit the Process Based BMS* (2nd edn), ASQ, Milwaukee, 2013.

British Standards Institute (BSI), BS EN ISO 9001:2008, *Quality Management Systems*, BSI, London, 2008.

Federal Information Processing Standards (FIPS), *Publications 183 and 194,* National Institute of Standards and Technology (NIST), Gaithesburg, 1993.

Hill, N. (ed), *Customer Satisfaction Measurement for ISO 9000:2000*, Butterworth-Heinemann, Oxford, 2002.

Hoyle, D., *ISO 9000 Quality Systems Handbook – updated for the ISO 9001:2008 standard: using the standards as a framework for business improvement*, Butterworth-Heinemann, Oxford, 2010.

International Standards Organization, ISO 9000-1: 1994, *Quality Management and Quality Assurance Standards – Part 1: Guidelines for Selection and Use.*

International Standards Organization, ISO 14001: 2008, *Environmental Management Systems – Specification with Guidance for Use.*

CHAPTER HIGHLIGHTS

Why a quality management system

- An appropriate quality management system will enable the objectives set out in the quality policy to be accomplished.
- The International Organization for Standardization (ISO) 9000 series sets out methods by which a system can be implemented to ensure that the specified customer requirements are met.
- A quality system may be defined as an assembly of components, such as the management responsibilities, process and resources.

Quality management system design and ISO 9000

- ISO 9000 links management system quality to product and process quality through a focus on demonstrating customer satisfaction and continuous improvement.
- Quality management systems should apply to and interact with processes in the organization. The activities are generally processing, communicating and controlling. These should be documented in the form of a quality manual.
- The system should follow the PLAN DO CHECK ACT cycle, through documentation, implementation, audit and review.
- The ISO 9000 family together form a coherent set of quality management system standards to facilitate mutual understanding across national and international trade.

Quality management system requirements

- The general categories of the ISO 9001:2000 standard on quality management systems include: management responsibility, resource management, product realization, measurement analysis and improvement, which are detailed in the standard.

Other management systems and models

- The International Standard ISO 14001 contains specifications for environmental management systems for ensuring and demonstrating compliance with the stated policies and objectives, and acting as a base for auditing and review schemes.
- ISO 14000 shares common principles with the ISO 9001 standard on quality management systems. The latter shows, by 'correspondence' between the two, under the main headings of both standards, that the quality standard has been aligned with the requirements of the environmental standard.
- ISO 9000 also makes comments on the relationship between quality management systems and excellence models. The two are based on the common principles of identifying strengths and weaknesses, evaluation, continuous improvement, external recognition.

Chapter 13

Continuous improvement – the basics

APPROACHES, METHODOLOGIES AND TOOLS

Continuous improvement is a term in common use throughout most industries and the public sector. It can become a meaningless term unless it is linked to organizational strategy, has a defined structure, a chosen approach, a methodology and an associated tool kit. Figure 13.1 provides a generic eight-stage structure that may be applied to most organizations. It begins unsurprisingly with leadership and the top-down cascade of objectives, particularly those related to achieving on-quality, on-time, on-cost delivery of products and services, chosen to boost customer confidence and drive efficiency improvements.

Unfortunately at this point the senior management in many organizations rush straight to the 'instruction manual' to implement approaches such as 'Lean' and 'Six Sigma' without establishing the much needed managerial infrastructure of stage 2. Good programme and systems management will be essential, as will communications with people inside and outside on what the chosen approach and even 'brand' means. The third stage requires careful choice of a pragmatic, fact-based improvement methodology which, hopefully, will yield rapid results using proven methodologies, supported by fact and data-based, hands-on tools and techniques to be recommended.

If this choice is based on the need to improve on-quality, on-time and on-cost delivery (Figure 13.2) then approaches that reduce the wastage of time and reduce variation to improve quality will be appropriate. Moreover, as these approaches also reduce costs, then overt cost-reduction programmes can be avoided since these often merely knock out capability and render the organization incapable of functioning properly – the equivalent of going to the doctor for help in reducing body weight only for him to chop off your left arm!

Lean approaches are designed to eliminate non-value added time and Six Sigma approaches are designed to reduce variation, so a combination of the two can provide an excellent ready-made solution for the CI approach – the so-called Lean Six Sigma

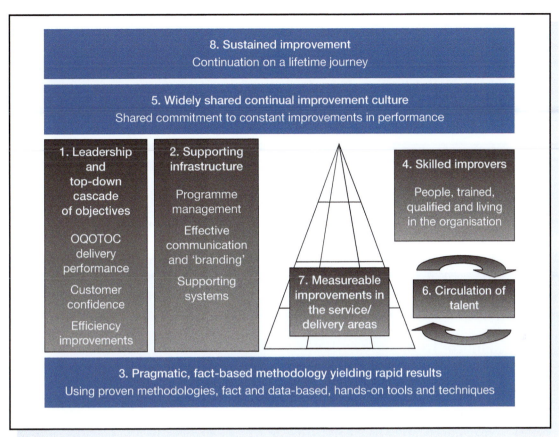

Figure 13.1
An overall approach structure for continuous improvement

or even Lean Sigma. To go with the overall approach a closed loop fact based improvement methodology is needed. This can be provided by the traditional Six Sigma method of DMAIC stages – Define, Measure, Analyse, Improve and Control (see later section in this chapter).

The fourth stage of the structure outlined in Figure 13.1 is skilled improvers – people, trained, qualified and 'living in the organization', as opposed to being part of a separate improvement task force of some kind. We are not all naturally born with the ability to make improvements in a structured and effective way and, as we need to be trained and developed to be engineers, chemists, dentists or lawyers, we need to be trained and developed to be improvers, whether we are engineers, chemists, dentists or lawyers. The best people to work on improving an engineering design process are the engineers who work in the process, who have been trained and developed how to make those improvements effectively, rather than an 'outside team' who are not fully conversant with how the process operates. That is not to say there is no place for an 'external task force', to tackle certain situations, but the norm should be that we expect the 'experts' who function in the process every day to make the changes required.

Continuous improvement – the basics

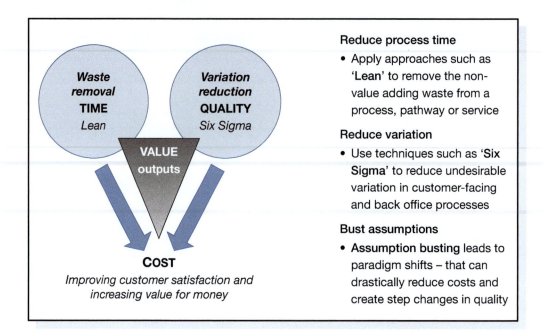

Reduce process time

- Apply approaches such as 'Lean' to remove the non-value adding waste from a process, pathway or service

Reduce variation

- Use techniques such as 'Six Sigma' to reduce undesirable variation in customer-facing and back office processes

Bust assumptions

- **Assumption busting** leads to paradigm shifts – that can drastically reduce costs and create step changes in quality

Figure 13.2
CI approach to delivering OQ OT OC

The development of a widely shared continual improvement culture, with shared commitment to constant improvements in performance is the fifth stage in Figure 13.1. The leadership plays a key role here, of course, establishing the 'norms' by asking for each and every improvement project to be properly defined and tackled using the chosen methodology, by properly trained improvers and specific expected outcomes – people soon change their behaviour when they get clear messages about what the top people want and expect.

The sixth stage of circulation of talent refers to situations where you do have specifically trained improvement task forces, such as the Black Belts trained in GE, and it is necessary to ensure they do not see their role as a dead-end job. Jack Welsh famously made the GE Six Sigma programme successful by insisting that the HR policies not only supported these trained and experienced improvers returning to line management, but stipulated that managers would not get promoted unless they had served time as a Black Belt in GE.

Only now are we ready to move to the seventh stage of making measurable improvements in the production, operations or service areas. Rushing into this without the establishment of the previous six stages usually results in disappointment and the claim that SPC is too statistical to work here, Six Sigma is Greek for us or Lean is too 'mean' for our organization.

Continuous improvement does mean sustained improvements on a lifetime journey. In the short-term world we live in this can be a challenge but do look around and accept that truly great organizations have developed a solid culture of doing the right things right every time. In a rapidly changing world, the only way this can be achieved is by having a culture of continually improving what is done for all customers and other stakeholders.

The 'DRIVER' framework for continuous improvement

The author and his colleagues have developed a fully closed loop improvement methodology which brings together the best of Lean, Six Sigma and Cost of Quality approaches – DRIVER with the stages of Define, Review, Investigate, Verify, Execute and Reinforce. This well-established approach, which has been used in literally hundreds of organizations, prevents people jumping from the problem to the solution without considering the improvement options:

- **D**EFINE – the scope and goals of the process to be improved, in terms of the customer requirements and the processes that deliver them
- **R**EVIEW – understand the 'as-is' processes and measure the current process performance to appreciate the 'value add' and 'waste'
- **I**NVESTIGATE – analyse the gap between the current and desired performance, prioritize problems and identify root causes
- **V**ERIFY – generate the improvement solutions, including the 'to-be' processes, to fix the problems and prevent them from recurring
- **E**XECUTE – implement the improved processes in the pathway in a way that 'holds the gains'
- **R**EINFORCE – capitalize on the improvement by 'learning the lessons' and establishing a re-assessment process for continuous improvement.

Figure 13.3 shows DRIVER, together with recommended tools and techniques taken from a range of established components of CI. Figures 13.4 to 13.9 give the detail of how these are applied throughout the various stages of an improvement project using DRIVER as the methodology. The next section and subsequent chapters provide further detail on the specific tools and techniques.

The need for data and some basic tools and techniques

In the never-ending quest for improvement in the ways processes are operated, numbers and information should form the basis for understanding, decisions and actions; and a thorough data gathering, recording and presentation system is essential:

- *Record* data – all processes can and should be measured – all measurements should be recorded.
- *Use* data – if data are recorded and not used they will be abused.
- *Analyse* data – data analysis should be carried out by means of some basic systematic tools.
- *Act* on the results – recording and analysis of data without action leads to frustration.

Continuous improvement – the basics

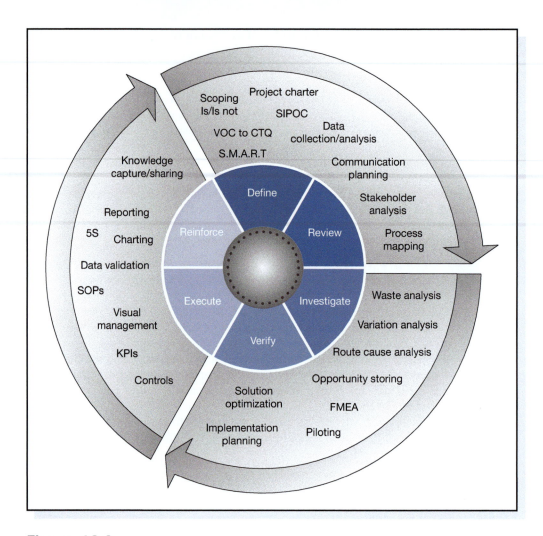

Figure 13.3
DRIVER – the dynamic improvement methodology (together with recommended tools & techniques)

In addition to the basic elements of a quality management system that provide a framework for recording, there exists a set of methods the Japanese quality guru Ishikawa has called the 'seven basic tools'. These should be used to interpret and derive the maximum use from data. The simple methods listed below, of which there are clearly more than seven, will offer any organization means of collecting, presenting and analysing most of its data:

- Process mapping or flowcharting – what is done?
- Check sheets/tally charts – how often is it done?
- Histograms – what do overall variations look like?
- Scatter diagrams – what are the relationships between factors?

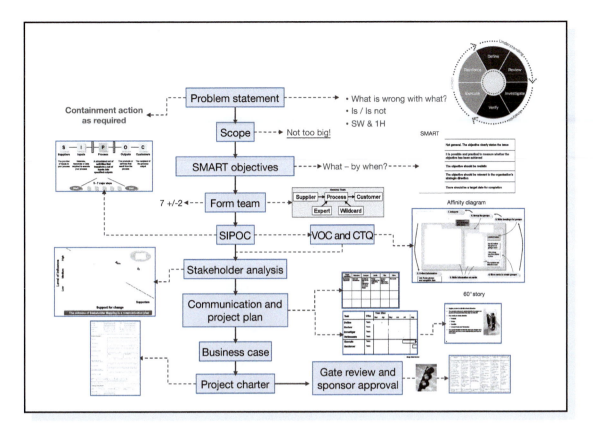

Figure 13.4
Define

- Stratification – how is the data made up?
- Pareto analysis – which are the big problems?
- Root cause and effect analysis and brainstorming (including CEDAC, NGT, RCA and the five whys) – what causes the problems?
- Force-field analysis – what will obstruct or help the change or solution?
- Emphasis curve – which are the most important factors?
- Control charts – which variations to control and how?

Sometimes more sophisticated techniques, such as analysis of variance, regression analysis and design of experiments, need to be employed.

The effective use of the tools requires their application by the people who actually work on the processes. Their commitment to this will be possible only if they are assured that management cares about improving quality. Managers must show they are serious by establishing a systematic approach and providing the training and implementation support required.

Improvements cannot be achieved without specific opportunities, commonly called problems, being identified or recognized. A focus on improvement opportunities leads to the creation of teams whose membership is determined by their work on and detailed knowledge of the process, and their ability to take improvement

Continuous improvement – the basics

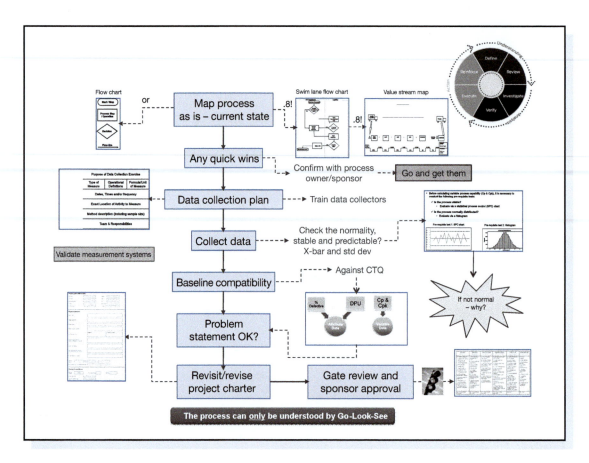

Figure 13.5
Review

action. The teams must then be provided with good leadership and the right tools to tackle the job.

The systematic approach of Figure 13.1 should lead to the use of factual information, collected and presented by means of proven techniques, to open a channel of communications not available to the many organizations that do not follow this or a similar structured approach to problem solving and improvement. Continuous improvements in the quality of products, services and processes can often be obtained without major capital investment, if an organization marshals its resources, through an understanding and breakdown of its processes in this way.

By using reliable methods, creating a favourable environment for team-based problem solving and continuing to improve using systematic techniques, the never-ending improvement helix (see Chapter 3) will be engaged. This approach demands the real time management of data, and actions focussed on processes and inputs rather than outputs. It will require a change in the language of many organizations from percentage defects, percentage 'prime' product and number of errors, to *process capability*. The climate must change from the traditional approach of 'If it meets the specification, there are no problems and no further improvements are necessary'. The

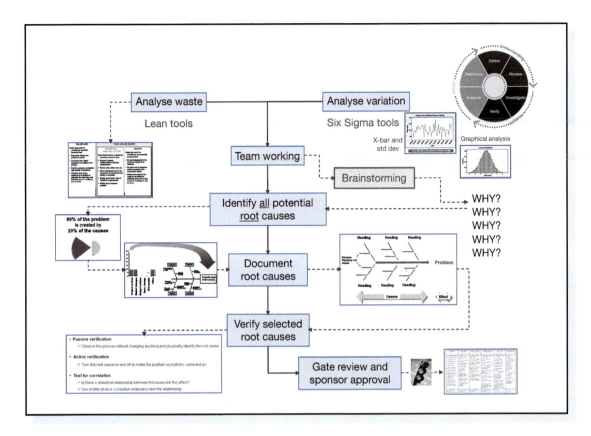

Figure 13.6
Investigate

driving force for this will be the need for better internal and external customer satisfaction levels, which will lead to the continuous improvement question, 'Could we do the job better?'

Basic tools and techniques

Understanding processes so that they can be improved by means of the systematic approach requires knowledge of a simple kit of tools or techniques. What follows is a brief description of each technique, but a full description and further examples of some of them may be found in *Statistical Process Control* by the author (6th edn, 2008).

Process mapping/flowcharting

The use of these techniques, which are described in Chapter 10, ensures a full understanding of the inputs, outputs and flow of the process. Without that understanding, it is not possible to draw the correct map or flowchart of the process. In flowcharting it is important to remember that in all but the smallest tasks no single person is able to complete a chart without help from others. This makes flowcharting a powerful improvement team forming exercise.

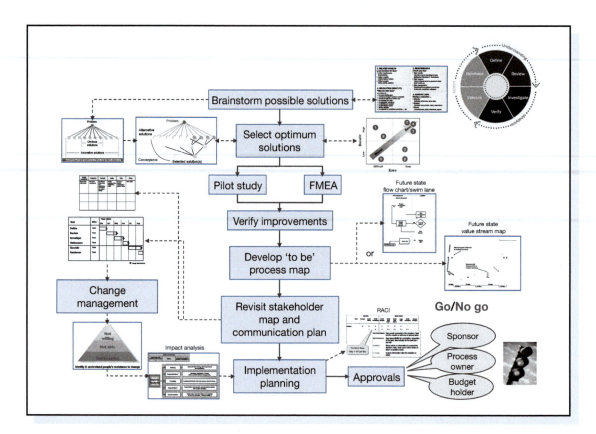

Figure 13.7
Verify

Check sheets or tally charts

A check sheet is a tool for data gathering, and a logical point to start in most process control or problem solving efforts. It is particularly useful for recording direct observations and helping to gather in facts rather than opinions about the process. In the recording process it is essential to understand the difference between data and numbers.

Data are pieces of information, including numerical, that are useful in solving problems or provide knowledge about the state of a process. Numbers alone often represent meaningless measurements or counts, which tend to confuse rather than to enlighten. Numerical data on quality will arise either from counting or measurement.

The use of simple check sheets or tally charts aids the collection of data of the right type, in the right form, at the right time. The objectives of the data collection will determine the design of the record sheet used.

Histograms

Histograms show, in a very clear pictorial way, the frequency with which a certain value or group of values occurs. They can be used to display both attribute and variable data, and are an effective means of letting the people who operate the process

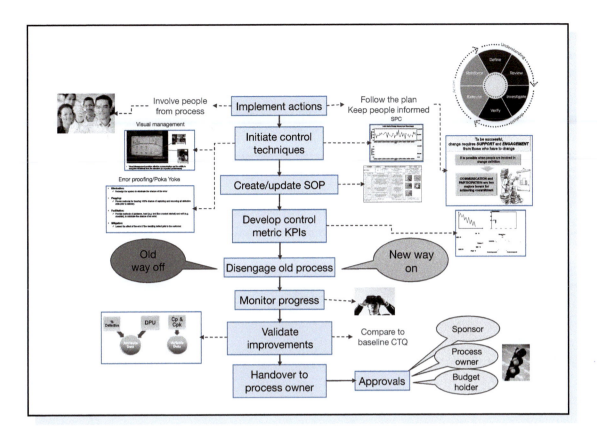

Figure 13.8
Execute

know the results of their efforts. Data gathered on truck turn-round times is drawn as a histogram in Figure 13.10.

Scatter diagrams

Depending on the technology, it is frequently useful to establish the association, if any, between two parameters or factors. A technique to begin such an analysis is a simple X-Y plot of the two sets of data. The resulting grouping of points on scatter diagrams (e.g. Figure 13.11) will reveal whether or not a strong or weak, positive or negative, correlation exists between the parameters. The diagrams are simple to construct and easy to interpret, and the absence of correlation can be as revealing as finding that a relationship exists.

Stratification

Stratification is simply dividing a set of data into meaningful groups. It can be used to great effect in combination with other techniques, including histograms and scatter diagrams. If, for example, three shift teams are responsible for a certain product output, 'stratifying' the data into the shift groups might produce histograms that indicate 'process adjustments' were taking place at shift changeovers.

Continuous improvement – the basics

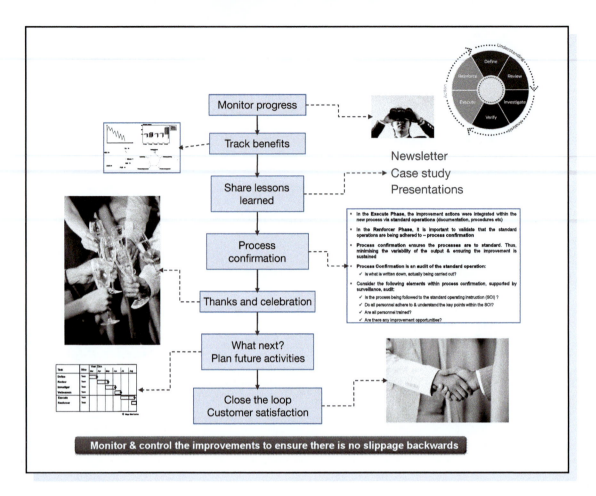

Figure 13.9
Reinforce

Pareto analysis

If the symptoms or causes of defective output or some other 'effect' are identified and recorded, it will be possible to determine what percentage can be attributed to any cause, and the probable results will be that the bulk (typically 80 per cent) of the errors, waste or 'effects', derive from a few of the causes (typically 20 per cent). For example, Figure 13.12 shows a *ranked frequency distribution* of incidents in the distribution of a certain product. To improve the performance of the distribution process, therefore, the major incidents (broken bags/drums, truck scheduling and temperature problems) should be tackled first. An analysis of data to identify the major problems is known as *Pareto analysis,* after the Italian economist who realized that approximately 90 per cent of the wealth in his country was owned by approximately 10 per cent of the people. Without an analysis of this sort, it is far too easy to devote resources to addressing one symptom only because its cause seems immediately apparent.

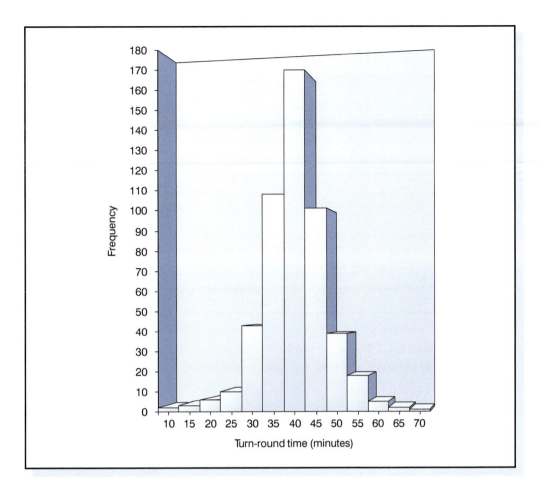

Figure 13.10
Frequency distribution for truck turn-round times (histogram)

Root cause and effect analysis and brainstorming

A useful way of mapping the inputs that affect quality is the *cause and effect diagram,* also known as the Ishikawa diagram (after its originator) or the fishbone diagram (after its appearance, Figure 13.13). The effect or incident being investigated is shown at the end of a horizontal arrow. Potential causes are then shown as labelled arrows entering the main cause arrow. Each arrow may have other arrows entering it as the principal factors or causes are reduced to their sub-causes and sub-sub-causes by *brainstorming.*

Brainstorming is a technique used to generate a large number of ideas quickly, and may be used in a variety of situations. Each member of a group, in turn, may be invited to put forward ideas concerning a problem under consideration. Wild ideas are safe to offer, as criticism or ridicule is not permitted during a brainstorming session. The people taking part do so with equal status to ensure this. The main objective is to create an atmosphere of enthusiasm and originality. All ideas offered

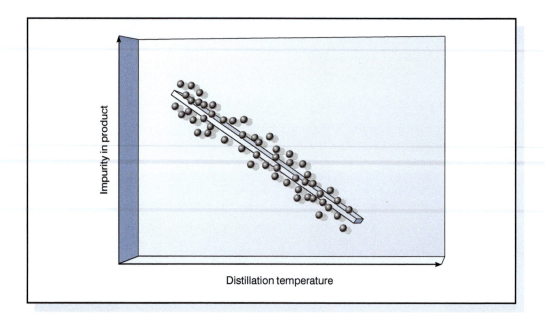

Figure 13.11
Scatter diagram showing a negative correlation between two variables

are recorded for subsequent analysis. The process is continued until all the conceivable causes have been included. The proportion of non-conforming output attributable to each cause, for example, is then measured or estimated, and a simple Pareto analysis identifies the causes that are most worth investigating.

A useful variant on the technique is negative brainstorming. Here the group brainstorms all the things that would need to be done to ensure a negative outcome. For example, in the implementation of TQM, it might be useful for the senior management team to brainstorm what would be needed to make sure TQM *was not* implemented well. Having identified in this way the potential roadblocks, it is easier to dismantle them.

CEDAC

A variation on the cause and effect approach, which was developed at Sumitomo Electric and now is used by many major corporations across the world, is the cause and effect diagram with addition of cards (CEDAC). The effect side of a CEDAC chart is a quantified description of the problem, with an agreed and visual quantified target and continually updated results on the progress of achieving it. The cause side of the CEDAC chart uses two different coloured cards for writing facts and ideas. This ensures that the facts are collected and organized before solutions are devised. The basic diagram for CEDAC has the classic fishbone appearance.

Nominal group technique (NGT)

The nominal group technique (NGT) is a particular form of team brainstorming used to prevent domination by particular individuals. It has specific application for

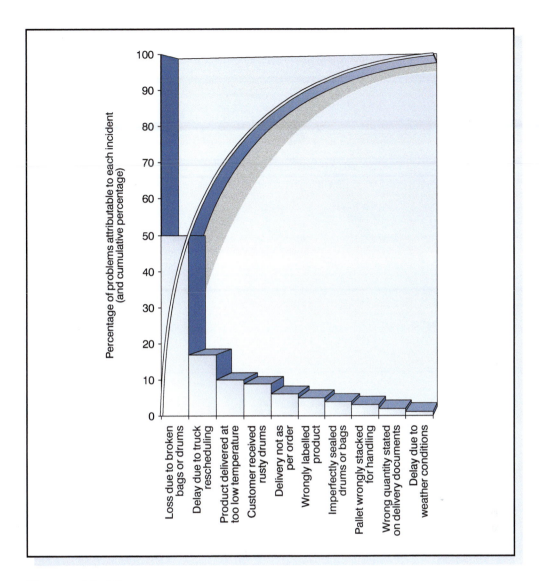

Figure 13.12
Incidents in the distribution of a chemical product

multi-level, multi-disciplined teams, where communication boundaries are potentially problematic.

In NGT a carefully prepared written statement of the problem to be tackled is read out by the facilitator (F). Clarification is obtained by questions and answers and then the individual participants (P) are asked to restate the problem in their own words. The group then discusses the problem until its formulation can be satisfactorily expressed by the team (T). The method is set out in Figure 13.14. NGT results in a set of ranked ideas that are close to a team consensus view obtained without domination by one or two individuals.

Continuous improvement – the basics

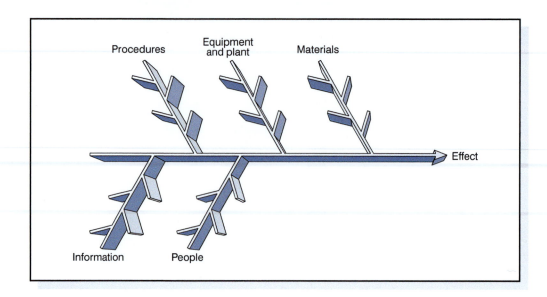

Figure 13.13
The cause and effect, Ishikawa or fishbone diagram

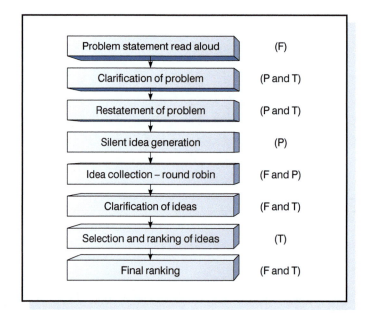

Figure 13.14
Nominal Group
Technique (NGT)

Force field analyses

Force field analysis is a technique used to identify the forces that either obstruct or help a change that needs to be made. It is similar to negative brainstorming-cause/effect analysis and helps to plan how to overcome the barriers to change or improvement. It may also provide a measure of the difficulty in achieving the change.

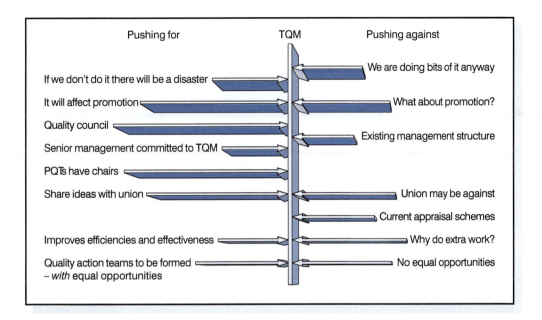

Figure 13.15
Force field analysis

The process begins with a team describing the desired change or improvement, and defining and objectives or solution. Having prepared the basic force field diagram, it identifies the favourable/positive/driving forces and the unfavourable/negative/restraining forces, by brainstorming. These forces are placed in opposition on the diagram and, if possible, rated for their potential influence on the ease of implementation. The results are evaluated. Then comes the preparation of an action plan to overcome some of the restraining forces, and increase the driving forces. Figure 13.15. shows a force field analysis produced by a senior management team considering the implementation of TQM in its organization.

The emphasis curve

This is a technique for ranking a number of factors, each of which cannot be readily quantified in terms of cost, frequency of occurrence, etc., in priority order. It is almost impossible for the human brain to make a judgement of the relative importance of more than three or four non-quantifiable factors. It is, however, relatively easy to judge which is the more important of two factors, using some predetermined criteria. The emphasis curve technique uses this fact by comparing only two factors at any one time.

The procedural steps for using the 'emphasis curve chart' are as follows:

1. List the factors for ranking under a heading 'Scope'.
2. Compare factor 1 with factor 2 and rank the most important. To assist in judging the relative importance of two factors, it may help to use weightings, e.g. degree of seriousness, capital investment, speed of completion etc., on a scale of 1 to 10.

3. Compare factor 1 with 3, 1 with 4, 1 with 5 and so on – ringing the most important number in the matrix.
4. Having compared factor 1 against the total scope, proceed to compare factor 2 with 3, 2 with 4 and so on.
5. Count the number of 'ringed' number 1s in the matrix and put the total in a right-hand column against Number 1. Next count the total number of 2s in the matrix and put total in column against Number 2 and so on.
6. Add up the numbers in the column and check the total, using the formula $\{n(n-1)\}/2$, where n is the number of entries in the column. This check ensures that all numbers have been 'ringed' in the matrix.
7. Proceed to rank the factors using the numbers in the column.
8. Generally the length of time to make a judgement between two factors does not significantly affect the outcome; therefore the rule is 'accept the first decision, record it and move quickly onto the next pair'.

Control charts

A control chart is a form of traffic signal whose operation is based on evidence from the small samples taken at random during a process. A green light is given when the process should be allowed to run. All too often processes are 'adjusted' on the basis of a single measurement, check or inspection, a practice that can make a process much more variable than it is already. The equivalent of an amber light appears when trouble is possibly imminent. The red light shows that there is practically no doubt that the process has changed in some way and that it must be investigated and corrected to prevent production of defective material or information. Clearly, such a scheme can be introduced only when the process is 'in control'. Since samples taken are usually small, there are risks of errors, but these are small, calculated risks and not blind ones. The risk calculations are based on various frequency distributions.

These charts should be made easy to understand and interpret and they can become, with experience, sensitive diagnostic tools to be used by operating staff and first-line supervision to prevent errors or defective output being produced. Time and effort spent to explain the working of the charts to all concerned are never wasted.

The most frequently used control charts are simple run charts, where the data is plotted on a graph against time or sample number. There are different types of control charts for variables and attribute data: for variables mean ($\bar{X}$) and range (**R**) charts are used together; number defective or **np** charts and proportion defective or **p** charts are the most common ones for attributes. Other charts found in use are moving average and range charts, numbers of defects (**c** and **u**) charts, and cumulative sum (cusum) charts. The latter offer very powerful management tools for the detection of trends or changes in attributes and variable data.

The cusum chart is a graph that takes a little longer to draw than the conventional control chart, but gives a lot more information. It is particularly useful for plotting the evolution of processes, because it presents data in a way that enables the eye to separate true changes from a background of random variation. Cusum charts can detect small changes in data very quickly, and may be used for the control of variables and attributes. In essence, a reference or 'target value' is subtracted from each successive sample observation, and the result accumulated. Values of this cumulative sum are plotted, and 'trend lines' may be drawn on the resulting graphs. If they are approximately horizontal, the value of the variable is about the same as the target

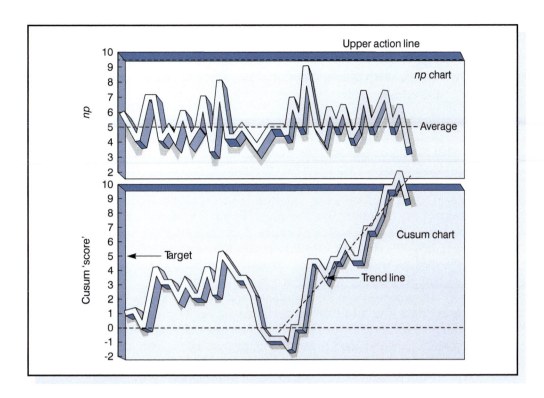

Figure 13.16
Comparison of cusum and *np* charts for the same data

value. A downward slope shows a value less than the target and an upward slope a value greater. The technique is very useful, for example, in comparing sales forecast with actual sales figures.

Figure 13.16 shows a comparison of an ordinary run chart and a cusum chart that have been plotted from the same data – errors in samples of 100 invoices. The change, which is immediately obvious on the cusum chart, is difficult to detect on the conventional control chart.

The range of type and use of control charts is now very wide, and within the present text it is not possible to indicate more than the basic principles underlying such charts. All of them can be generated electronically using the various software tools available. A full treatment of all control chart methods is given in *Statistical Process Control* (6th edn 2008).

STATISTICAL PROCESS CONTROL (SPC)

The responsibility for quality in any transformation process must lie with the operators of that process. To fulfil this responsibility, however, people must be provided with the tools necessary to:

- Know whether the process is capable of meeting the requirements.
- Know whether the process is meeting the requirements at any point in time.
- Make correct adjustment to the process or its inputs when it is not meeting the requirements.

The techniques of statistical process control (SPC) will greatly assist in these stages. To begin to monitor and analyse any process, it is necessary first of all to identify what the process is, and what the inputs and outputs are. Many processes are easily understood and relate to known procedures, e.g. drilling a hole, compressing tablets, filling cans with paint, polymerising a chemical using catalysts. Others are less easily identifiable, e.g. servicing a customer, delivering a lecture, storing a product in a warehouse, inputting to a computer. In many situations it can be extremely difficult to define the process. For example, if the process is inputting data into a computer terminal, it is vital to know if the scope of the process includes obtaining and refining the data, as well as inputting. Process definition is so important because the inputs and outputs change with the scope of the process.

All processes can be monitored and brought 'under control' by gathering and using data – to measure the performance of the process and provide the feedback required for corrective action, where necessary. SPC methods, backed by management commitment and good organization, provide objective means of *controlling* quality in any transformation process, whether used in the manufacture of artefacts, the provision of services or the transfer of information.

SPC is not only a tool kit, it is a strategy for reducing variability, the cause of most quality problems: variation in products, in times of deliveries, in ways of doing things, in materials, in people's attitudes, in equipment and its use, in maintenance practices, in everything. Control by itself is not sufficient. Total quality management requires that the processes should be improved continually by reducing variability. This is brought about by studying all aspects of the process, using the basic question: 'Could we do this job more consistently and on target?' The answer drives the search for improvements. This significant feature of SPC means that it is not constrained to measuring conformance, and that it is intended to lead to action on processes that are operating within the 'specification' to minimize variability.

Process control is essential, and SPC forms a vital part of the TQM strategy. Incapable and inconsistent processes render the best design impotent and make supplier quality assurance irrelevant. Whatever process is being operated, it must be reliable and consistent. SPC can be used to achieve this objective.

In the application of SPC there is often an emphasis on techniques rather than on the implied wider managerial strategies. It is worth repeating that SPC is not only about plotting charts on the walls of a plant or office, it must become part of the company-wide adoption of TQM and act as the focal point of never-ending improvement. Changing an organization's environment into one in which SPC can operate properly may take several years rather than months. For many companies SPC will bring a new approach, a new 'philosophy' and the importance of the numbers and information should not be disguised. Simple presentation of data using diagrams, graphs and charts should become the means of communication concerning the state of control of processes. It is on this understanding that improvements will be based.

The SPC system

A systematic study of any process through answering the questions:

- Are we capable of doing the job correctly?
- Do we continue to do the job correctly?
- Have we done the job correctly?
- Could we do the job more consistently and on target?

provides knowledge of the *process capability* and the sources of non-conforming outputs. This information can then be fed back quickly to marketing, design and the 'technology' functions. Knowledge of the current state of a process also enables a more balanced judgement of equipment, both with regard to the tasks within its capability and its rational utilization.

Statistical process control, procedures exist because there is variation in the characteristics of all material, articles, services and people. The inherent variability in each transformation process causes the output from it to vary over a period of time. If this variability is considerable, it is impossible to predict the value of a characteristic of any single item or at any point in time. Using statistical methods, however, it is possible to take meagre knowledge of the output and turn it into meaningful statements that may then be used to describe the process itself. Hence, statistically based process control procedures are designed to divert attention from individual pieces of data and focus it on the process as a whole. SPC techniques may be used to measure and control the degree of variation of any purchased materials, services, processes and products, and to compare this, if required, to previously agreed specifications. In essence, SPC techniques select a representative, simple, random sample from the 'population', which can be an input to or an output from a process. From an analysis of the sample it is possible to make decisions regarding the current performance of the process.

Organizations that embrace the TQM concepts should recognize the value of SPC techniques in areas such as sales, purchasing, invoicing, finance, distribution, training and in the service sector generally. These are outside the traditional areas for SPC use, but it needs to be seen as an organization-wide approach to reducing variation with the specific techniques integrated into a programme of change throughout. A Pareto analysis, a histogram, a flowchart or a control chart is a vehicle for communication. Data are data and, whether the numbers represent defects or invoice errors, weights or delivery times, or the information relates to machine settings, process variables, prices, quantities, discounts, sales or supply points, is irrelevant – the techniques can always be used.

In the author's experience, some of the most exciting applications of SPC have emerged from organizations and departments which, when first introduced to the methods, could see little relevance in them to their own activities. Following appropriate training, however, they have learned how to, for example:

- *Pareto analyse* errors on invoices to customers and industry injury data.
- *Brainstorm and cause and effect analyse* reasons for late payment and poor purchase invoice matching.
- *Histogram* defects in invoice matching and arrival of trucks at certain times during the day.
- *Control chart* the weekly demand of a product.

Distribution staff have used control charts to monitor the proportion of late deliveries, and Pareto analysis and force field analysis to look at complaints about the distribution system. Bank operators have been seen using cause and effect analysis, NGT and histograms to represent errors in the output from their services. Moving average and cusum charts have immense potential for improving processes in the marketing area.

Those organizations that have made most progress in implementing continuous improvement have recognized at an early stage that SPC is for the whole organization. Restricting it to traditional manufacturing or operational activities means that a window of opportunity for improvement has been closed. Applying the methods and techniques outside manufacturing will make it easier, not harder, to gain maximum benefit from SPC.

BIBLIOGRAPHY

Bauer, J.E., Duffy, G.Z. and Westcott, R.T. (eds), *The Quality Improvement Handbook*, Quality Press, Milwaukee, 2002.

Carlzon, J., *Moments of Truth,* Ballinger, Cambridge, Mass., 1987.

Joiner, B., *Fourth Generation Management,* McGraw-Hill, New York, 1994.

Neave, H., *The Deming Dimension,* SPC Press, Knoxville, 1990.

Oakland, J.S., *Statistical Process Control: A Practical Guide* (6th edn), Butterworth-Heinemann, Oxford, 2008.

Ohno, T. and Rosen, C.B., 1988, *Toyota Production System: Beyond Large-scale Production*, Productivity Press, TOWN, 1988.

Price, F., *Right First Time*, Gower, London, 1985.

Ryuka Fukuda, *CEDAC – A Tool for Continuous Systematic Improvement*, Productivity Press, Cambridge, Mass., 1990.

Scherkenbach, W.W., *Deming's Road to Continual Improvement*, SPC Press, Knoxville, 1991.

Shingo, S., *Zero Quality Control: Source Inspection and the Poka-yoke System*, Productivity Press, Stamford, Conn., 1986.

Wheeler, D.J., *Understanding Statistical Process Control* (2nd edn), SPC Press, Knoxville, 1992.

Wheeler, D.J., *Understanding Variation*, SPC Press, Knoxville, 1993.

CHAPTER HIGHLIGHTS

Approaches, methodologies and tools

- Continuous improvement must be linked to organizational strategy, have a defined structure, a chosen approach, a methodology and an associated tool kit.
- A generic eight stage structure for CI may be applied to most organizations; it involves: leadership and top down cascade of objectives, supporting infrastructure, pragmatic, fact-based methodology yielding rapid results (e.g. Lean Six Sigma), skilled improvers, widely shared continual improvement culture, circulation of talent, measurable improvements in the service/ delivery areas, sustained improvement.

The 'DRIVER' framework for continuous improvement

- A fully closed loop improvement methodology, which brings together the best of Lean, Six Sigma and Cost of Quality approaches is 'DRIVER' with the stages of Define, Review, Investigate, Verify, Execute and Reinforce.

The need for data and some basic tools and techniques

- Numbers and information will form the basis for understanding, decisions and actions in never-ending improvement – record data, use/analyse data, act on results.
- A set of simple tools is needed to interpret fully and derive maximum use from data. More sophisticated techniques may need to be employed occasionally; the effective use of the tools requires the commitment of the people who work on the processes, which in turn needs management support and the provision of training.
- The basic tools and the questions answered are:
  - Process flowcharting – what is done?
  - Check/tally charts – how often is it done?
  - Histograms – what do variations look like?
  - Scatter diagrams – what are the relationships between factors?
  - Stratification – how is the data made up?
  - Pareto analysis – which are the big problems?
  - Cause and effect analysis and brainstorming (also CEDAC and NGT) – what causes the problem?
  - Force-field analysis – what will obstruct or help the change or solution?
  - Emphasis curve – which are the most important factors?
  - Control charts (including cusum) – which variations to control and how?

Statistical process control

- People operating a process must know whether it is capable of meeting the requirements, know whether it is actually doing so at any time and make correct adjustments when it is not. SPC techniques will help here.
- Before using SPC, it is necessary to identify what the process is, what the inputs/outputs are, and how the suppliers and customers and their requirements are defined. The most difficult areas for this can be in non-manufacturing.
- All processes can be monitored and brought 'under control' by gathering and using data. SPC methods, with management commitment, provide objective means of controlling quality in any transformation process.
- SPC is not only a tool kit, it is a strategy for reducing variability, part of never-ending improvement. This is achieved by answering the following questions:
  - Are we capable of doing the job correctly?
  - Do we continue to do the job correctly?
  - Have we done the job correctly?
  - Could we do the job more consistently and on target?
- SPC provides knowledge and control of process capability.
- SPC techniques have value in the service sector and in the non-manufacturing areas, such as marketing and sales, purchasing, invoicing, finance, distribution, training and personnel.

Chapter 14

Continuous improvement – more advanced, including Taguchi and Six Sigma

SOME ADDITIONAL TECHNIQUES FOR PROCESS DESIGN AND IMPROVEMENT

Seven additional qualitative methods tools may be used as part of quality function deployment (see Chapter 6) or to improve processes. These do not replace the basic systematic tools described in the previous chapter, neither are they extensions of these; rather they are systems and documentation methods used to achieve success in design by identifying objectives and intermediate steps in the finest detail. The seven 'new tools' are:

1. Affinity diagram.
2. Interrelationship diagraph.
3. Tree diagram.
4. Matrix diagram or quality table.
5. Matrix data analysis.
6. Process decision programme chart (PDPC).
7. Arrow diagram.

The tools are interrelated, as shown in Figure 14.1 and are summarized below.

1. Affinity diagram

This is used to gather large amounts of language data (ideas, issues, opinions) and organizes them into groupings based on the natural relationship between the items. In other words, it is a form of brainstorming.

288

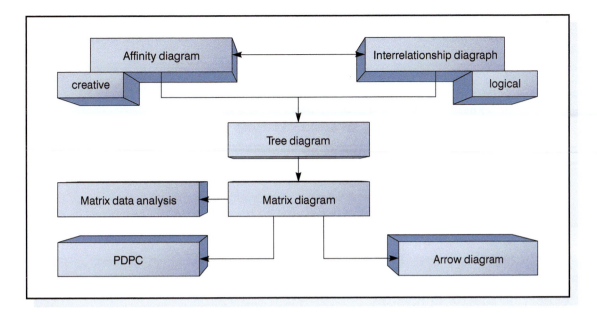

Figure 14.1
The seven 'new tools' of quality design

The steps for generating an affinity diagram are as follows:

1. Assemble a group of people familiar with the problem of interest. Six to eight members in the group works best.
2. Phrase the issue to be considered. It should be vaguely stated so as not to prejudice the responses in a predetermined direction.
3. Give each member of the group a stack of cards and allow 5–10 minutes for everyone individually in the group to record ideas on the cards, writing down as many ideas as possible.
4. At the end of the 5–10 minutes each member of the group, in turn, reads out one of his/her ideas and places it on the table for everyone to see, without criticism or justification.
5. When all ideas are presented, members of the group place together all cards with related ideas repeating the process until the ideas are in a few groups.
6. Look for one card in each group that captures the meaning of that group.

The output of this exercise is a compilation of a maximum number of ideas under a limited number of major headings. This data can then be used with other tools to define areas for attack. One of these tools is the interrelationship diagraph.

2. Interrelationship diagraph

This tool is designed to take a central idea, issue or problem, and map out the logical or sequential links among related factors. While this still requires a very creative process, the interrelationship diagraph begins to draw the logical connections that surface in the affinity diagram. In designing, planning and problem solving it is obviously not enough to just create an explosion of ideas. The affinity diagram allows some organized creative patterns to emerge but the interrelationship diagraph lets *logical* patterns become apparent.

Figure 14.2 gives an example of a simple interrelationship diagraph.

3. Systems flow/tree diagram

The systems flow/tree diagram (usually referred to as a tree diagram) is used to systematically map out the full range of activities that must be accomplished in order to reach a desired goal. It may also be used to identify all the factors contributing to a problem under consideration. One of the strengths of this method is that it forces the user to examine the logical and chronological link between tasks.

Depending on the type of issue being addressed, the tree diagram will be similar to either a cause and effect diagram or a flowchart, although it may be easier to interpret because of its clear linear layout. If a problem is being considered, each branch of the tree diagram will be similar to a cause and effect diagram.

4. Matrix diagrams

The matrix diagram is the heart of the seven additional tools and the house of quality described in Chapter 6. The purpose of the matrix diagram is to outline the interrelationships and correlations between tasks, functions or characteristics, and to show their relative importance. There are many versions of the matrix diagram, but the most widely used is a simple L-shaped matrix known as the *quality table* in which

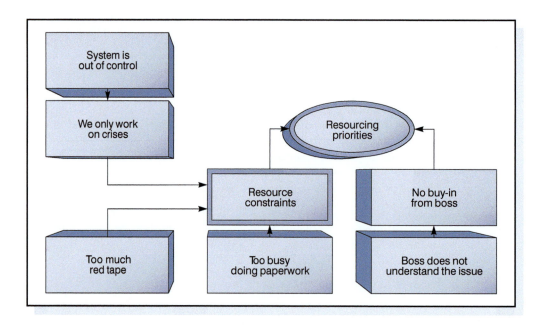

Figure 14.2
Example of the interrelationship digraph

customer demands (the whats) are analysed with respect to substitute quality characteristics (the hows). See Figure 14.3. Correlations between the two are categorized as strong, moderate and possible. The customer demands shown on the left of the matrix are determined in co-operation with the customer in a joint meeting if possible.

The right side of the chart is often used to compare current performance to competitors' performance, company plan and potential sales points with reference to the customer demands. Weights are given to these items to obtain a 'relative quality weight', which can be used to identify the key customer demands. The relative quality weight is then used with the correlations identified on the matrix to determine the key quality characteristics.

T-shaped matrix diagram

A T-shaped matrix is nothing more than the combination of two L-shaped matrix diagrams. Figure 14.4 shows one application, the relationship between a set of courses in a curriculum and two important sets of considerations: who should do the training for each course and which would be the most appropriate functions to attend each of the courses?

There are other matrices that deal with ideas such as product or service function, cost, failure modes, capabilities, etc., and there are at least forty different types of matrix diagrams available.

5. Matrix data analysis

Matrix data analysis is used to take data displayed in a matrix diagram and arrange them so that they can be more easily viewed and show the strength of the relationship between variables. It is used most often in marketing and product research. The concept behind matrix data analysis is fairly simple, but its execution (including data gathering) is complex.

		MFR	Ash	Importance	Current	Best competitor	Plan	IR	SP	RQW
						Substitute quality characteristics				
Customer demands	No film breaks	◉ 17	▲ 6	4	4	4	4	1	◯	5.6
	High rates	◉ 23		3	3	4	4	1.3		4.6
	Low gauge variability	◉ 37	▲ 7	4	3	4	4	1.3	◯	7.3

◉ Strong correlation
◯ Some correlation
▲ Possible correlation
IR Improvement ratio
SP Sales point
RQW Relative quality weight

Figure 14.3
Example of the matrix diagram

Who trains?

	SQC	7 Old tools	7 New tools	Reliability	Design review	QC basics	QCC facilitator	Diagnostic tools	Problem solving	Communication skills	Organize for quality	Design of experiments	Company mission	Quality planning	Just-in-time	New superv. training	Company TQM system	Group dynamics skills	SQC course/execs.
Human resources dept.																			
Managers																			
Operators*																			
Consultants																			
Production operator																			
Craft foreman																			
GLSPC co-ordinator																			
Plant SPC co-ordinator																			
University																			
Technology specialists																			
Engineers																			

* Need to tailor to groups

X = Full
O = Overview

Courses

Who attends?

	SQC	7 Old tools	7 New tools	Reliability	Design review	QC basics	QCC facilitator	Diagnostic tools	Problem solving	Communication skills	Organize for quality	Design of experiments	Company mission	Quality planning	Just-in-time	New superv. training	Company TQM system	Group dynamics skills	SQC course/execs.
Executives																			
Top management																			
Middle management																			
Production supervisors																			
Supervisor functional																			
Staff																			
Marketing																			
Sales																			
Engineers																			
Clerical																			
Production worker																			
Quality professional																			
Project team																			
Employee involvement																			
Suppliers																			
Maintenance																			

Figure 14.4
T-matrix on company-wide training

6. Process decision programme chart

A process decision programme chart (PDPC) is used to map out each event and contingency that can occur when progressing from a problem statement to its solution. The PDPC is used to anticipate the unexpected and plan for it. It includes plans for counter-measures on deviations. The PDPC is related to a failure mode and effect analysis and its structure is similar to that of a tree diagram.

7. Arrow diagram

The arrow diagram is used to plan or schedule a task. To use it, one must know the sub-task sequence and duration. This tool is essentially the same as the standard Gantt chart used in project planning. Although it is a simple and well-known tool for planning work, it is surprising how often it is ignored. The arrow diagram is useful in analysing a repetitive job in order to make it more efficient.

Summary

What has been described in this section is a system for improving the design of products, processes and services by means of seven additional qualitative tools. For the most part the seven tools are neither new nor revolutionary, but rather a compilation and modification of some methods that have been around for a long time. These tools do not replace statistical methods or other techniques, but they are meant to be used together as part of continuous improvement.

The tools work best when representatives from all parts of an organization are involved in their use and execution of the results. Besides the structure that the tools provide, the co-operation between functions or departments that is required will help break down barriers within the organizations.

While design, marketing and operations people will see the most direct applications for these tools, proper use of the 'philosophy' behind them requires participation from all parts of an organization. In addition, some of the seven additional tools can be used in direct problem solving activities.

TAGUCHI METHODS FOR PROCESS IMPROVEMENT

Genichi Taguchi was a noted Japanese engineering specialist who advanced 'quality engineering' as a technology to reduce costs and improve quality simultaneously. The popularity of Taguchi methods testifies to the merit of his philosophies on quality. The basic elements of Taguchi's ideas, which have been extended here to all aspects of product, service and process quality, may be considered under four main headings.

1. Total loss function

An important aspect of the quality of a product or service is the total loss to society that it generates. Taguchi's definition of product quality as 'the loss imparted to society from the time a product is shipped', is rather strange, since the word *loss* denotes the very opposite of what is normally conveyed by using the word *quality*. The essence of his definition is that the smaller the loss generated by a product or service from the time it is transferred to the customer, the more desirable it is.

The main advantage of this idea is that it encourages a new way of thinking about investment in quality improvement projects, which become attractive when the resulting savings to customers are greater than the cost of improvements.

Taguchi claimed, with some justification, that any variation about a target value for a product or process parameter causes loss to the customer. The loss may be some simple inconvenience, but it can represent actual cash losses, owing to rework or badly fitting parts, and it may well appear as loss of customer goodwill and eventually market share. The loss (or cost) increases exponentially as the parameter value moves away from the target, and is at a minimum when the product or service is at the target value.

2. Design of products, services and processes

In any product or service development three stages may be identified: product or service design, process design, and production or operations. Each of these overlapping stages has many steps, the output of one often being the input to others. The output/input transfer points between steps clearly affect the quality and cost of the final product or service. The complexity of many modern products and services demands that the crucial role of design be recognized. Indeed the performance of the quality products from the Japanese automotive, camera and machine tool industries can be traced to the robustness of their product and process designs.

The prevention of problems in using products or services under varying operations and environmental conditions must be built in at the design stage. Equally, the costs during production or operation are determined very much by the actual manufacturing or operating process. Controls, including SPC methods, added to processes to reduce imperfections at the operational stage are expensive, and the need for controls *and* the production of non-conformance can be reduced by correct initial designs of the process itself.

Taguchi distinguished between *off-line* and *on-line* quality control methods, 'quality control' being used here in the very broad sense to include quality planning, analysis and improvement. Off-line QC uses technical aids in the *design* of products and processes, whereas on-line methods are technical aids for controlling quality and costs in the *production* of products or services. Too often the off-line QC methods focus on evaluation rather than improvement. The belief by some people (often based on experience!) that it is unwise to buy a new model or a motor car 'until the problems have been sorted out' testifies to the fact that insufficient attention is given to improvement at the product and process design stages. In other words, the bugs should be removed *before* not after product launch. This may be achieved in some organizations by replacing detailed quality and reliability evaluation methods with approximate estimates, and using the liberated resources to make improvements.

3. Reduction of variation

The objective of a continuous quality improvement programme is to reduce the variation of key products' performance characteristics about their target values. The widespread practice of setting specifications in terms of simple upper and lower limits conveys the wrong idea that the customer is satisfied with all values inside the specification band, but is suddenly not satisfied when a value slips outside one of the limits. The practice of stating specifications as tolerance intervals only can lead manufacturers to produce and despatch goods whose parameters are just inside the specification band. Owing to the interdependence of many parameters of component parts and assemblies, this is likely to lead to quality problems.

The target value should be stated and specified as the ideal, with known variability about the mean. For those performance characteristics that cannot be measured on the continuous scale, the next best thing is an ordered categorical scale such as excellent, very good, good, fair, unsatisfactory, very poor, rather than the binary classification of 'good' or 'bad' that provides meagre information with which the variation reduction process can operate.

Taguchi introduced a three-step approach to assigning nominal values and tolerances for product and process parameters:

a) System design – the application of scientific engineering and technical knowledge to produce a basic functional prototype design. This requires a fundamental understanding of the needs of the customer *and* the production environment.
b) Parameter design – the identification of the settings of product or process parameters that reduce the sensitivity of the designs to sources of variation. This requires a study of the whole process system design to achieve the most robust operational settings, in terms of tolerance to ranges of the input variables.
c) Tolerance design – the determination of tolerances around the nominal settings identified by parameter design. This requires a trade-off between the customer's loss due to performance variation and the increase in production or operational costs.

4. Statistically planned experiments

Taguchi pointed out that statistically planned experiments should be used to identify the settings of product and process parameters that will reduce variation in performance. He classified the variables that affect the performance into two categories: design parameters and sources of 'noise'. As we have seen earlier, the nominal settings of the *design parameters* define the specification for the product or process. The *sources of noise* are all the variables that cause the performance characteristics to deviate from the target values. The *key* noise factors are those that represent the major sources of variability, and these should be identified and included in the experiments to design the parameters at which the effect of the noise factors on the performance is minimum. This is done by systematically varying the design parameter settings and comparing the effect of the noise factors for each experimental run.

Statistically planned experiments may be used to identify:

a) The design parameters that have a large influence on the product or performance characteristic.
b) The design parameters that have no influence on the performance characteristics (the tolerances of these parameters may be relaxed).
c) The settings of design parameters at which the effect of the sources of noise on the performance characteristic is minimal.
d) The settings of design parameters that will reduce cost without adversely affecting quality.

Taguchi methods stimulated a great deal of interest in the application of statistically planned experiments to product and process designs. The use of 'design of experiments' to improve industrial products and processes is not new – Tippett used these techniques in the textile industry more than 50 years ago. What Taguchi did, however, is to acquaint us with the scope of these techniques in off-line quality control.

Taguchi's methods, like all others, should not be used in isolation, but be an integral part of continuous improvement.

SIX SIGMA

Since the early 1980s, most of the world has been in what the author has called a 'quality revolution'. Based on the simple premise that organizations of all kinds exist mainly to serve the needs of the customers of their products or services, good quality management has assumed great importance. Competitive pressures on companies and governments' demands on the public sector have driven the need to find more effective and efficient approaches to managing businesses and non-profit-making organizations.

In the early days of the realization that improved quality was vital to the survival of many companies, especially in manufacturing, senior managers were made aware, through national campaigns and award programmes, that the basic elements had to be right. They learned through adoption of quality management systems, the involvement of improvement teams and the use of quality tools that improved business performance could be achieved only through better planning, capable processes and the involvement of people. These are the basic elements of a TQM approach and this has not changed no matter how many sophisticated approaches and techniques come along.

The development of TQM has seen the introduction and adoption of many dialects and components, including quality circles, international systems and standards, statistical process control, business process re-engineering, lean manufacturing, continuous improvement, benchmarking and business excellence.

An approach finding favour in some companies is Six Sigma, most famously used in Motorola, General Electric and Allied Signal. This operationalized TQM into a project-based system, based on delivering tangible business benefits, often directly to the bottom line. Strange combinations of the various approaches have led to Lean Six Sigma, Lean Sigma and other company specific acronyms such as 'Statistically Based Continuous Improvement (SBCI)'!

The Six Sigma improvement model

There are five fundamental phases or stages in applying the Six Sigma approach to improving performance in a process: Define, Measure, Analyse, Improve and Control (DMAIC). These form an improvement cycle grounded in Deming's original Plan, Do, Check, Act (PDCA) (Figure 14.5). In the Six Sigma approach, DMAIC provides a breakthrough strategy and disciplined methods of using rigorous data gathering and statistically based analysis to identify sources of errors and ways of eliminating them. It has become increasingly common in so-called 'six sigma organizations' for people to refer to 'DMAIC Projects'. These revolve around the three major strategies for

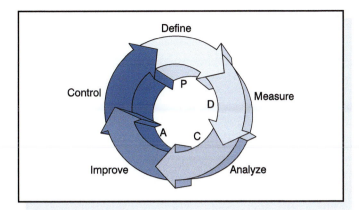

Figure 14.5
The Six Sigma improvement model – DMAIC

Table 14.1 The DMAIC steps

D	Define the scope and goals of the improvement project in terms of customer requirements and the process that delivers these requirements – inputs, outputs, controls and resources.
M	Measure the current process performance – input, output and process – and calculate the short- and longer-term process capability – the sigma value.
A	Analyze the gap between the current and desired performance, prioritize problems and identify root causes of problems. Benchmarking the process outputs, products or services against recognized benchmark standards of performance may also be carried out.
I	Generate the improvement solutions to fix the problems and prevent them from recurring so that the required financial and other performance goals are met.
C	This phase involves implementing the improved process in a way that 'holds the gains'. Standards of operation will be documented in systems such as ISO 9000 and standards of performance will be established using techniques such as statistical process control (SPC).

processes to bring about rapid bottom-line achievements – design/redesign, management and improvement.

Table 14.1 shows the outline of the DMAIC steps

Building a Six Sigma organization and culture

Six Sigma approaches question many aspects of business, including its organization and the cultures created. The goal of most commercial organizations is to make money through the production of saleable goods or services and, in many, the traditional measures used are capacity or throughput based. As people tend to respond to the way they are being measured, the management of an organization tends to get what it measures. Hence, throughput measures may create work-in-progress and finished goods inventory thus draining the business of cash and working capital. Clearly, supreme care is needed when defining what and how to measure.

Six Sigma organizations focus on:

- understanding their customers' requirements
- identifying and focusing on core-critical processes that add value to customers
- driving continuous improvement by involving all employees
- being very responsive to change
- basing managing on factual data and appropriate metrics
- obtaining outstanding results, both internally and externally.

The key is to identify and eliminate variation in processes. Every process can be viewed as a chain of independent events and, with each event subject to variation, variation accumulates in the finished product or service. Because of this, research suggests that most businesses operate somewhere between the 3 and 4 sigma level. At this level of performance, the real cost of quality is about 25–40 per cent of sales revenue. Companies that adopt a six sigma strategy can readily reach the 5 sigma level and reduce the cost of quality to 10 per cent of sales. They often reach a plateau here and to improve to six sigma performance and 1 per cent cost of quality takes a major rethink.

Properly implemented six sigma strategies involve:

- leadership involvement and sponsorship
- whole organization training
- project selection tools and analysis
- improvement methods and tools for implementation
- measurement of financial benefits
- communication
- control and sustained improvement.

One highly publicised aspect of the six sigma movement, especially in its application in companies such as General Electric (GE), Motorola, Allied Signal and GE Capital in Europe, is the establishment of process improvement experts, known variously as 'Master Black Belts', 'Black Belts' and 'Green Belts'. In addition to these martial arts related characters, who perform the training, lead teams and do the improvements, are other roles which the organization may consider, depending on the seriousness with which they adopt the six sigma discipline. These include the:

- Leadership Group or Council/Steering Committee
- Sponsors and/or Champions/Process Owners
- Implementation Leaders or Directors – often Master Black Belts
- Six Sigma Coaches – Master Black Belts or Black belts
- Team Leaders or Project Leaders – Black Belts or Green Belts
- Team Members – usually Green Belts.

Many of these terms will be familiar from TQM and continuous improvement activities. The 'Black Belts' reflect the finely honed skill and discipline associated with the six sigma approaches and techniques. The different levels of Green, Black and Master Black Belts recognize the depth of training and expertise.

Mature six sigma programmes, such as at GE, Johnson & Johnson and Allied Signal, have about 1 per cent of the workforce as full-time Black Belts. There is typically one Master Black Belt to every ten Black Belts or about one to every 1,000 employees.

A Black Belt typically oversees/completes 5–7 projects per year, which are led by Green Belts who are not employed full-time on six sigma projects (Figure 14.6).

The means of achieving six sigma capability are, of course, the key. At Motorola and GE this included millions of dollars spent on a company-wide education programmes, documented quality systems linked to quality goals, formal processes for planning and achieving continuous improvements, individual QA organizations acting as the customer's advocate in all areas of the business, a Corporate 'Quality Council' for co-ordination, promotion, rigorous measurement and review of the various quality systems/programmes to facilitate achievement of the policy.

Ensuring the financial success of Six Sigma projects

Six sigma approaches are not looking for incremental or 'virtual' improvements, but breakthroughs. This is where Six Sigma has the potential to outperform other improvement initiatives. An intrinsic part of implementation is to connect improvement to bottom line benefits and projects should not be started unless they plan to deliver significantly to the bottom line.

Estimated cost savings vary from project to project, but reported average results range from $100–150,000 per project, which typically last four months. The average Black Belt will generate $500,000–1,000,000 benefits per annum, and large savings are claimed by the leading exponents of Six Sigma. For example, GE has claimed returns of $1.2bn from its initial investment of $450m.

Six sigma project selection takes on different faces in different organizations. While the overall goal of any six sigma project should be to improve customer results and business results, some projects will focus on production/service delivery processes and others will focus on business/commercial processes. Whichever they are, all six sigma projects must be linked to the highest levels of strategy in the

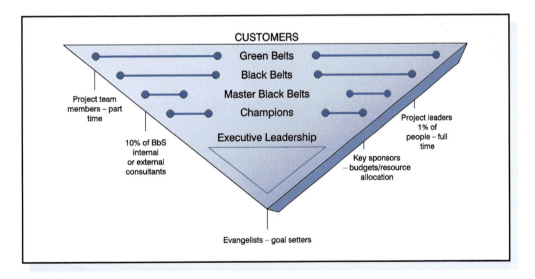

Figure 14.6
A Six Sigma company

organization and be in direct support of specific business objectives. The projects selected to improve business performance must be agreed upon by both the business and operational leadership, and someone must be assigned to 'own' or be accountable for the projects, as well as someone to execute them.

At the business level, projects should be selected based on the organization's strategic goals and direction. Specific projects should be aimed at improving such things as customer results, non-value add, growth, cost and cash flow. At the operations level, six sigma projects should still tie to the overall strategic goals and direction but directly involve the process/operational management. Projects at this level then should focus on key operational and technical problems that link to strategic goals and objectives.

When it comes to selecting six sigma projects, key questions which must be addressed include:

- what is the nature of the projects being considered?
- what is the scope of the projects being considered?
- how many projects should be identified?
- what are the criteria for selecting projects?
- what types of results may be expected from six sigma projects?

Project selection can rely on a 'top-down' or 'bottom-up' approach. The top-down approach considers a company's major business issues and objectives and then assigns a champion – a senior manager most affected by these business issues – to broadly define the improvement objectives, establish performance measures and propose strategic improvement projects with specific and measurable goals that can be met in a given time period. Following this, teams identify processes and critical-to-quality characteristics, conduct process baselining and identify opportunities for improvement. This is the favoured approach and the best way to align 'localized' business needs with corporate goals.

At the process level, six sigma projects should focus on those processes and critical-to-quality characteristics that offer the greatest financial and customer results potential. Each project should address at least one element of the organization's key business objectives, and be properly planned.

Concluding observations and links with TQM, SPC, Excellence, etc.

Six Sigma is not a new or different technique, its roots can be found in Total Quality Management (TQM) and Statistical Process Control (SPC) but it is more than TQM or SPC re-badged. It is a framework within which powerful TQM and SPC tools can be allowed to flourish and reach their full improvement potential. With the TQM philosophy, many practitioners promised long-term benefits over 5–10 years, as the programmes began to change hearts and minds. Six Sigma is about delivering breakthrough benefits in the short term and is distinguished from TQM by the intensity of the intervention and pace of change.

Excellence approaches such as the EFQM Excellence Model, Six Sigma, Lean or Lean Sigma are complementary vehicles for achieving better organizational performance. The Excellence Model can play a key role in the baselining phase of strategic improvement, whilst the six sigma breakthrough strategy is a delivery vehicle for achieving excellence through:

1. Committed leadership
2. Integration with top level strategy
3. A cadre of change agents – Black Belts
4. Customer and market focus
5. Bottom line impact
6. Business process focus
7. Obsession with measurement
8. Continuous innovation
9. Organizational learning
10. Continuous reinforcement.

These are 'mapped' onto the Excellence model in Figure 14.7.

There is a whole literature and many conferences have been held on the subject of Six Sigma, Lean and Lean Sigma and it is not possible here to do justice to the great deal of thought that has gone into the structure of these approaches. As with Taguchi methods the major contribution of Six Sigma/Lean Sigma has not been in the creation of new technology or methodologies, but in bringing to the attention of senior management the need for a disciplined structured approach and their commitment, if real performance and bottom line improvements are to be achieved. See also the next chapter on Lean systems.

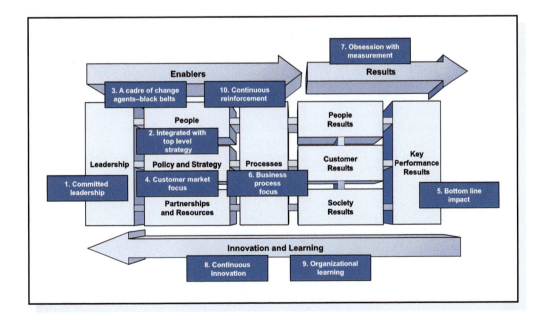

Figure 14.7
The Excellence Model and Six Sigma

Technical note

Sigma is a statistical unit of measurement that describes the distribution about the mean of any process or procedure. A process or procedure that can achieve plus or minus *six sigma* capability can be expected to have a defect rate of no more than a few parts per million, even allowing for some shift in the mean. In statistical terms, this approaches *zero defects*.

In a process in which the characteristic of interest is a variable, defects are usually defined as the values which fall outside the specification limits. Assuming and using a normal distribution of the variable, the percentage and/or parts per million defects can be found. For example, in a centred process with a specification set average ± 3 there will be 0.27 per cent or 2700 ppm defects. This may be referred to as 'an unshifted ± 3 sigma process' and the quality called '± 3 sigma quality'. In an 'unshifted ± 6 sigma process', the specification range is average 6σ and it produces only 0.002 ppm defects.

It is difficult in the real world, however, to control a process so that the mean is always set at the nominal target value – in the centre of the specification. Some shift in the process mean is expected. The P̲pm defects produced by such a 'shifted process' are the sum of the ppm outside each specification limit, which can be obtained from the normal distribution.

BIBLIOGRAPHY

Bendell, T., Wilson, G. and Millar, R.M.G., *Taguchi Methodology with Total Quality*, IFS, Bedford, 1990.

Bhote, K.R., *World Class Quality – Using Design of Experiments to Make It Happen* (2nd edn), AMACOM, New York, 1999.

Caulcutt, R., *Statistics in Research and Development* (2nd edn), Chapman and Hall, London, 1991.

Harry, M. and Schroeder, R., *Six-Sigma – The Breakthrough Management Strategy Revolutionising the World's Top Corporations*, Doubleday, New York, 2000.

Neave, H., *The Deming Dimension*, SPC Press, Knoxville, 1990.

Oakland, J.S., *Statistical Process Control: A Practical Guide* (6th edn), Butterworth-Heinemann, Oxford, 2008.

Pande, P.S., Neumann, R.P. and Cavanagh, R.R., *The Six-Sigma Way – How GE, Motorola and Other Top Companies Are Honing Their Performance*, McGraw-Hill, New York, 2000.

Pyzdek, T. and Keller, P.A., *The Six Sigma Handbook* (3rd edn), ASQ Press, Milwaukee, 2009.

Ranjit, Roy, *A Primer on the Taguchi Method*, Van Nostrand Reinhold, New York, 1990.

Shingo, S., *Zero Quality Control: Source Inspection and the Poka-yoke System*, Productivity Press, Stamford, Conn., 1986.

Taguchi, G., *Systems of Experimental Design*, Vols 1&2, Unipub/Kraus Int., New York, 1987.

Wheeler, D.J., *Understanding Statistical Process Control* (2nd edn), SPC Press, Knoxville, 1992.

Wheeler, D.J., *Understanding Variation*, SPC Press, Knoxville, 1993.

CHAPTER HIGHLIGHTS

Some additional techniques for process design and improvement

- Seven 'new tools' may be used as part of quality function deployment (QFD, see Chapter 6) or to improve processes. These are systems and documentation methods for identifying objectives and intermediate steps in the finest detail.
- The seven new tools are: affinity diagram, interrelationship digraph, tree diagram, matrix diagrams or quality table, matrix data analysis, process decision programme chart (PDPC) and arrow diagram.
- The tools are interrelated and their promotion and use should lead to better designs in less time. They work best when people from all parts of an organization are using them. Some of the tools can be used in activities related to problem solving and design.

Taguchi methods for process improvement

- Genichi Taguchi has advanced 'quality engineering' as a technology to reduce costs and make improvements.
- Taguchi's approach may be classified under four headings; total loss function; design of products, services and processes; reduction in variation; and statistically planned experiments.
- Taguchi methods, like all others, should not be used in isolation, but as an integral part of continuous improvement.

Six Sigma

- Six Sigma is not a new technique – its origins may be found in TQM and SPC. It is a framework through which powerful TQM and SPC tools flourish and reach their full potential. It delivers breakthrough benefits in the short term through the intensity and speed of change. The Excellence Model is a useful framework for mapping the key Six Sigma breakthrough strategies.
- A process that can achieve six sigma capability (where sigma is the statistical measure of variation) can be expected to have a defect rate of a few parts per million, even allowing for some drift in the process setting.
- Six Sigma is a disciplined approach for improving performance by focusing on enhancing value for the customer and eliminating costs which add no value.
- There are five fundamental phases/stages in applying the six sigma approach: Define, Measure, Analyse, Improve and Control (DMAIC). These form an improvement cycle similar to Deming's Plan, Do, Check, Act (PDCA), to deliver the strategies of process design/redesign, management and improvement, leading to bottom line achievements.
- Six sigma approaches question organizational cultures and the measures used. Six sigma organizations, in addition to focusing on understanding customer requirements, identify core processes, involve all employees in continuous improvement, are responsive to change, base management on fact and metrics and obtain outstanding results.

Continuous improvement – more advanced

- Properly implemented six sigma strategies involve: leadership involvement and sponsorship, organization-wide training, project selection tools and analysis, improvement methods and tools for implementation, measurement of financial benefits, communication, control and sustained improvement.
- Six sigma process improvement experts, named after martial arts – Master Black Belts, Black Belts and Green Belts – perform the training, lead teams and carry out the improvements. Mature six sigma programmes have about 1 percent of the workforce as Black Belts.

Chapter 15

Continuous improvement – Lean systems

INTRODUCTION TO LEAN THINKING

As described in Chapter 11 on process redesign, Womack, Jones and Roos first introduced the concept of Lean Manufacturing in 1990 by describing an approach to quality and operational management that had been adopted by the Japanese car industry, led primarily by Toyota. They were in fact describing an approach that, although inspired by Henry Ford's production line at the River Rouge Model-T Ford assembly plant, had evolved continuously from the 1950s through into the 1980s, but which had been consolidated at Toyota into an organized 'system', the so-called Toyota Production System, or *TPS*. The pioneers that laid the foundations for this were Taiichi Ohno, who became an Executive Vice President in 1975, and his consultant and advisor Dr Shigeo Shingo. Whereas Ohno's career remained with Toyota, Shingo went on to successfully introduce these concepts to many other industries, not only in Japan, but also in the US and in Europe.

The fact that other industries adopted the concept of 'lean' early on in its development is important. Womack's book was based on research that specifically focused on the automotive industry but the concepts and most of the tools used were not only transferred and adapted across other manufacturing industries as lean manufacturing, but have also become established practices as part of the evolving nature of what has continued as Total Quality Management.

The evolving nature of the lean approach is an important issue as the practices usually needed tailoring to the specific context of the organizations adopting them. Over recent times many industries have discovered that the concepts and practices embodied in lean are indeed transferable and adaptable, for example to the service environment. There has also developed much interest in the applicability of lean in the public sector, and a number of those organizations have already adopted lean initiatives with varying degrees of success.

Table 15.1 Lean Thinking – some myths and facts

Myth	Fact
Lean is a manufacturing concept	Lean is a concept that applies to every organizational setting
Lean tools and techniques are not transferable and/or are not relevant to the public sector	Every organization, whether in manufacturing, service or public sector has its own unique context and therefore some tools and techniques will not transfer well – however, had 'lean' started in the public sector, the private sector would have found that not all the tools and techniques worked for them either. There has always been a need for sector and even organization specific tools and techniques
Lean is about cutting back on people resource	Lean is about eliminating non value-adding waste, not laying off employees
Lean is another management fad	The concepts and practices of lean have been in operation since (at least) the 1950s and are still with us
Lean is resource hungry to implement	Lean is most successful when the skills and knowledge required are embedded in the organization
Lean is expensive to implement	Lean potentially saves far more than it costs to implement

With this background in mind, it is worthwhile considering a few 'myths' about lean (Table 15.1).

There are four key issues that come out of this.

1. Lean is transferable

The fundamental concept behind lean is that of stripping out non-value adding activity so that an organization can truly focus on delivering what its customers actually want. In other words, it is not a good idea to expend our far from limitless resources on things that do not achieve what we are trying to deliver.

This tenet holds true for the private sector because it aims to reduce operating costs but not at the expense of diminishing customer requirements or value. The overriding goal may be profit maximization, but that is all about increasing the efficiency of how resources are utilized in order to provide something customers want, in order to provide future 'business'. In the public sector, this relates to getting more for less. In other words, delivery capability increases whilst internal costs come down. Whichever the sector, effort should be focussed on delivering what is needed whilst reducing effort expended on what is not.

Specific 'lean' tools/practices that work in manufacturing may not work so well in other sectors but one would expect this. It is not the tools that are the issue as these can be adapted and/or developed to suit a specific organizational context – many already have. It is very easy to look at specific practices (often given as manufacturing examples) and dismiss them as 'not appropriate' to another context. Even as service and public sector lean examples start to emerge, the response can still be the same – 'yes, but I couldn't do this – it wouldn't work in my organization'.

The key lean principal is removing non-value adding activities so the challenge is about developing (or adapting) tools that will work, just as the pioneers of lean thinking did within their own context of the automotive industry.

2. Lean does not necessarily mean losing people

Language and words used in organizations are very powerful and therefore important. The simplest things that are taken for granted on a day-to-day basis within one context can become powerful symbols and cause divisions in ideology in others.

'Lean' can become a powerful word in a particular context and it may not necessarily communicate the intentions of the people who collectively developed the lean thinking. This was about freeing up resources to get on with the important things and delivering value add, rather than – say – reducing headcount. The pioneers of 'lean' thinking were trying to reduce the time people spent on non-value adding activities so they could focus more on what would add value. It has become an empowering philosophy in many lean organizations, including those hit with hard financial pressures, that they have built their capability by reducing waste.

3. Lean is not just a fashionable idea that will go away eventually

The concept of lean has been with us at least since the Japanese took on board these principals in manufacturing. Many years on we are still discussing it, so it is probably not a fad, although terms and emphases may come and go. What is certainly true is that these concepts, tools, techniques and philosophies have grown, adapted and evolved over time and have delivered sometimes quite breathtaking efficiency gains for those who have embraced them properly, even in the most unlikely environments.

If the aim is to adapt, or even transform, the way we deliver products and services, including in the public sector, and to deliver more for less, whilst developing and maintaining our focus on the customer or citizen and meeting the goals and objectives of other stakeholders, then 'lean' thinking will help us by ensuring that we have the internal capability required.

4. Implementing lean is an investment

It is very rare to find examples where significant change has not had a concomitant cost and there are generally major resource issues when impacting any significant change. Lean is about reducing waste and cost – it is about doing more for less – so what really needs to be considered here is the fact that lean provides an opportunity to measure savings and actively demonstrate the savings made. Clearly, any cost of change can be set against the cost savings of the change.

It is worth noting however that in the manufacturing sector, the aim of introducing these techniques was to reduce cost whilst at the same time not losing value added to the customer. As these techniques are adapted to different situations, this aspect of delivering a quantifiable benefit to both customers and the organization could potentially be lost. It is therefore critical that this aspect is not lost in any translation.

Lean and Six Sigma

Six Sigma was an improvement approach pioneered by Motorola and subsequently GE (see Chapter 14). It has its roots in the work of the early 'TQM gurus' such as Deming and Ishikawa, who emphasized the need to understand and control variation in processes to improve the management of quality. Although six sigma is geared more to reducing process variation, and as such has again been criticized for its manufacturing bias, it has nevertheless been found to be of enormous benefit in other sectors, especially where there are transactional processes. There has also been a lot of interest and indeed success in combining the two approaches of Lean and Six Sigma into 'Lean Six Sigma' or 'Lean Sigma' programmes, where the lean element addresses waste and lead times and the six sigma element addresses process variation (Figure 15.1).

This combined view produces a more holistic approach and provides a broader set of improvement tools and techniques. However, it has been argued that many service and public sector processes are not transactional in nature and attempts to reduce variation may actually be counterproductive, if not impossible to achieve. It therefore becomes easy for people to dismiss the approach as not being relevant to these sectors, whereas many of the elements contained within the overall approach are very relevant indeed. There is a need to understand how this thinking can be adapted and tailored to the specific context and needs of any specific organization.

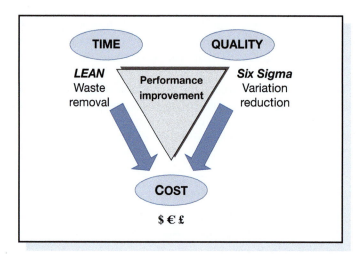

Figure 15.1
Lean Six Sigma

Approaches to Lean interventions

As we have seen, Lean is not new and has been around as an improvement approach since Ohno and Shingo first introduced the concept in relation to the Toyota Production System. Many of these ideas have also been described in other documented approaches to improvement, including TQM, and a number of authors and practitioners have started to distinguish between lean in manufacturing, lean in services and lean in public services, etc. In principle there are many similarities between these approaches.

Having described lean manufacturing in *The Machine that Changed the World* in 1990, Womack and Jones continued to develop their ideas and in their follow-up book *Lean Thinking*, published in 1996, they identified five key principles to guide an organization's implementation of lean:

1. Provide the *Value* actually desired by customers
2. Identify the *Value Stream* for each product
3. Line up the remaining steps in a *Continuous Flow*
4. Let the customer *Pull* value from the firm
5. Endlessly search for *Perfection*.

The emphasis placed on lean was that of understanding the 'core value-adding processes' and the stripping out of all non-value adding activity from these. To ensure that these then run smoothly, all supplying and support processes need to be designed and run to deliver as a continuous flow so that, as activity is *pulled* through the system by customer demand, things get done only when they are required to be done, so eliminating waste activity, unnecessary inventory and time delays.

If this was the basic concept of lean, a number of tools and techniques were developed in order to enable this including, for example, Just In Time supply (supplying only *what* is needed, *when* it is needed), Kanban systems (ensuring a constant supply of materials, but only as needed) and Poke-Yoke (mistake proofing). Although many of the tools in the 'lean tool-kit' come from a manufacturing origin, the principles behind them are applicable to any delivery process, including in the services and public sectors.

The concept of the value stream is fundamental to lean. In many manufacturing industries, this can be seen as an internal core process of value-adding activity that ultimately delivers a value proposition (in the form of a product) to the customer. Customers very rarely get involved in the delivery of that product, be it a car or a computer (Figure 15.2).

However, in services, as in many support processes in manufacturing, the customer does interact at various 'touch-points' along the value delivery chain (Figure 15.3).

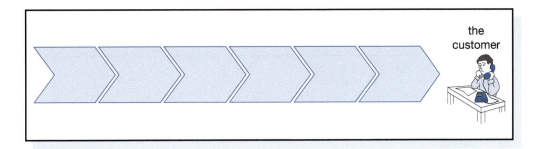

Figure 15.2
Value stream 1

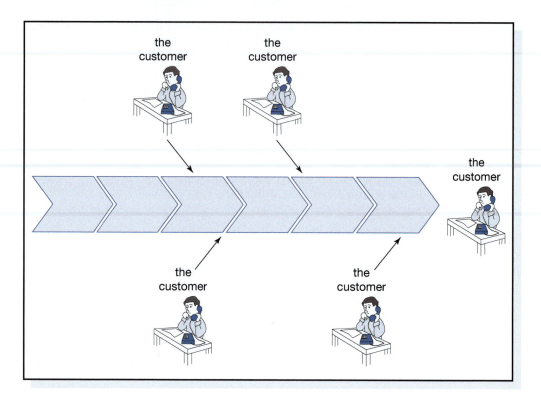

Figure 15.3
Value stream 2

In *Lean Solutions*, Womack and Jones expanded their approach again to give a more thorough view of the customer, stressing the need to understand 'consumption', as in the customer's demand requirements, and 'provision', in terms of the organization's capability to deliver these requirements.

They advocated the need for a 9-step approach (Table 15.2).

In terms of tools and techniques, the primary methodology presented was the Value-Stream Map, to identify materials and information flows and understand how and where value was created (and destroyed/wasted).

Table 15.2 Womack and Jones' 9-step approach

1. **Learning to see** *Consumption*
2. **Learning to see** *Provision*
3. **Solve my Problem** *Completely*
4. **Don't Waste my** *Time*
5. **Get me Exactly** *What* **I Want**
6. **Provide Value** *Where* **I Want it**
7. **Solve my Problem** *When* **I Want**
8. **Get me the Solution I** *Really Want*
9. **Solve my** *Complete* **Problem** *Permanently*

In this approach, lean is seen as a transformational with three critical success factors:

- Leadership
- process understanding and process thinking
- understanding both lean consumption and lean provision.

Fundamental to the approach was the concept of providing what the customer *actually wants* through the application of a lean approach, i.e.:

- Identifying the *value stream* providing the value and removing all wasted steps
- putting the remaining steps into *continuous flow*
- so the consumer can *pull* value from the system
- while pursuing *perfection*.

Whereas six sigma programmes have always been geared towards achieving significant cost savings by measuring their impact on the 'bottom line', lean has tended to focus more on the concepts of customer value and elimination of waste, rather than ROI. Bringing the two approaches together brings more of a focus on the need to prioritize projects for the impact they will have in terms of cost reduction.

The six sigma methodology popularized by Motorolla (in manufacturing) and then GE Capital (in services) has become the standard for six-sigma improvement processes and many advocate using the same approach for lean six-sigma interventions. This five-phase framework is known as DMAIC: Define Measure, Analyse, Improve and Control.

Whereas Womack and Jones did not get into specific tools and techniques other than value-stream mapping, others have adopted more pragmatic approaches and suggested specific tools and techniques that are applicable in each of the five phases. The adapted DMAIC framework to show the tools and techniques applicable to lean, as opposed to six sigma, is shown in Figure 15.4.

Although DMAIC has substantial acceptance as a six-sigma and lean six-sigma methodology, there are some who struggle with its use as a purely lean approach,

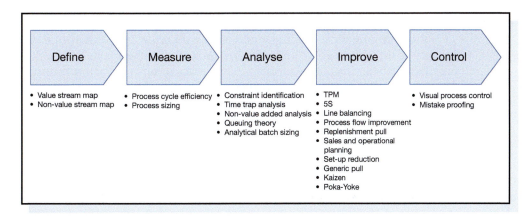

Figure 15.4
DMAIC and the lean toolkit

especially in service and public sector environments, although there are similarities with the Deming Plan, Do, Check, Act (PDCA) cycle.

VALUE STREAM MAPPING

Value stream mapping (VSM) studies the set of specific actions required to bring a product family from raw material to finished goods as per customer demand, concentrating on information management and physical transformation tasks.

The outputs of a VSM based study are a current state map, future state map and implementation plan for getting from the current to the future state. Using VSM it should be possible to bring the lead time closer and closer to the actual value added processing time by attacking the identified bottlenecks and constraints. Bottlenecks addressed could include long setup times, unreliable equipment, unacceptable first pass yield, or high work or process inventories.

The VSM technique has been central to the approach advocated by Womack and Jones from their original work with Toyota, and is still the mainstay of lean interventions. The principle is to describe the current process, looking at both physical and materials flows and information flows, in a highly visual format and to apply measures to each process step to identify the time taken and the cost involved. By also identifying which activities add value and which do not, it is possible to analyse the process from a value creating perspective and determine the potential gains from eliminating non-value adding activity.

Whilst this fits well with the **Define** phase of DMAIC, value stream mapping (VSM) is also used in the generation of a **Target Operating Model (TOM)**, and should come after analysis and investigative work to identify what can be done to reduce non-value adding activities. Figure 15.5 shows a value stream map from a manufacturing context where the original time taken has been identified and then altered to show what has been made possible through understanding what can be changed.

In this example, the time taken for information to get from the customer to the manufacturer prior to the start of manufacture was 66 days and amounted to 1360 minutes of actual process time. Manufacturing time took 21 days (1075 minutes actual process time).

By identifying these timings and then challenging how much of that time was actually spent adding value and what was not, it was possible to reduce the total lead time from 87 (66 + 21) days to 20 (15 +5) – a saving of 67 days throughput time. Although this is from a manufacturing environment, the overall process (in terms of process steps and flow) is very similar to many service processes in that much of the non-value added time is spent 'in transit' waiting for things to be done – often on administrative tasks.

VSM has been used successfully in the health sector in reducing patient wait times, where process delays are often due to a non-alignment of the process steps – in the language of the Womack and Jones' model they are not arranged in 'continuous flow' and therefore are not adding value and in fact are increasing waste and cost.

Although there are some standardized approaches to VSM, there are also different approaches in operation. Figure 15.6 shows a value stream map for out-patient heart failure services at a hospital. The map takes the form of a horizontal timeline showing

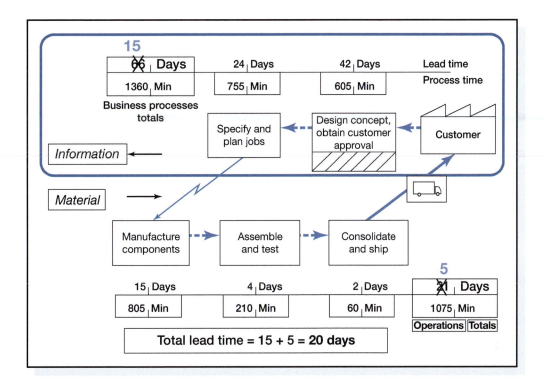

Figure 15.5
Value stream map (example 1)

the sequential process steps, but also plotted on a vertical axis showing the cost of value-adding activity (above the timeline) and the cost of non-value-adding activity (below the timeline). The addition of staff costs provides a powerful diagnostic tool by which to address how non-value adding activities might be eliminated.

Although VSM would appear to be a simplistic tool/technique, it is important that how the various activities make up the process is understood. In complex service environments there may be some apparent non-value adding steps that are essential to another process that are in some way linked to the process being investigated. It is important that these dependencies are understood and it is therefore essential that value stream maps are not created by individuals but by teams of people working in the process who know what is going on and can challenge each others' perspectives. Lean should be a cross-functional, team-based approach. A development of VSM by the author and his colleagues is 'Carbon Stream Mapping' (CSM) in which processes are studied to identify carbon emission at various stages; Figure 15.7 is an example of a CSM.

Process families

The manufacturing based example in Figure 15.5 may share similarities with some service processes, at least in part. In all organizations there are processes that, whilst

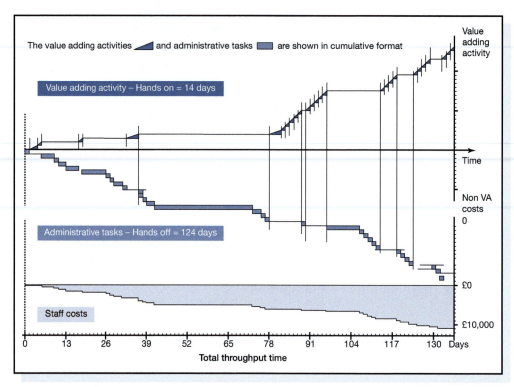

Figure 15.6
Value stream map (example 2, out patient heart failure services)

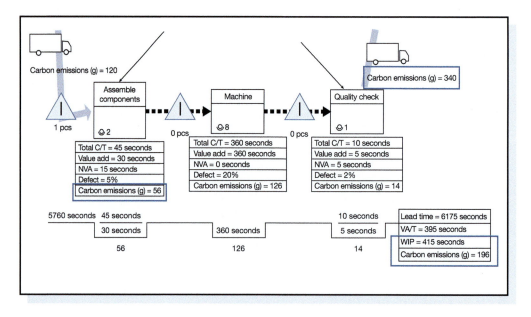

Figure 15.7
Carbon stream mapping (CSM) example

at first sight may seem to have nothing in common, when they are broken down into their component steps they actually pass through the same stages, often worked on by the same departments and/or people. When different processes share the same process structure, they can be described as being in the same 'process family'.

This is important from a VSM perspective, because working on a process family rather than a single process often yields far more information about the value adding and non-value adding steps across the family of processes. As long as the needs of each individual process are taken into account, and one process does not suffer as a result of changes to the 'family', it is possible to leverage the gains by applying them to a broader set of activities.

Value stream scopes (end-to-end processes – field to fork)

When Hammer and Champy brought Business Process Reengineering to the world's attention in 1993, they highlighted the fact that major gains can be made in process improvement by looking at processes from an 'end-to-end' perspective. This principle applies equally to VSM and lean. Value streams build up in the connection of the different activities that stem from initial 'provision' to final 'consumption' (using the language of Womack and Jones).

If we think of the food that we eat, this is the end result of a value stream that ultimately begins with animals and/or plants in a field (Figure 15.8). The series of sequential steps that lead ultimately to that birthday cake or Chicken Tikka Masala comprise an end-to-end value chain. Of course we could take it further still, but the bigger the scope of the Value Stream, the greater the potential for real gains.

Scoping the value stream is a key decision in VSM as it will directly dictate the levels of improvement that are possible.

An insurance company was trying to take time out of its new business process in order to compete with internet-based competitors, as they could not turn around

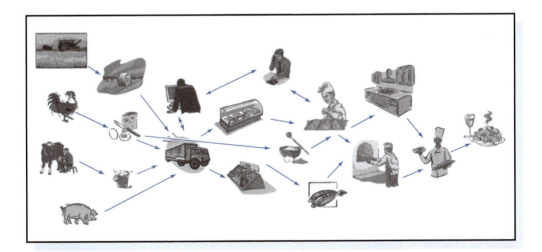

Figure 15.8
Field to fork

insurance proposals as quickly. Their information systems started tracking times on receipt of the proposals and finished when the documentation was sent to print. They scoped the intervention around their information systems, as anything outside was in theory not measurable. Whilst trying to save minutes on internal processes, nobody was looking at the days lost in the print room, in transit, proposals sat on the desks of brokers, etc. (Figure 15.9).

Four steps for VSM

Value stream mapping is a powerful technique that is at the heart of a lean approach. As a visual imaging tool, it allows the visualization of other concepts, such as understanding what the customer sees as value and how that value is created by the internal processes of provision. It allows a quantification of that value in terms of time and cost and allows an understanding of the concept of continuous flow as compared to bottlenecks and constraints that add to time and cost and destroy or lessen customer value. If we add to this the concept of customer touch-points – identifying where the customer interacts with the elements within the value stream, which is so important in the service context – we can more clearly understand the concepts of customer value.

By looking at current methodologies of VSM it is possible to discern a four step process (Figure 15.10).

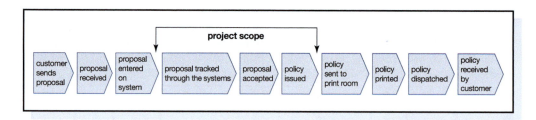

Figure 15.9
Scoping the value stream

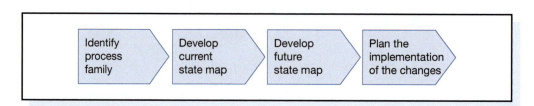

Figure 15.10
The 4 steps of VSM

THE BUILDING BLOCKS OF LEAN

It is beyond the scope of this chapter to describe all the individual tools and techniques that comprise lean. VSM is undoubtedly central to a lean approach in that it is the primary visualization technique for understanding how value is created (or destroyed) on the 'provision' side of the value stream.

Although the literature, and indeed practice, includes tools and techniques such as Just in Time, Kanban, Total Productive Maintenance (TPM), Cellular Production and Flow, etc (Figure 15.11), many of these tools are specific to the workplace and some, for example TPM, may be less immediately applicable to the public sector. However these tools are designed to implement the required changes and whereas some are undoubtedly adaptable to service setting, new tools, more specific to a service environment, will emerge as lean becomes more widespread in this environment.

Two other tools/techniques worth noting at this stage, however, are 5S and Kaizen.

5S

5S has been described as a system that creates a disciplined, clean and well-ordered work environment. Its original role in the manufacturing context was that of ensuring

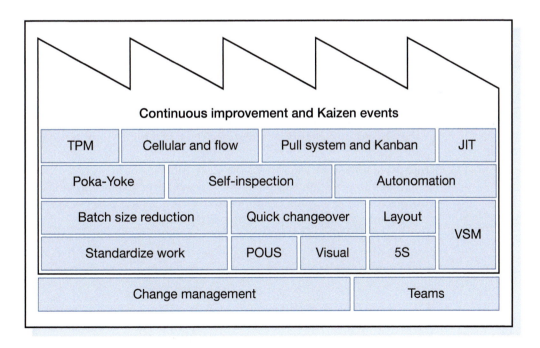

Figure 15.11
The building blocks of Lean

that work areas were free from clutter, clean and laid out in such a way that tools weren't lost and time was not wasted on non-value adding activities to do with general house-keeping. The secret to this was contained in five simple steps:

1. Sort

The first step is to free up the workplace by getting rid of everything that is not required for the work – this reduces problems and decreases lead times.

2. Set in order

What is left is then organized so that it is easily accessible and time is not lost looking for things that are misplaced.

3. Shine

This step relates to cleanliness and keeping things in good working order. Although this is easy to visualize in a shop floor manufacturing environment, the basic concepts can still apply in services.

4. Standardize

In this step, the above three steps are standardized so that they become a routine part of day-to-day work. Management should create the time resources for people to perform the first three steps.

5. Sustain

The fifth and last step is about embedding the practice in business as usual by involving people at all levels, through peer and leadership behaviours, 5S audits, 5S goals and providing feedback on performance.

In the health sector in the UK, 5S has been successfully adapted to a 'CANDO' approach:

C = **Clean up** – sorting out necessary from unnecessary items for that area that are required to support flow of the service or product.

A = **Arranging** – the ergonomics of the work area are looked at and the handling of tools or supplies to assess how they should be most effectively arranged and stored for ease of work and safe working practices.

N = **Neatness** – this initially involves a large scale clean up and the establishment of standard practices for the organization of items in the area, followed by a regular maintenance clean taking 5–10 minutes per shift.

D = **Discipline** – the maintenance of the C, A, N stages and requires the team in that area to maintain compliance to the set agreed standards for the workplace. This discipline is supported by an auditing process to sustain and improve workplace organization performance.

O = **Ongoing improvement** – once CANDO has initially been carried out the team are then encouraged to continue to make improvements that will assist them in operating in the most efficient manner with high levels of safety.

5S is a way of engaging everyone in the organization in lean activities and does present an opportunity for tying into a cultural approach in which individuals take personal responsibility for their own workplace and their attitudes to how they work.

Some authors have noted that in the final step of this approach, sustain (or ongoing improvement) another approach is often useful – Kaizen.

The role of Kaizen

The word Kaizen comes from the Japanese: Kai (meaning change or action toward) and Zen (the good or better) – see also Chapter 17. It is heavily associated with the Toyota Production System (TPS) and is therefore inevitably associated with the lean approach. Although the word has generally become associated with a philosophy of continuous improvement, it is also associated with the concept of rapid, short-term improvement activity – the Kaizen 'blitz'.

In a Kaizen blitz, typically a cross-functional, multi-level team of 6 to 12 members is brought together to focus on solving a specific work problem (i.e. not a factory or organization wide process), and is characterized by a typical 3–5 day time period. The aim is to focus on achieving a rapid, breakthrough result by rapidly developing, testing and refining solutions to leave a new process in place. It is not about planning, but about doing, and keeping things simple. Although the team needs to be trained in Kaizen before the project, tools are kept very simple and few in number. Budgets are typically low but savings can be substantial and achieved very quickly. If a process owner exists, he or she usually takes responsibility and provides leadership, but the core team members are from the area being tackled – the actual people who do the real work.

The key principles of Kaizen Blitz are:

- Being open minded
- Maintaining a positive attitude
- Rejecting excuses and seeking solutions
- Asking Why? Why? Why? Why? and Why again? – there is no such thing as a stupid question
- Taking action – not seeking perfection, but implementing now with the resources at hand
- Using all the team's knowledge
- Disregarding rank – everyone is equal.

It is interesting to note that VSM is typically a three-day activity and it is not surprising that the two elements are often seen as being complementary parts of a lean intervention.

The Kaizen method has been successfully implemented in many service environments and there are many case studies documented. In general, however, it is possible to identify a generalized 10 step approach (Table 15.3).

Table 15.3 A 10-step service Kaizen methodology

1	Define the Problem
2	Define the Goal
3	Define the Baseline
4	Define the Process
5	Explore Possible Sources of Variation
6	Identify Corrective Actions to Eliminate Waste
7	Develop an Action Plan to Take Action
8	Review Results
9	Replicate and Make Additional Improvements
10	Celebrate

Many organizations make mistakes in implementing lean because they focus on specific tools and techniques that are not well adapted to their organizational setting. When this happens, not only does the lean initiative fail to deliver what was expected, but this becomes another example of 'how lean does not work in our environment'. However, those organizations that have *succeeded* in implementing lean approaches have done so by adapting or developing approaches based on the underlying principles of lean, i.e. by focusing on what the customer sees as being of value, and then identifying activities that do not add value and taking steps to eliminate them.

Organizations that have repeatedly delivered major sustainable benefits from lean implementations have tended to follow a structured approach to improvement that has focused on what will work for them in their context, rather than trying to follow a set 'recipe book' based on specific tools and techniques. By adopting a pragmatic approach it becomes possible to tailor specific tools and interventions from the vast armoury that is available to deliver what is required within that specific organizational setting.

The importance of a pragmatic approach (rather than slavishly following a theoretical model) needs to be emphasized, and it is possible to discern from the evidence available a pattern in terms of what approach an organization could follow to ensure that they consider the main concepts underpinning lean without being tied to specific tools and techniques that will not work well in their context.

Organizations that take this path to lean implementation tend to focus on six key phases (Table 15.4) – see also Chapter 13.

DRIVER (Figure 15.12) has been successfully used in many private and public sector organizations as an improvement methodology and is tried and tested. It is clear that this approach embraces the concepts of lean whilst allowing sufficient flexibility for the detailed tools and techniques that could and should be used. It is therefore presented as a pragmatic improvement approach for lean interventions.

Define

At the start of any lean intervention, it is important to go through a number of key steps. The goals of the project should be articulated in terms of desired outcomes, and

Table 15.4 DRIVER – a six-phase approach to Lean

1	Define	Define the scope of the improvement project in terms of customer and organizational goals
2	Review	Map the current process and measure its performance to understand how it adds value and when it doesn't
3	Investigate	Analyse the gap between current and desired performance identifying problems and prioritizing opportunities
4	Verify	Generate and validate improvement solutions to realize the opportunities
5	Execute	Implement the improved process to sustain performance improvement
6	Reinforce	Learn the lessons and continuously improve

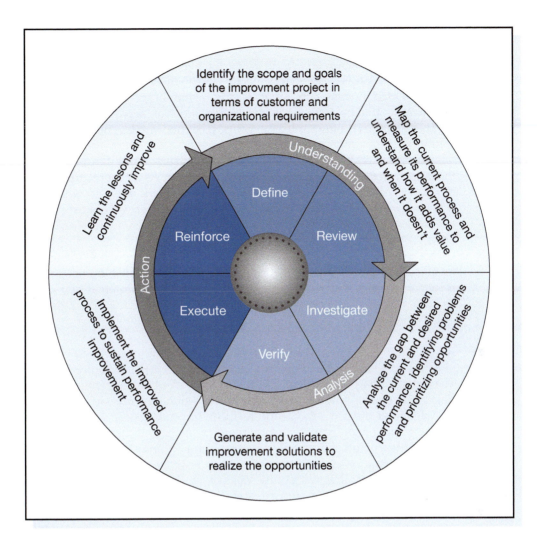

Figure 15.12
DRIVER: a pragmatic improvement approach

these should ideally be expressed in terms of the requirements of relevant stakeholders. For any one project the individual stakeholders may vary. For public sector organizations, for example, one would expect them to include at the least citizen, government and organizational stakeholders.

Stakeholder requirements also include two specific inputs to the lean intervention:

- The voice of the customer, in terms of what value means to the customer
- The voice of the business, in terms of driving out non-value adding activities from the value stream so as to reduce time and waste and thereby cost.

The complex multiple-stakeholder perspective in many organizations means that gaining a clear understanding of the value proposition for all stakeholders will be challenging but requirements tend to be easier to articulate when expressed in terms of value.

Other approaches have highlighted the need to identify first the process that the lean intervention will be conducted on and then to understand the process family within which it sits – in other words, what other processes exist within the organization that have the same underlying series of steps and activities. The process family perspective allows multiple gains to be made by leveraging the benefits of any solution across a number of different processes.

The scope of the process is critical to the outcomes and it must be very clear from the outset what is in scope and what is out of scope. As a general guideline one should always scope the process at its largest and then work down to what is feasible and achievable. It is important to look at end-to-end processes (field to fork) if possible to avoid the dangers of scoping at too low a level and thereby missing the most critical areas for removal of non-value adding activity.

Other key considerations at this stage include project management, team membership, governance at the intervention level, establishing performance metrics related to stakeholder requirements, communications strategies and a direct, visible and clearly articulated link between this lean intervention and the vision and values of the organization.

Review

The Review phase sees a description and analysis of the current state ('as-is'). At this stage, it is appropriate to start to build the value stream map, based on the team's knowledge of the process. Clearly team involvement is crucial to get a realistic and meaningful view of 'reality' and, as the view is built, it may need to be checked and refined. Measures need to collected and mapped onto the process steps to identify time and costs.

It is important that voice of the business and voice of the customer data continue to be developed during this phase and as the view is built up, as this will challenge the current state analysis. The author and his colleagues have found in service environments, where the customer has direct contact within the value stream, it is important to identify these customer 'touch-points' and, likewise, understand the value propositions that exist at each touch-point. It is also highly likely that supporting processes will be identified as impacting on the value stream and these will need to be explored to provide the cause and effect analysis that will take place in the next phase.

Investigate

Once the current state picture has been built and metrics have been added to the value stream map to provide a clear view of the time taken for each process step and the costs incurred to deliver both value adding and non-value adding activities, we move into the 'Investigate' phase. The aim in this phase is to find ways by which we can reduce the time taken and reduce non-value adding costs whilst still delivering the value add to customers and even increasing it, if required. This phase requires an in-depth challenging of the status quo and any assumptions that might be present in order to generate possible solutions to eliminate non-value adding activity and 'dead time' when nothing is actually being done.

If the customer interacts with the process, touch-points need to be explored in detail to understand what is adding value, what is not adding value and what is actually destroying value in the customer's eyes.

Any other processes interacting with this process, such as supplying process, will also need to be investigated for impacts and mapped as appropriate. Similarly, any processes that the process under investigation impacts upon should also be identified, as any changes made could have effects elsewhere in the 'system'.

This phase is really about understanding where and how changes might be made and what the effects of those changes will be. A range of possible solutions may be developed and all possibilities should be considered in order to generate potential 'future states'.

Verify

The investigation comes to an end when all possibilities have been explored and further ideas are exhausted. We now need to decide on the solutions that will be implemented by verifying the impacts of the various ideas and determining which are the most appropriate for the specific context we are in.

In multiple stakeholder environments, we need to be sure of the impact of any changes on different stakeholder groups and some of the ideas generated previously may need to be modified or even discarded if they present problems to key stakeholder groups. It is also important to ensure that the proposed solutions fit with the organization's strategy and values.

The aim of this phase is to generate the definitive 'future state' map that will then be implemented, and in doing so, there may be substantial testing and retesting of assumptions. In some cases there may be a need for pilot studies to test the feasibility of some of the proposed changes. The proposed 'future state' process design then should generate confidence it can be successfully implemented and will deliver the potential gains identified.

Execute

Once the new process has been determined the changes will need to be implemented, which requires a robust change management plan. Implementing new ways of working is always a challenge and there are a number of documented methodologies specifically relating to this – see the figure of 8 framework in Chapter 9. Implementation needs to be tailored to the organization's culture and specific requirements and is usually best carried out as a 'phased' approach.

Senior management commitment to the proposed changes is critical, as is a broader buy-in from key stakeholders. Staff affected by the changes also need to be on board with the proposed changes and, therefore, it is essential that confidence that the required results can and will be achieved by the changes is established at all levels and with all constituencies involved. Clearly excellent two-way communications will be an essential component of the implementation.

The implementation will require appropriate documentation to be developed. Training requirements will need to be determined and appropriate interventions developed and delivered. These will undoubtedly include training staff in lean tools and techniques and, if consultants have been involved, a full skills transfer programme should be in place, including mentoring and coaching as appropriate. If a pilot has not been conducted in the Verify stage, it is advisable to establish a 'safe test' facility before full roll out.

Some lean tools described originally by Ohno and now part of the lean 'toolkit' also now play a key role. For example, mistake proofing systems can be built in and tools identified in the Investigate and Verify phases now need to be implemented along with the new process.

Reinforce

Having made changes that will create a better value proposition for both the organization and its customers (and other stakeholders), it is important that the changes are held and that the process does not 'slip back' to its previous state. It is here that the concepts of Kaizen as a continuous improvement philosophy come in. Techniques such as 5S can considerably impact the culture of the organization, and people should be encouraged to 'think lean' in their everyday operations.

Whereas initial capability should have been built in the Execute phase, there will be a need for ongoing capability development to ensure that knowledge is not lost. Many successful lean initiatives have included processes for communicating learning points and establishing communities of lean 'champions' and/or 'practitioners,' who are skilled and knowledgeable in lean, and take a lead in continuing lean interventions. The development of 'Lean Academies' is also popular in some industries.

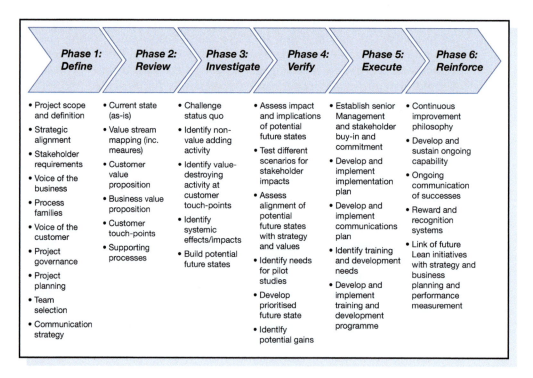

Figure 15.13
DRIVER: a six phase approach to Lean

Ongoing communication of successes is a good way of ensuring that the initial momentum and key messages are not lost, as is building in an appropriate reward and recognition system to ensure that positive lean behaviours are encouraged.

Companies and organizations that succeeded in establishing an ongoing commitment to lean have embedded the thinking in the day-to-day culture and have ensured that there is a direct link between lean initiatives and the organizational strategy and performance measurement framework.

Figure 15.13 shows a summary of the DRIVER framework for lean.

BIBLIOGRAPHY

Bercaw, R.G., *Lean Leadership for Healthcare,* ASQ, Milwaukee, 2013.

Womack, J.P. and Jones, D.T., 'From Lean Production to the Lean Enterprise', *Harvard Business Review,* Vol.72, No. 2, March–April, pp. 93–103, 1994.

Womack, J.P. and Jones, D.T., 'Beyond Toyota: how to root out waste and pursue perfection', *Harvard Business Review,* Vol.74, No. 5, September–October, pp. 140–158, 1996.

Womack, J. P. and Jones D. T., *Lean Thinking: Banish Waste and Create Wealth in Your Corporation,* Simon & Schuster, New York, 1996.

Womack, J. P. and Jones D. T., *Lean Solutions,* Simon & Schuster UK Ltd, London, 2005.

Womack, J.P., Jones, D.T. and Roos, D., *The Machine that Changed the World,* New York: Rawson MacMillan, 1990.

CHAPTER HIGHLIGHTS

Introduction to lean thinking

- Womack, Jones and Roos first introduced the concept of Lean Manufacturing by describing an approach that had been adopted by the Japanese car industry, led primarily by Toyota – the so-called Toyota Production System.
- The evolving nature of the lean approach means that the practices need tailoring to the specific context of the organizations adopting them; many industries and sectors have discovered that lean concepts and practices are transferable and adaptable, including to the public sector.
- There are certain myths and facts about lean which lead to four issues to consider: lean is transferable, lean does not necessarily mean losing people, lean is not a fashionable idea that will go away eventually, and lean is an investment.

Lean and Six Sigma

- There has been a lot of interest and success in combining the two approaches of Lean and Six Sigma into 'Lean Six -Sigma' or 'Lean Sigma' programmes, where the lean element addresses waste and lead times and the six sigma element addresses process variation and quality; this combined view produces a holistic approach and provides a broad set of improvement tools and techniques.

Approaches to lean interventions

- Womack and Jones identified five key principles to guide an organization's implementation of lean: provide the *Value* actually desired by customers; identify the *Value Stream* for each product; line up the remaining steps in a *Continuous Flow*; let the customer *Pull* value from the firm; endlessly search for *Perfection*.
- The emphasis placed on lean is that of understanding the 'core value-adding processes' and the stripping out of all non-value adding activity; all supplying and support processes need to be designed and run to deliver as a continuous flow so that, as activity is *pulled* through the system by customer demand, things get done only when they are required to be done, so eliminating waste activity, unnecessary inventory and time delays.

Value stream mapping

- Value stream mapping (VSM) studies the set of specific actions required to bring a product family from raw material to finished goods, as per customer demand, concentrating on information management and physical transformation tasks.
- The outputs of a VSM based study are a current state map, future state map and implementation plan for getting from the current to the future state.
- A development of VSM by the author and his colleagues is 'Carbon Stream Mapping' (CSM) in which processes are studied to identify carbon emission at various stages.
- In all organizations there are processes that, when broken down into the component steps, actually pass through the same stages, often worked on by the same departments and/or people; when different processes share the same process structure, they can be described as being in the same 'process family.'

The building blocks of lean

- The building blocks of lean include tools and techniques such as Just in Time, Kanban, Total Productive Maintenance (TPM), Cellular Production and Flow, Poka-Yoke and 5S (CANDO in health sector); these are designed to implement the required change; no doubt new tools more specific to service environments will emerge as lean becomes more widespread in its application.

DRIVER: a context-dependant process view of lean

- Organizations that have succeeded in implementing lean approaches have done so by adapting or developing approaches based on the underlying principles of lean, i.e. by focusing on what the customer sees as being of value, and then identifying activities that do not add value and taking steps to eliminate them.
- The 'DRIVER' improvement methodology has been successfully tried and tested in many private and public sector organizations. It is clear that the DRIVER approach embraces the concepts of lean whilst allowing sufficient flexibility for the detailed tools and techniques that could and should be used; it is therefore presented as a pragmatic improvement approach for lean interventions.

Part IV Discussion questions

1. Explain what is meant by taking a business process management (BPM) approach to running an organization outlining the main advantages of adopting BPM successfully. What would be the key components of an implementation plan for BPM?

2. Develop a high level process framework for an organization of your choice identifying the 'value adding' processes and the main support processes. Give a breakdown to the first level sub-processes of one value adding process and one support process.

3. Using an appropriate process modelling technique show the core processes for a company manufacturing and selling fast moving consumer goods. Identify the key inputs and outputs for the processes and explain how you would engage the senior management of the company in the development of the process framework for the business.

4. Explain the basic philosophy behind quality management systems such as those specified in ISO 9000 series. How can an effective quality management system contribute to continuous improvement in an international banking operation?

5. Explain what is meant by independent third part certification to a standard such as ISO 9000 and discuss the merits of such a scheme for an organization.

6. Compare and contrast the role of quality management systems in the following organizations:
   a) a private hospital;
   b) a medium-sized engineering company;
   c) a branch of a major bank.

7. English Aerospace is concerned about its poor quality and delivery performance with the EA847. Considerable penalty costs are incurred if the company fails to meet agreed specifications or delivery dates. As the company's new Chief Quality and Business Improvement Officer you have been asked to lead an improvement programme. Describe the approaches and methodologies you would adopt and list some of the tools and techniques that might be used in a systematic approach.

8. The marketing department of a large chemical company is reviewing its sales forecasting activities. Over the last three years the sales forecasts have been grossly inaccurate. As a result, a process improvement team has been formed to look at this problem. Give an account of how you would advise that team in this situation and outline a programme of work for them to consider.

9. It has been suggested by Deming and Ishikawa that statistical techniques can be used by staff at all levels within an organization. Comment on this view and explain how such techniques could help:
   a) Senior managers to assess performance
   b) Sales staff to demonstrate process capability to customers
   c) Process teams to achieve quality improvement.

10. 'Lean thinking and systems' have been used widely in many sectors to bring about performance improvement. Prepare a presentation on 'Lean' for the senior management team of an organization of your choice, so that they may understand the concept and its building blocks. Recommend an appropriate systematic approach and tool kit. Make proposals on how they should go about implementing Lean in their organization.

Part V

People

The people are the masters.

Edmund Burke, 1729–1797

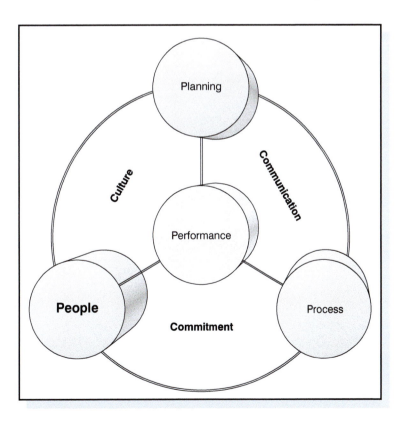

Chapter 16

Human resource management

STRATEGIC ALIGNMENT OF HRM POLICIES

Technological advances and variation in demand for products and services worldwide since the turn of the century has created relative instability, cyclic hiring and downsizing in many organizations. However, during these times the way in which people are managed and developed at work has become recognized as one of the primary keys to improved and sustained organizational performance. This is reflected by popular idioms such as 'people are our most important asset' or 'people make the difference'. Indeed, such axioms now appear in the media and on corporate public relations documents with such regularity that the accuracy and integrity of such assertions has begun to be questioned. This chapter draws on some of the research undertaken by the European Centre *for* Business Excellence (EC*for*BE), the research and education division of Oakland Consulting, which focused on world-class, successful and, in many cases, award-winning organizations. It describes the main people management activities that are currently being used in these leading edge organizations.

There is an overwhelming amount of evidence that successful organizations pay much more than lip service to the claim that people are their most important resource. This is consistent with the recognition that intellectual capital reflects a significant part of any company's value; and that knowledge management (Chapter 17) is a key strategic activity, especially if the tacit knowledge within an organization is to be properly leveraged for the benefit of the company.

On a general level, successful organizations share a fundamental philosophy to value and invest in their employees. More specifically, world-class organizations value and invest in their people through the following activities:

- Strategic alignment of human resource management (HRM) policies
- Effective communication
- Employee empowerment and involvement
- Motivation through recognition of excellence

- Training and development
- Teams and teamwork
- Review and continuous improvement
- The fostering of social cohesion.

It is clear that leading edge organizations adopt a common approach or plan, illustrated in Figure 16.1, to align their HR policies to the overall business strategy.

Key elements of the HR strategy (e.g. skills, recruitment and selection, health and safety, appraisal, employee benefits, remuneration, training etc.) are first identified, usually by the HR director, who then reports regularly to the board and the HR plan, typically spanning three years, is aligned with the overall business objectives and is an integral part of company strategy. For example, if a business objective is to expand at a particular site, then the HR plan provides the necessary additional manpower with the appropriate skills profile and training support. The HR plan is revised as part of the overall strategic planning process. Divisional boards then liaise with the HR director to ensure that the HR plan supports and is aligned with overall policy.

In addition, the HR director holds regular meetings with key personnel from employee relations, health and safety, training and recruitment etc. to review and monitor the HR plan, drawing upon published data and benchmarking activities in all relevant areas of policy and practice. Divisional managing directors and the HR director report progress on how the HR plan is supporting the business to the quality council or board. An overview of this human resource process is illustrated in Figure 16.2.

Although it is beyond the scope of this chapter to make a detailed examination of HR policy, it is prudent to outline briefly some of the common practices that emerged from the identified best practice relating to selection and recruitment, skills and competencies, appraisal, and employee reward, recognition and benefits.

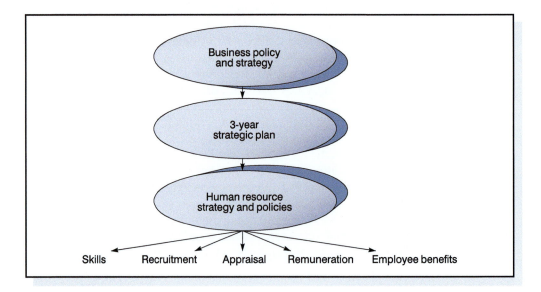

Figure 16.1
Strategic alignment of HRM policies

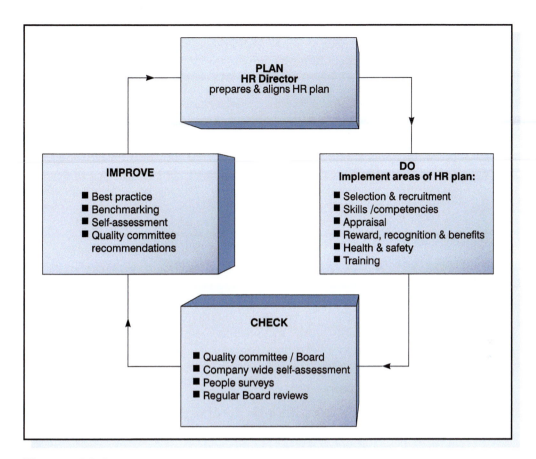

Figure 16.2
Human resource process

Selection and recruitment

The following practices are common amongst the organizations studied regarding selection and recruitment:

1. Ensure fairness by using standard tools and practices for job descriptions and job evaluations.
2. Enhance 'transparency' and communication through jargon-free booklets that provide detailed information to new recruits about performance, appraisal, job conditions and so on.
3. Ensure that job descriptions are responsibility rather than task oriented.
4. Train all managers and supervisors in interviewing and other selection techniques.
5. Align job descriptions and competencies so that people with the appropriate skills and attributes for the job are identified.
6. Compare the organization's employment terms and conditions (on a regular basis) with published data on best practice and documents to ensure the highest standards are being met.

7. Review HR policies regularly to ensure that they fully reflect legislative and regulatory changes together with known best practice.
8. In the recruitment of new graduates start early, at least six months before graduation, as the best students make selections early to pick the best available jobs.

Skills/competencies

Since the publication of *The Competent Manager* (Boyatsis 1982), the terms competence and competency have been widely used and underpin the work of the specific bodies in any country associated with vocational qualifications and occupational standards. In line with this, good organizations have skills / competence-based human resource management policies underpinning selection and recruitment, training and development, promotion and appraisal.

Although numerous lists of generic management competencies have been published, in essence they are all very similar and are closely allied to the core management competencies underpinning HR policies: leadership, motivation, people management skills, team working skills, comprehensive job knowledge, planning and organizational skills, customer focus, commercial and business awareness, effective communication skills – oral and written – and change management skills, coupled with a drive for continuous improvement.

Appraisal process

As with other HR policies, the main thrust of the appraisal process is alignment – of personal, team and corporate goals – coupled with appraisals to help individuals achieve their full potential (see Figure 16.3).

Without exception, the appraisal systems described in world-class organizations are based on objectives. Agreed objectives are also time-based so that completion dates

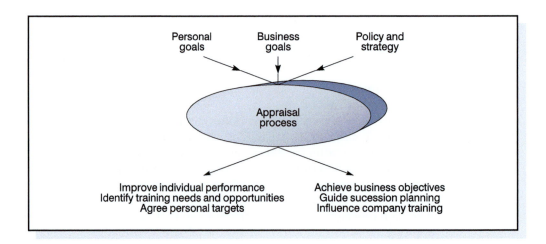

Figure 16.3
The appraisal process

provided the opportunity for automatic review processes. Typically, employees are appraised annually and the managers conducting appraisals attend training in appraisal skills. Before each appraisal, the appraisee and appraiser each complete preparation forms thus making the interview a two-way discussion on performance against objectives during – say – the previous twelve months. Training and development work to achieve the objectives is agreed and, if necessary, additional help is available in the form of advice and counselling.

Employee reward, recognition and benefits

Although an in-depth study of the policies and practices relating to financial reward and recognition was beyond the scope of the research, it is possible to highlight the following activities that were common amongst the organizations:

- Rewards are based on consistent, quality-based performance.
- Awards are given to employees and also to customers, suppliers, universities, colleges, students etc.
- Financial incentives are offered for company-wide suggestions and new idea schemes.
- Internal promotion, for example, from non-supervisory roles to divisional managing directors encourage a highly motivated workforce and enhance job security.
- Commendations include ad hoc recognition for length of service, outstanding contributions, etc.
- Recognition is given through performance feedback mechanisms, development opportunities, pay progressions and bonuses.
- Recognition systems operate at all levels of the organization but with particular emphasis on informal recognition ranging from a personal 'thank you' to recognition at team meetings and events.

With regard to employee benefits, it is well documented that benefits are seen as a tangible expression of the psychological bond between employers and employees. However, to maximize effectiveness benefits packages should be selected on the basis of what is good for the employee as well as the employer. Moreover, when employees can design their own benefits package both they and the company benefit.

Leading edge organizations favour a 'cafeteria' approach to employee benefits and in recent years there has been increasing interest in this idea to maximize flexibility and choice, particularly in the area of fringe benefits, which can make up a high proportion of the total remuneration package. Under such schemes, the company provides a core package of benefits to all employees (including salary) and a 'menu' of other costed benefits (e.g. personal medical care, dental care, company car, health insurance, etc.) from which the employee can select their personal package.

Some of the ideas underpinning cafeteria benefits sit well with the literature on motivation – to emphasize that different individuals have different needs and expectations from work. Moreover, through communicating the benefits package and providing employees with benefit flexibility, the positive impact is further increased; not only are employees more likely to get what benefits they want, but also, communication makes them more aware of the benefits they are gaining thus informing and increasing morale.

Effective communication emerges from the research as an essential facet of people management – be it communication of the organization's goals, vision, strategy and policies or the communication of facts, information and data. For business success, regular, two-way communication, particularly face-to-face with employees, is an important factor in establishing trust and a feeling of being valued. Two-way communication is regarded as both a core management competency and as a key management responsibility. For example, a typical list of management responsibilities for effective communication is to:

- Regularly meet all their people
- Ensure people are briefed on key issues in language free of technical jargon
- Communicate honestly and as fully as possible on all issues which affect their people
- Encourage team members to discuss company issues and give upward feedback
- Ensure issues from team members are fed back to senior managers and timely replies given.

Regular two-way communication also involves customers, shareholders, financial communities and the general public.

Communications process

Successful organizations follow a systematic process for ensuring effective communications as shown in Figure 16.4.

Plan

Typically, the HR function, e.g. the HR director, is responsible for the communication process. He/she assesses the communication needs of the organization and liaises with divisional directors, managers or local management teams to ensure that the communication plans are in alignment with overall policy and strategy. A communication programme accompanies any major changes on organization policy or objectives.

Do

A comprehensive mix of diverse media are used to support effective communication throughout their organization. These include:

Videos	Posters	Open-door policies
Surveys	Campaigns	E-mail
Magazines	Briefings/Blogs	Notice boards
Newsletters	Conferences	Internet/Intranet
Appraisals	Meetings	Focus groups

It is evident that the introduction of electronic systems has brought about radical changes in communications. Typically employees are able to access databases,

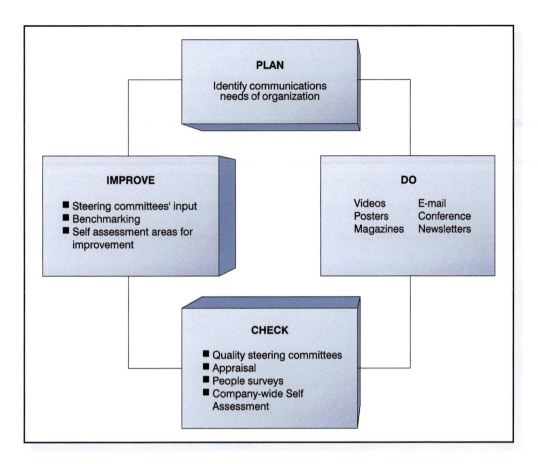

Figure 16.4
Best practice communications process

spreadsheets, word processing, e-mail and diary facilities. Information on business performance, market intelligence and quality issues can also be easily and quickly cascaded. Further, video conferencing is used to facilitate internal face-to-face communications with major customers across the world, resulting in substantial savings on travel and associated costs. Furthermore, provision is made for depots, units, regions, divisions, departments, etc. to hold 'virtual' meetings and conferences. Feedback questionnaires then check that events are valuable and help the planning of future events.

Check

Quality steering or review committees, people surveys, appraisal and company wide self-assessment are used to review the effectiveness of the communications process. Appraisal and staff survey data are analysed to ensure that the communications process is continuing to deliver effective upward, downward and lateral communications. Reports are then made on a quarterly, six-monthly and/or annual basis to the chief executive and/or the most senior team on the effectiveness and relevance of the communications process. The people survey data are also used to ascertain employee perceptions and to keep in touch with current opinion.

Improve

The results of the various review processes highlight areas for improvement and results are verified by benchmarking against, for instance, a national survey. Quality steering committees then put forward recommendations for future planning and continuous improvements.

Communications structure

Successful organizations place great emphasis on communication channels that enable people at all levels in the organization to feel able to talk to each other. Consequently, managers are not only trained but 'are committed to being open-minded, honest, more visible and approachable'. Many formal and informal communication mechanisms exist, all designed to foster an environment of open dialogue, shared knowledge, information and trust in an effective upward, downward, lateral and cross-functional structure such as the one illustrated in Figure 16.5.

(See also Chapter 18 for more detail on the communication process.)

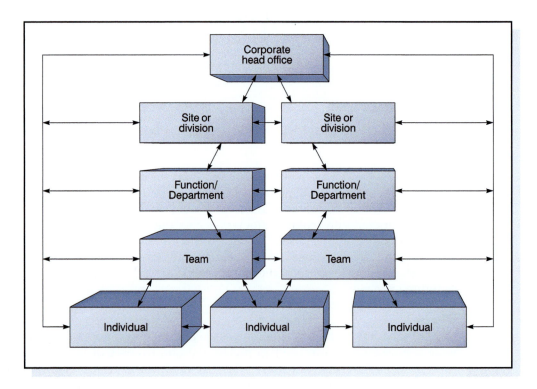

Figure 16.5
Multi-directional communications structure

To encourage employee commitment and involvement, successful organizations place great importance on empowering their employees. The positive effects of employee empowerment are well documented but the notion has been challenged with some writers claiming that it is not possible to empower people – rather, it is possible only to create a climate and a structure in which people will take responsibility. Nonetheless, it is clear that the organizations studied in the research considered empowerment to be a key issue *and* make efforts to create a working environment that is conducive to the employees taking responsibility.

Many companies with impressive customer satisfaction scores, for example, subscribe to the importance of employee empowerment by encouraging employees to:

- Set their own goals
- Judge their own performance
- Take ownership of their actions
- Identify with the company (e.g. to become stock/shareholders).

Similarly, TNT report that 'all employees are empowered to respond to normal and extraordinary situations without further recourse' and that they have 'worked hard to create a no blame culture where our people are empowered to take decisions to achieve their objectives'. Along the same lines, Hewlett-Packard advocate teamwork and high levels of empowerment combined with a strong setting of objectives and freedom for employees to achieve them. To address these issues, management map out processes to provide employees with the necessary authority and skills. In addressing the issue of not caring, employee surveys reveal that appraisal systems can be a major roadblock and the appraisal process may need to be revised.

Common initiatives

There are three common initiatives which successful organizations place great store by:

1. *Corporate employee suggestion schemes* – these provide a formalized mechanism for promoting participative management, empowerment and employee involvement.
2. *Company wide culture change programmes* – in the form of workshops, ceremonies and events used to raise awareness and to empower individuals and teams to practice continuous improvement.
3. *Measurement of key performance indicators (KPIs)* – whereby the effectiveness of staff involvement and empowerment is measured by improvements in human resource key performance indicators (KPIs), such as labour turnover, absenteeism, accident rates and lost time through accidents. Typically KPI measurements, coupled with appraisal feedback and survey results, are regularly reviewed by the HR director who uses the information as the basis for reports and suggestions for improvements to the board.

On a more general level, successful organizations increase commitment by empowering and involving more and more of their employees in formulating plans that shape the business vision. As more people understand the business and where it is planned to go, the more they become involved in and committed to developing the organization's goals and objectives.

Motivation through recognition of excellence

Publicly recognising excellent contributions by individuals and teams is an essential part of a HR program; it is key to motivating people and engendering an ongoing commitment to continuous improvement. This, however, requires that processes be put in place to identify and recognize excellence and that the forums and media for recognizing outstanding achievements be planned rather than ad hoc.

Line managers have the responsibility for mentoring those immediately under them: setting training and development goals and reviewing performance. This also places them in the ideal position to identify outstanding achievements. Guidance needs to be provided by senior HR management, however, and the process of identification has to be regular and ongoing.

TRAINING AND DEVELOPMENT

The training and development of people at work has increasingly come to be recognized as an important part of successful management. Major changes in an organization, the introduction of new technology and widening of ranges of tasks all require training provision. The approaches and systems set down in ISO 9000, benchmarking and self-assessment against frameworks such as the EFQM Excellence Model have further highlighted the need for properly trained employees.

It is widely acknowledged that many writers and practising managers sing the praises of training, saying it is a 'symbol of the employers' commitment to staff', or that it shows an organization's strategy is based on 'adding value rather than lowering costs'. However, others claim that a lack of effective training can still be found in many organizations today and that serious doubts remain as to whether or not management actually *does* invest in the training of their human resources.

It is perhaps not surprising then that research on successful and award winning organizations revealed an on-going commitment to investing in the provision of planned, relevant and appropriate training. In such organizations training was found to be carefully planned through training needs analysis processes that linked the training needs with those of the organization, groups, departments, divisions and individuals. To maintain training relevancy and currency, databases of training courses are widely available and, to encourage diversification, employees are able to realize their full potential by training in quality, job skills, general education, health and safety and so on, through exams, qualifications, assessor training, etc. Typically, training strategies in these organizations require managers to:

- Play an active role in training delivery (cascade training) and support (including quality improvement tools and techniques)

- Receive training and development based on personal development plans
- Fund training and improvement activities to allow autonomy at 'local' levels for short payback investments
- Co-ordinate discussions and peer assessments to develop tailored training plans for individuals.

As a result of their investments these companies boast business benefits such as:

- increases in sales volumes
- not losing customers to competitors
- low employee turnover.

What is particularly noteworthy about the training activities identified is that they are almost identical to those processes and activities commonly found in the management literature on the theory of training. Many writers have developed models of the training process which can be summarized into the four phases shown in Figure 16.6.

The Assessment Phase identifies what is needed (the content of the training) at the organizational, group and individual levels. This may involve some overlap, e.g. an individual's poor sales performance may be a symptom of production problems at the group level, as well as a need for more product innovation at the organizational level. The Assessment Phase thus involves identifying training needs by assessing the gaps between future requirements of a job and the current skills, knowledge or attitudes of the person in the job. So the organization looks at what is presently happening and what should or could be happening. Any differences between the two will give some indications of training needs.

The Planning/Design Phase identifies where and when the training will take place and involves such questions as:

- Who needs to be trained?
- What competencies are required?
- How long will training take?
- What are the expected benefits of training?

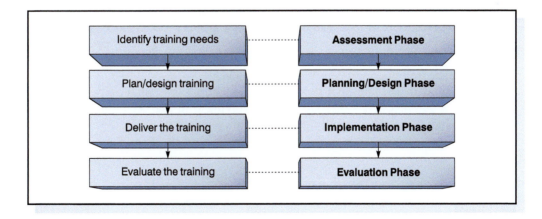

Figure 16.6
A systematic model of training

- Who/How many will undertake the training?
- What resources are needed e.g. money, equipment, accommodation etc.?

Typically, the organizations involved in the studies planned their training programmes, e.g. annually, according to the needs of the business and circulated lists of available courses well in advance of the training dates. The lists ensure all managers are aware of what is provided so that they are able to schedule attendance by staff. The strategic training plan is supported by an annual budgeting and planning system with quarterly meetings to monitor and review performance. The budgeting and planning process with its integral HR element is cascaded throughout the organization to teams at all locations.

The Implementation Phase involves the actual delivery of the training. This might be on site or away from the premises and will include training techniques such as: simulators; business games; case studies; coaching and mentoring; planned experience; computer assisted instruction. Demonstration or 'sitting next to Nellie' is another commonly used training technique.

Induction and devolved training form an integral part of the training implementation phase. New employees attend induction courses and are issued with personal development documents giving details of what training and assessment will take place in the first few months of employment, as well as copies of the vision and mission statements. At regular intervals throughout the induction period, e.g. every three or four weeks, new employees are then reviewed to identify training and development needs for the remainder of the year.

In addition, much of the training is devolved to line managers through facilitation and facilitator packs. This requires the training and development of all levels of managers and supervisors in facilitation skills. Line managers then identify team members to be trained as facilitators. Adopting this approach is said to create an environment in which everyone is aware of training and development issues for themselves and their colleagues.

The Evaluation Phase is widely acknowledged as one of the most critical steps in the training process and can take many forms such as observation, questionnaires, interviews etc. For example, in this phase the overall effectiveness of training is evaluated and this provides feedback for the trainers, for future improvements to the programme, for senior managers and the trainees themselves. Providing trainees with a set of training objectives will help them know what they need to learn and give them feedback on their progress. This will then influence their attitudes towards future training and even the company itself.

In sum, it seems that successful organizations approach training and development in a planned and systematic way involving training needs analysis, assessment of training content, carefully planned implementation and continuous evaluation and review – a convincing argument for the value of theory when it is put into practice.

TEAMS AND TEAMWORK

It is clear that leading edge organizations place great emphasis on the value of people working together in teams. This is hardly surprising as a great deal of theory and research indicates that people are motivated and work better when they are part of a team. Teams can also achieve more through integrated efforts and problem solving.

Teams are a management tool and are most effective when team activity is clearly linked to organizational strategy. For this, the strategy must be communicated to influence team direction, which then links to the production of team mission statements and the use of team agendas and scorecards. Importantly though, many people emphasize the value of cross-functional teams, which proved to be a common feature in many of the organizations studied. Here, teams that have originally evolved out of the old functional departments or units within an organization gain experience and benefit from team building and become cross functional. Each team is required to identify its customers, the customer requirements and what measures need to be used to ensure that those requirements are being satisfied.

The fostering of social cohesion

We live at a time of high employee mobility and low levels of corporate loyalty. It is said that modern employees are increasingly self-interested and that they develop their competencies and CVs so that they can attract the highest remuneration in the market. While there is evidence to support that many employees do have such a global view, there is a counter cyclical trend among employees in world-class companies. These people identify with their work organizations and enjoy being with their colleagues. They do not seek to move at the first opportunity because they enjoy where they are; they receive recognition and they are given challenges.

All the factors in this section combine to form a highly satisfying work environment; however, social cohesion is of equal importance to those already discussed. World-class companies take the same planned and structured approach to social activity as they do to everything else they do. (See also Chapter 17 for more detail on teams and teamwork).

ORGANIZING PEOPLE FOR QUALITY

In some organizations management systems are still viewed in terms of the internal dynamics between marketing, design, sales, production/operations, distribution, accounting, etc. A change is required from this to a larger process-based system that encompasses and integrates the business interests of customers and suppliers. Management needs to develop an in-depth understanding of these relationships and how they may be used to cement the partnership concept. A quality function can be the organization's focal point in this respect, and should be equipped to gauge internal and external customers' expectations and degree of satisfaction. It should also identify deficiencies in all business functions and processes, and promote improvements.

The role of the quality function is to make quality an inseparable aspect of every employee's performance and responsibility. The transition in many companies from quality departments with line functions will require careful planning, direction and monitoring. Quality professionals have developed numerous techniques and skills, focused on product or service quality. In many cases there is a need to adapt these to broader, process applications. The first objectives for many 'quality managers' will be to gradually disengage themselves from line activities, which will then need to be dispersed throughout the appropriate operating departments. This should allow quality to be understood as a 'process' at a senior level and to be concerned with the following throughout the organization:

- Encouraging and facilitating improvement
- Monitoring and evaluating the progress of improvement
- Promoting 'partnerships' in relationships with customers and suppliers
- Designing, planning, managing, auditing and reviewing quality management systems
- Planning and providing training and counselling or consultancy
- Giving advice to management on:
  a) Establishment of process management and control
  b) Relevant statutory/legislation requirements with respect to quality
  c) Quality and process improvement programmes
  d) Inclusion of quality elements in all processes, job instructions and procedures.

Quality directors and managers may have an initial task, however, to help those who control the means to implement this concept – the leaders of industry and commerce – to really believe that quality must become an integral part of all the organization's operations.

The author has a vision of quality as a strategic business management function that will help organizations to change their cultures. To make this vision a reality, quality professionals must expand the application of quality concepts and techniques to all business processes and functions, and develop new forms of providing assurance of quality at every supplier-customer interface. They will need to know the entire cycle of products or services, from concept to the *ultimate* end user. An example of this was observed in the case of a company manufacturing pharmaceutical seals, whose customer expressed concern about excess aluminium projecting below and round a particular type of seal. This was considered a cosmetic defect by the immediate customer, the Health Service, but a safety hazard by a blind patient – the *customer's customer*. The prevention of this 'curling' of excess metal meant changing practices at the mill that rolled the aluminium – at the *supplier's supplier*. Clearly, the quality professional dealing with this problem needed to understand the supplier's processes and the ultimate customer's needs, in order to judge whether the product was indeed capable of meeting the requirements.

The shifts in 'philosophy' will require considerable staff education in many organizations. Not only must people in other functions acquire quality management skills, but quality personnel must change old attitudes and acquire new skills – replacing the inspection, calibration, specification-writing mentality with knowledge of defect prevention, wide-ranging process-based quality management systems design, audit and improvement. Clearly, the challenge for many quality professionals is not so much making changes in their organization, as recognising the changes required in themselves. It is more than an overnight job to change the attitudes of an 'inspection police' force into those of a consultative, team-oriented improvement resource. This emphasis on prevention and improvement-based systems elevates the role of quality professionals from a technical one to that of general management. A narrow departmental view of quality is totally out of place in an organization aspiring to

TQM, and many quality directors and managers will need to widen their perspective and increase their general knowledge of the business to encompass all facets of the organization.

To introduce the concepts of process management will require not only a determination to implement change but sensitivity and skills in inter-personal relations. This will depend very much of course on the climate within the organization. Those whose management is truly concerned with co-operation and concerned for the people will engage strong employee support for the quality manager or director in his catalytic role in the improvement process. Those with aggressive, confrontational management will create for the quality professional impossible difficulties in obtaining support from the 'rank and file'.

Quality appointments

Perhaps the most interesting initial findings from a recent research project by the author and his colleagues on 'Quality in the 21st Century' (Oakland 2011) is to do with the 'Quality Leaders' perceived to be required:

- They will need first and foremost to be **business people**, with multifunctional experience, who understand and speak the language of business and can relate to those running the business.
- The lack of that type of person in Quality is the root cause of complaints about top management from most quality people – but it is because they are **not the right people**!

Many organizations have realized the importance of the contribution a senior, qualified director of quality, senior vice president of quality and operational excellence, chief quality (and business improvement) officer, or other similar titles, can make to the prevention strategy. Smaller organizations may well feel that the cost of employing a full-time quality manager is not justified, other than in certain very high-risk areas. In these cases a member of the management team may be appointed to operate on a part-time basis, performing the quality management function in addition to his/her other duties. To obtain the best results from a quality director/manager, he/she should be given sufficient authority to take necessary action to secure the implementation of the organization's quality policy, and must have the personality to be able to communicate the message to all employees, including staff, management and directors. Occasionally the quality director/manager may require some guidance and help on specific technical quality matters, and one of the major attributes required is the knowledge and wherewithal to acquire the necessary information and assistance.

In large organizations, then, it may be necessary to make several specific appointments or to assign details to certain managers. The following actions may be deemed to be necessary.

Assign a quality director, manager or co-ordinator

This person will be responsible for the planning and implementation of TQM. He or she will be chosen first for process, project and people management abilities rather than detailed knowledge of quality assurance matters. Depending on the size and complexity of the organization, and its previous activities in quality management, the position may be either full or part-time, but it must report directly to the chief executive.

Appoint a quality manager adviser

A professional expert on quality management may be required to advise on the 'technical' aspects of planning and implementing TQM. This is a consultancy role, and may be provided from within or without the organization, full or part-time. This person needs to be a persuader, philosopher, teacher, adviser, facilitator, reporter and motivator. He or she must clearly understand the organization, its processes and interfaces, be conversant with the key functional languages used in the business, and be comfortable operating at many organizational levels. On a more general level this person must fully understand and be an effective advocate and teacher of TQM, be flexible and become an efficient agent of change.

Steering committees and teams

Devising and implementing total quality management in an organization takes considerable time and ability. It must be given the status of a senior executive project. The creation of cost effective performance improvement is difficult, because of the need for full integration with the organization's strategy, operating philosophy and management systems. It may require an extensive review and substantial revision of existing systems of management and ways of operating. Fundamental questions may have to be asked, such as 'do the managers have the necessary authority, capability and time to carry this through?'

Any review of existing management and operating systems will inevitably 'open many cans of worms' and uncover problems that have been successfully buried and smoothed over – perhaps for years. Authority must be given to those charged with following TQM through with actions that they consider necessary to achieve the goals. The commitment will be continually questioned and will be weakened, perhaps destroyed, by failure to delegate authoritatively.

The following steps are suggested in general terms. Clearly, different types of organization will need to make adjustments to the detail, but the component parts are the basic requirements.

A disciplined and systematic approach to continuous improvement may be established in a Quality or Business Excellence 'Steering Committee or Council' (Figure 16.7). The committee/council should meet at least monthly to review strategy, implementation progress and improvement. It should be chaired by the chief executive, who must attend every meeting – only death or serious illness should prevent him/her being there. Clearly, postponement may be necessary occasionally, but the council should not carry on meeting without the chief executive present. The council members should include the top management team and the chairmen of any 'site' steering committees or process management teams, depending on the size of the organization. The objectives of the council are to:

- Provide strategic direction on quality for the organization.
- Establish plans for quality on each 'site'.

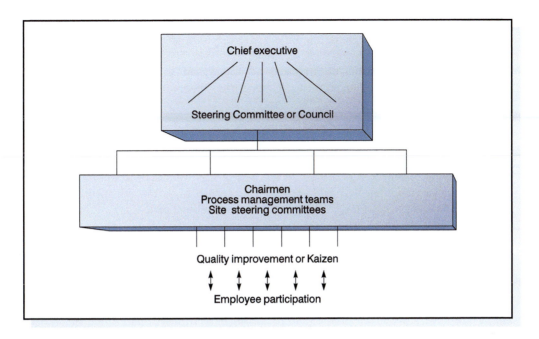

Figure 16.7
Employee participation through the team structure

- Set up and review the process teams that will own the key or critical business processes.
- Review and revise quality plans for implementation.

The process management teams and any site steering committees should also meet monthly, shortly before the senior steering committee/council meetings. Every senior manager should be a member of at least one such team. This system provides the 'top-down' support for employee participation in process management and development. It also ensures that the commitment to quality at the top is communicated effectively through the organization.

The three-tier approach of steering committee, process management teams and quality improvement teams allows the first to concentrate on quality strategy, rather than become a senior problem-solving group. Progress is assured if the team chairmen are required to present a status report at each meeting.

The process management teams or steering committees control all the quality improvement teams and have responsibility for:

- The selection of projects for the teams
- Providing an outline and scope for each project to give to the teams
- The appointment of team members and leaders
- Monitoring and reviewing the progress and results from each team project.

As the focus of this work will be the selection of projects, some attention will need to be given to the sources of nominations. Projects may be suggested by:

a) Steering Committee/Council members representing their own departments, process management teams, their suppliers or their customers, internal and external
b) Quality improvement teams
c) Kaizen teams or quality circles (if in existence)
d) Suppliers
e) Customers.

The process team members must be given the responsibility and authority to represent their part of the organization in the process. The members must also feel that they represent the team to the rest of the organization. In this way the team will gain knowledge and respect and be seen to have the authority to act in the best interests of the organization, with respect to their process.

QUALITY CIRCLES OR KAIZEN TEAMS

No book on TQM would be complete without a mention of Kaizen teams and quality circles. Kaizen is a philosophy of continuous improvement of all the employees in an organization so that they perform their tasks a little better each day. It is a never-ending journey centred on the concept of starting anew each day with the principle that methods can always be improved.

Kaizen Teian is a Japanese system for generating and implementing employee ideas. Japanese suggestion schemes have helped companies to improve quality and productivity, and reduced prices to increase market share. They concentrate on participation and the rates of implementation, rather than on the 'quality' or value of the suggestion. The emphasis is on encouraging everyone to make improvements quickly – there and then.

Kaizen Teian improvements are usually small-scale quick 'hot' solutions, in the worker's own area, and are easy and cheap to implement. Key points are that the objectives are clear, and implementation is rapid, which results in many small improvements that can accumulate to massive total savings and improvements.

One of the most publicized aspects of the Japanese approach to quality has been these quality circles or Kaizen teams. The quality circle may be defined then as a group of workers doing similar work who meet:

• Voluntarily
• Regularly
• In normal working time
• Under the leadership of their 'supervisor'
• To identify, analyse and solve work related problems
• To recommend solutions to management.

Where possible quality circle or Kaizen team members should implement the solutions themselves.

The quality circle concept first originated in Japan in the early 1960s, following a postwar reconstruction period during which the Japanese placed a great deal of emphasis on improving and perfecting their quality control techniques. As a direct result of work carried out to train foremen during that period, the first quality circles were conceived, and the first three circles registered with the Japanese Union of Scientists and Engineers (JUSE) in 1962. Since that time the growth rate has been phenomenal. The concept has spread to Taiwan, the USA and Europe, and circles in many countries have been successful. Many others have failed.

In the early days it was very easy to regard quality circles as the magic ointment to be rubbed on the affected spot, and unfortunately many managers in the West first saw them as a panacea for all ills. There are no panaceas, and to place this concept into perspective, Juran, who was an important influence in Japan's improvement in quality, stated that quality circles represented only 5–10 per cent of the canvas of the Japanese success. The rest is concerned with understanding quality, its related costs and the organization, systems and techniques necessary for achieving customer satisfaction.

Given the right sort of commitment by top management, introduction and environment in which to operate, quality circles can produce the 'shop floor' motivation to achieve quality performance at that level. Circles should develop out of an understanding and knowledge of quality on the part of senior management. They must not be introduced as a desperate attempt to do something about poor quality. The term 'quality circle' may be replaced with a number of acronyms but the basic concepts and operational aspects may be found in many organizations.

The structure of a quality circle or Kaizen organization

The unique feature about quality circles or Kaizen teams is that people are asked to join and not told to do so. Consequently, it is difficult to be specific about the structure of such a concept. It is, however, possible to identify four elements in a circle organization:

- Members
- Leaders
- Facilitators or co-ordinators
- Management.

Members form the prime element of the concept. They will have been taught the basic problem solving and process control approaches and techniques and, hence, possess the ability to identify and solve work related problems.

Leaders are usually the immediate supervisors or foremen of the members. They will have been trained to lead a circle or Kaizen team and bear the responsibility of its success. A good leader, one who develops the abilities of the circle members, will benefit directly by receiving valuable assistance in tackling nagging problems.

Facilitators are the managers of the quality circle or Kaizen programmes. They, more than anyone else, will be responsible for the success of the concept, particularly within an organization. The facilitators must co-ordinate the meetings, the training and energies of the leaders and members, and form the link between the circles and the rest of the organization. Ideally the facilitator will be an innovative industrial teacher, capable of communicating with all levels and with all departments within the organization.

Management support and commitment are necessary to Kaizen and quality circles or, like any other concept, they will not succeed. Management must retain its prerogatives, particularly regarding acceptance or non-acceptance of recommendations, but the quickest way to kill a programme is to ignore a proposal arising from it. One of the most difficult facts for management to accept, and yet one forming the cornerstone of the Kaizen/quality circle philosophy, is that the real 'experts' on performing a task are those who do it day after day.

Training Kaizen teams and quality circles

The training of circle/Kaizen leaders and members is the foundation of all successful programmes. The whole basis of the training component is that the ideas must be easy to take in and be put across in a way that facilitates understanding. Simplicity must be the key word, with emphasis being given to the basic techniques. Essentially there are eight segments of training:

1. Introduction to quality circles or the Kaizen approach, including the 'Blitz'
2. Brainstorming
3. Data gathering and histograms
4. Cause and effect analysis
5. Pareto analysis
6. Sampling and stratification
7. Control charts
8. Presentation techniques.

Managers should also be exposed to some training in the part they are required to play in the Kaizen/quality circle philosophy. Such a programme can be effective only if management believes in it and is supportive and, since changes in management style may be necessary, managers' training is essential.

Operation of quality circles/Kaizen teams

There are no formal rules governing the size of a quality circle/Kaizen team. Membership usually varies from three to fifteen people, with an average of seven to eight. It is worth remembering that, as the circle becomes larger than this it becomes increasingly difficult for all members to participate.

Meetings can be held in the work area or away from it so that members are free from interruptions, and are mentally and physically at ease. If away from the work space, the room should be arranged in a manner conducive to open discussion, and any situation that physically emphasizes the leader's position should be avoided. To a large extent the nature of the problems selected will determine the nature of the meetings, the interval between them and the venue.

Great care is needed to ensure that every meeting is productive, no matter how long it lasts or how frequently is it held. Any of the following activities may take place during a circle meeting:

- Training – initial or refresher
- Problem identification
- Problem analysis
- Preparation and recommendation for problem solution
- Management presentations
- Quality circle/Kaizen team administration.

It is sometimes necessary for quality circles to contact experts in a particular field, e.g. engineers, quality experts, safety officers, maintenance personnel. This communication should be strongly encouraged, and the normal company channels should be used to invite specialists to attend meetings and offer advice. The experts may be considered to be 'consultants', the quality circle/Kaizen team retaining responsibility for improving a process or solving the particular problem. The overriding purpose of quality circles or Kaizen teams is to provide the powerful motivation of allowing people to take some part in deciding their own actions and futures.

Kaizen Blitz events

Rapid benefits realization and effective employee engagement are clear features of the so-called 'Kaizen Blitz' approach which uses 1–2 week duration events, with multi-functional teams focused on single pieces of plant. 'World Class Manufacturing' (WCM) Kaizen methodologies have been developed and these should be tailored to focus on specific problem needs, such as:

- Asset Care & Total Productive Maintenance (TPM)
- Changeover time and quality (perhaps requiring numerous engineering call outs after each changeover).

The first day of Kaizen Blitz events usually covers training on all aspects of WCM, but with emphasis on the priority areas relevant to the task in hand, such as plant reliability. The bulk of the events should be very hands-on; predominantly on the shop floor or in analysis/problem-solving workshops. Using the insights and experience of cross-functional teams is key to gaining insight to solving problems but also to engage the team in the improvement process; breaking them out of the low morale fire-fighting culture so they actually see some benefit being realized through their ideas. Individuals are identified from the outset that should be trained in the Kaizen Blitz methodology so as to quickly remove dependence on external support and build capability in the business in the Kaizen approaches. The focus during the Kaizen Blitz events is implementation of improvement actions, but anything not finished is added to an implementation plan to be managed post-event.

Typical achievements from such Kaizen Blitz events include:

- Improved plant reliability:
  - For example, the fraction of unplanned downtime due to breakdowns was reduced from 16.5 per cent to 6.5 per cent in 4 months in one pharmaceutical company
  - Planned maintenance procedures updated using FMEA risk based approaches including the frequency and depth of maintenance
  - Production procedures improved, including changeover procedures, fault-finding matrices, pre-production check lists and deep clean procedures
  - Technical improvements identified through the Kaizen events.
- Skills and methods transfer:
  - The methodology is usually packaged to become a permanent method in the organization, to be used as required to support future improvement initiatives

- Individuals are fully trained in running the Kaizen events for the site
- Methodology ownership is transferred to manager with supporting improvement engineers or workers
- Plans for rolling out to wider groups.

REVIEW, CONTINUOUS IMPROVEMENT AND CONCLUSIONS

In organizations that achieve outstanding performance and deliver real improvement, processes for reviewing performance and continuous improvement exist at the individual, team, departmental / divisional and organizational levels. These include such processes as:

- Annual staff surveys and subsequent actions, which are viewed as the cornerstones of continuous improvement. The people surveys are also critically reviewed against data from other world-class organizations and benchmarks to determine best practice and feed into the continuous improvement processes.
- Quality committees, the HR department and cross-functional teams drawn from sites, depots, regions, units, divisions review feedback from surveys as well as the format of the surveys.
- Ongoing performance feedback and development through on the job coaching plus regular one-to-one individual and team reviews.

This chapter has highlighted the main people management activities that are currently being used in some world-class organizations. A general conclusion from the research supporting this is that successful organizations pay much more than lip service to the popular idiom 'people are our most important asset'. Indeed successful organizations value and invest in their people in a never-ending quest for effective management and development of their employees. This involves rigorous planning of processes, skilful implementation, regular review of processes and continuous improvement practices.

From a theoretical viewpoint, these findings about people management activities in successful organizations are hardly surprising, since the management literature is strewn with examples of the benefits of systematic planning, followed by strategic implementation, regular review and continuous improvement. Nonetheless, from a practical viewpoint, the real value of the findings is that they flesh out in some detail those people management activities that are being used to good effect in some world-class organizations which are reaping the benefits of putting theory into practice.

ACKNOWLEDGEMENT

The author is grateful to Dr Susan Oakland for the contribution she made to this chapter.

BIBLIOGRAPHY

Blanchard, K. and Herrsey, P., *Management of Organizational Behaviour: Utilizing Human Resources* (4th edn), Prentice-Hall, Englewood Cliffs, NJ, 1982.

Boyatsis, R., *The Competent Manager: A Model for Effective Performance*, Wiley, New York, 1982.

Choppin, J., *Quality Through People: A Blueprint for Proactive Total Quality Management*, Rushmere Wynne, Bedford, 1997.

Collins, J.C. and Porras, J.I., *Built to Last: Successful Habits of Visionary Companies*, Random House, New York, 1998.

Dale, B.G., Cooper, C. and Wilkinson, A., *Total Quality and Human Resources – A Guide to Continuous Improvement*, Blackwell, Oxford, 1992.

Imai, M., *Gemba Kaizen: A Common Sense, Low Cost Approach to Management*, Quality Press, Milwaukee, 1997.

Imai, M., *Kaizen*, McGraw-Hill, New York, 1986.

Kotter, J.P. and Heskett, J.L., *Corporate Culture and Performance*, The Free Press, New York, 1992.

Marchington, M. and Wilkinson, A., *Core Personnel and Development*, Institute of Personnel and Development, London, 1997.

Oakland, J.S., 'Leadership and Policy Deployment: the backbone of TQM', *Total Quality Management & Business Excellence*, Vol. 22, No.5, pp. 517–534, 2011.

Oakland, S. and Oakland, J.S., 'Current people management activities in world-class organizations', *Total Quality Management*, Vol. 12, No. 6, pp. 25–31, 2001.

Ouchi, W., *Theory Z: How American Business can meet the Japanese Challenge*, Addison Wesley, New York, 1985.

Stone, R.J., *Human Resource Management* (4th edn), Wiley, London, 2002.

CHAPTER HIGHLIGHTS

Strategic alignment of HRM policies

- In recent years the way people are managed has been recognized as a key to improving performance. Recent research (ECforBE) on world-class, award winning organizations has identified the main people management activities used in leading edge organizations.
- World-class organizations value and invest in people through: strategic alignment of HRM policies, effective communications, employee empowerment and involvement, training and development, teams and teamwork and review and continuous improvement.
- Leading edge organizations adopt a common approach to aligning HR policies with business strategy. Key elements of policy such as skills, recruitment and selection, training health and safety, appraisal, employee benefits and remuneration, are first identified. The HR plan is then devised as part of the strategic planning process, following a plan, do, check, improve (PDCI) cycle.

Effective communication

- Regular two-way communication, particularly face-to-face, is essential for success.
- Again the PDCI cycle provides a systematic process for ensuring effective

communications, which uses benchmarking and self-assessment as part of the improvement effort.

Employee empowerment and involvement

- To encourage employee commitment and involvement, successful organizations place great importance on empowering employees. This can include people setting own goals, judging own performance, taking ownership of actions and identifying with the organization itself (perhaps as shareholders).
- Common initiatives include: employee suggestion schemes, culture change programmes and measurement of KPIs. Generally commitment is increased by involving more employees in planning and shaping the vision.
- Publicly recognizing excellent contributions by individuals and teams is an essential part of a HR programme.

Training and development

- Training and development has been highlighted by many initiatives as a critical success factor, although lack of effective training still predominates in many organizations.
- In successful organizations, training is planned through needs analysis, use of databases, training delivery at local levels and peer assessments for evaluation.

Teams and teamwork

- Leading edge organizations place great value in people working in teams, because this motivates and causes them to work better.
- Teams are most effective when their activities are clearly linked to the strategy, which in turn is communicated to influence direction. Cross-functional teams are particularly important to address end-to-end processes.
- World-class companies take the same planned and structured approach to social activity as they do to everything else they do.

Organizing people for quality

- The quality function should be the organization's focal point of the integration of the business interests of customers and suppliers into the internal dynamics of the organization.
- Its role is to encourage and facilitate quality and process improvement; monitor and evaluate progress; promote the quality chains; plan, manage, audit and review systems; plan and provide quality training, counselling and consultancy; and give advice to management.
- In larger organizations a quality director will contribute to the prevention strategy. Smaller organizations may appoint a member of the management team to this task on a part-time basis. An external TQM adviser is usually required.
- In devising and implementing TQM for an organization, it may be useful to ask first if the managers have the authority, capability and time to carry it through.

- A disciplined and systematic approach to continuous improvement may be established in a steering committee/council, whose members are the senior management team.
- Reporting to the steering committee are the process management teams or any site steering committees, which in turn control the quality improvement or Kaizen teams and quality circles.

Quality circles or Kaizen teams

- Kaizen is a philosophy of small step continuous improvement, by all employees. In Kaizen teams the suggestions and rewards are small but the implementation is rapid.
- A quality circle or Kaizen team is a group of people who do similar work meeting voluntarily, regularly, in normal working time, to identify, analyse and solve work-related problems, under the leadership of their supervisor. They make recommendations to management. Alternative names may be given to the teams, other than 'quality circles.'
- Rapid benefits realization and effective employee engagement are clear features of the so-called 'Kaizen Blitz' approach which uses short duration events, with multi-functional teams, focused on single problems or pieces of plant.

Review, continuous improvement and conclusions

- Effective organizations use processes for reviewing performance and continuous improvement at the individual, team, divisional/departmental and organizational levels. These include surveys of staff, committees/teams and ongoing performance feedback.

Chapter 17

Culture change through teamwork

THE NEED FOR TEAMWORK

The complexity of most of the processes that are operated in industry, commerce and the services places them beyond the control of any one individual. Furthermore, as any supply chain becomes more fragmented, the interface between organizations gives rise to many of the more significant problems and opportunities. Developing effective solutions requires the involvement of all the stakeholders in the problem areas. The only really efficient way to tackle process management and improvement is through the use of some form of teamwork which has many advantages over allowing individuals to work separately:

- A greater variety of complex processes and problems may be tackled – those beyond the capability of any one individual or even one department or organization– by the pooling of expertise and resources.
- Processes and problems are exposed to a greater diversity of knowledge, skill, experience, and are solved more efficiently.
- The approach is more satisfying to team members, and boosts morale and ownership through participation in process management, problem solving and decision making.
- Processes and problems that cross departmental or functional boundaries can be dealt with more easily, and the potential/actual conflicts are more likely to be identified and solved.
- The recommendations are more likely to be implemented than individual suggestions, as the quality of decision making in *good teams,* is high.
- Processes and problems that cross organizational boundaries can only be effectively dealt with by inter-organizational teams.

Most of these factors rely on the premise that people are willing to support any effort in which they have taken part or helped to develop.

When properly managed and developed, teams improve the process of problem solving, producing results quickly and economically. Teamwork throughout any organization is an essential component of the implementation of TQM and process management, for it builds trust, improves communications and develops interdependence. Much of what has been taught previously in management has led to a culture in the West of independence, with little sharing of ideas and information. Knowledge is very much like organic manure – if it is spread around it will fertilize and encourage growth, if it is kept closed in, it will eventually fester and rot.

Good teamwork changes the independence to interdependence through improved communications, trust and the free exchange of ideas, knowledge, data and information (Figure 17.1). The use of the face-to-face interaction method of communication, with a common goal, develops over time the sense of dependence on each other. This forms a key part of any quality improvement process, and provides a methodology for employee recognition and participation, through active encouragement of group activities.

Teamwork provides an environment in which people can grow and use all the resources effectively and efficiently to make continuous improvements. As individuals grow, the organization grows. It is worth pointing out, however, that employees will not be motivated towards continual improvement in the absence of:

- Commitment from top management
- The right organizational 'climate'
- A mechanism for enabling individual contributions to be effective.

All these are focused essentially at enabling people to feel, accept and discharge responsibility. More than one organization has made this part of their strategy – to

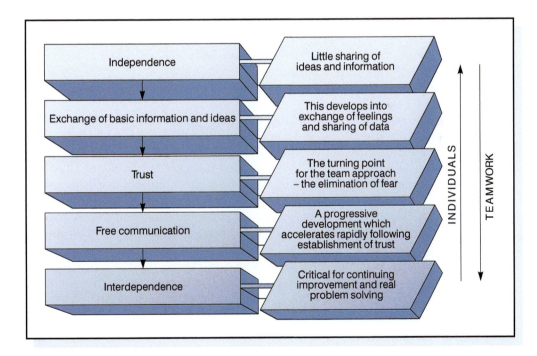

Figure 17.1
Independence to interdependence through teamwork

Culture change through teamwork

'empower people to act'. If one hears from employees comments such as 'We know this is not the best way to do this job, but if that is the way management want us to do it, that is the way we will do it', then it is clear that the expertise existing at the point of operation has not been harnessed and the people do not feel responsible for the outcome of their actions. Responsibility and accountability foster pride, job satisfaction and better work.

Empowerment to act is very easy to express conceptually, but it requires real effort and commitment on the part of all managers and supervisors to put into practice. Recognition that only partially successful but good ideas or attempts are to be applauded and not criticized is a good way to start. Encouragement of ideas and suggestions from the workforce, particularly through their part in team or group activities, requires investment. The rewards are total commitment, both inside the organization and outside through the supplier and customer chains.

Teamwork to support process management and improvement has several components. It is driven by a strategy, needs a structure and must be implemented thoughtfully and effectively. The strategy that drives the improvement comprises the:

- vision and mission of the organization
- critical success factors
- core process framework.

These components have been dealt with in other chapters. The structural and implementation aspects of teamwork are the subject of the remainder of this chapter, to bring together a fully functional *people – process – technology* triangle.

RUNNING PROCESS MANAGEMENT AND IMPROVEMENT TEAMS

Process management and improvement teams are groups of people with the appropriate knowledge, skills and experience who are brought together specifically by management to improve processes and/or tackle and solve particular problems, usually on a project basis. They are cross-functional and often multi-disciplinary.

The 'task force' has long been a part of the culture of many organizations at the 'technology' and management levels. But process teams go a step further; they expand the traditional definition of 'process' to cover the entire end-to-end operating system. This includes technology, software, communication and other units, operating procedures and the process equipment itself. By taking this broader view, the teams can address new problems. The actual running of process teams calls several factors into play:

- Team selection and leadership
- Team objectives
- Team meetings
- Team assignments
- Team dynamics
- Team results and reviews.

Team selection and leadership

The most important element of a process team is its members. People with knowledge and experience relevant to the process or solving the problem are clearly required. However, there should be a limit of five to ten members to keep the team small enough to be manageable but allow a good exchange of ideas. Membership should include appropriate people from groups outside the operational and technical areas directly 'responsible' for the process, if their presence is relevant or essential. In the selection of team members it is often useful to start with just one or two people who are clearly concerned directly with the process. If they try to draw maps or flowcharts (see Chapter 10) of the relevant processes, the requirement to include other people, in order to understand the process and complete the charts, will aid the team selection. This method will also ensure that all those who can make a significant contribution to the process and its improvement are represented. It will be important often to have representation from other companies in the supply chain – selecting the right individual is crucial.

The process owner has a primary responsibility for team leadership, management and maintenance, and his/her selection and training is crucial to success. The leader need not be the highest ranking person in the team, but must be concerned about accomplishing the team objectives (this is sometimes described as 'task concern') and the needs of the members (often termed 'people concern'). Weakness in either of these areas will lessen the effectiveness of the team in solving problems or making breakthroughs. The need for team leadership training is often overlooked; never assume that just because people have been elevated to supervisory or project management roles, they necessarily know how to lead a team – many companies do not train in these basic skills. Skill development may be needed in areas such as facilitation, meeting management and motivation. Needs should be identified and training directed at correcting deficiencies in these crucial aspects.

Team objectives

At the beginning of any process improvement project it is important that the objective should be clearly defined and agreed. This may be in problem or performance improvement terms and it may take some time to define – but agreement is important. Also at the start of every meeting the objectives should be stated as clearly as possible by the leader. This can take a simple form: 'This meeting is to continue the discussion from last Tuesday on the development of our design manual and its trial and adoption throughout the company. Last week we agreed on the overall structure of the manual and today we will look in detail at the structure of the first section.' Project and/or meeting objectives enable the team members to focus thoughts and efforts on the aims, which may need to be restated if the team becomes distracted by other issues.

Team meetings

Meetings need to be seen as a part of a process working towards a longer-term goal – and hence, planning for each meeting and maintaining the continuity between meetings is important. An agenda should be prepared by the leader and distributed to each team member before every meeting. It should include the following information:

- Meeting place, time and how long it will be.
- A list of members (and co-opted members) expected to attend.
- Any preparatory assignments for individual members or groups.
- Any supporting material to be discussed at the meeting.

Early in a project the leader should orient the team members in terms of the approach, methods and techniques they will use to solve the problem. This may require a review of the:

1. Systematic approach, such as *DRIVER* (Chapters 13 and 15)
2. Procedures and rules for using some of the basic tools, e.g. brainstorming – no judgement of initial ideas.
3. Role of the team in the continuous improvement process.
4. Authority of the team.

To make sure that the meeting process is used to maximum advantage it is important that the team leader manages the meeting process; there are several important aspects to this. First of all bear in mind the overall meeting plan (including an approximate timeframe), then for each topic that is addressed:

- maintain the participation of everyone
- maintain focus on the topic being considered
- maintain momentum, keep the process moving forward
- achieve closure, before moving on, capture where the group is up to and where and how it will proceed.

A team secretary should be appointed to take the minutes of the meeting and distribute them to members as soon as possible after each meeting. The minutes should not be overly formal, but reflect decisions and carry a clear statement of the action plans, together with assignments of tasks. They may be hand-written initially, copied and given to team members at the end of the meeting, to be followed later by a more formal document that will be seen by any member of staff interested in knowing the outcome of the meeting. In this way the minutes form an important part of the communication system, supplying information to other teams or people needing to know what is going on and any resulting assignments.

Team assignments

It is never possible to solve problems by meetings alone. What must come out of those meetings is a series of action plans that assign specific tasks to team members. This is the responsibility of the team leader. Agreement must be reached regarding the responsibilities for individual assignments, together with the time scale, and this must be made clear in the minutes. Task assignments must be decided while the team is together and not by separate individuals in after-meeting discussions. Make sure that task assignments are realistic to the time frame and resources available. This may need the allocation of additional resources and the team leader may need to negotiate for this with senior management. The use of RACI – who's *Responsible*, *Accountable*, *Communicated* with, *Involved* – provides a good structure and discipline for meeting outcomes.

Team dynamics

In any team activity the interactions between the members are vital to success. If solutions to problems are to be found, the meetings and ensuing assignments should assist and harness the creative thinking process. This is easier said than done, because many people have either not learned or been encouraged to be innovative. The team leader clearly has a role here to:

- Create a 'climate' for creativity.
- Encourage all team members to speak out and contribute their own ideas or build on others.
- Allow differing points of view and ideas to emerge.
- Remove barriers to idea generation, e.g. incorrect pre-conceptions that are usually destroyed by asking 'Why?'
- Support all team members in their attempts to become creative.
  In addition to the team leader's responsibilities, the members should:
  a) Prepare themselves well for meetings, by collecting appropriate data or information (*facts*) pertaining to a particular problem.
  b) Share ideas and opinions.
  c) Encourage other points of view.
  d) Listen 'openly' for alternative approaches to a problem or issue.
  e) Help the team determine the best solutions.
  f) Reserve judgement until all the arguments have been heard *and* fully understood.
  g) Accept individual responsibility for assignments and group responsibility for the efforts of the team.

Team results and reviews

A process approach to improvement and problem solving is most effective when the results of the work are communicated and acted upon. Regular feedback to the teams, via their leaders, will assist them to focus on objectives, and review progress.

Reviews also help to deal with certain problems that may arise in teamwork. For example, certain members may be concerned more with their own personal objectives than those of the team. This may result in some manipulation of the problem solving process to achieve different goals, resulting in the team splitting apart through self-interest. If recognized, the review can correct this effect and demand greater openness and honesty.

A different type of problem is the failure of certain members to contribute and take their share of individual and group responsibility. Allowing other people to do their work results in an uneven distribution of effort, and leads to bitterness. The review should make sure that all members have assigned and specific tasks, and perhaps lead to the documentation of duties in the minutes. A team roster may even help. If some members of a team are not contributing and cannot be induced to do so, consideration should be given to their replacement. However this can become a more complex issue, if the team leader does not manage team processes well and people believe they are wasting their time. There may be a high level of frustration that could lead to some members withdrawing support.

A third area of difficulty, which may be improved by reviewing progress, is the ready-fire-aim syndrome of action before analysis. This often results from team leaders being too anxious to deal with a problem. A review should allow the problem to be redefined adequately and expose the real cause(s). This will release the trap the team may be in of trying do something before they really know what should be done. The review will provide the opportunity to rehearse the steps in the systematic approach.

TEAMWORK AND ACTION-CENTRED LEADERSHIP

Over the years there has been much academic work on the psychology of teams and on the leadership of teams. Three points on which all authors are in agreement are that teams develop a personality and culture of their own, respond to leadership and are motivated according to criteria usually applied to individuals.

Key figures in the field of human relations, like Douglas McGregor (Theories X & Y), Abraham Maslow (Hierarchy of Needs) and Fred Hertzberg (Motivators and Hygiene Factors), all changed their opinions on group dynamics over time as they came to realize that groups are not the democratic entity that everyone would like them to be, but respond to individual, strong, well-directed leadership, both from without and within the group, just like individuals.

Action Centred Leadership

Several decades ago John Adair, senior lecturer in Military History and the Leadership Training Adviser at the Military Academy, Sandhurst and later assistant director of the Industrial Society, developed what he called the action-centred leadership model, based on his experiences at Sandhurst, where he had the responsibility to ensure that results in the cadet training did not fall below a certain standard. He had observed that some instructors frequently achieved well above average results, owing to their own natural ability with groups and their enthusiasm. He developed this further into a team model, which is the basis of the approach of the author and his colleagues to this subject.

In developing his simple model for teamwork and leadership, which is still one of the best around today, Adair brought out clearly that for any group or team, big or small, to respond to leadership, they need a clearly defined *task*, and the response and achievement of that task are interrelated to the needs of the *team* and the separate needs of the *individual members* of the team (Figure 17.2).

The value of the overlapping circles is that it emphasizes the unity of leadership and the interdependence and multifunctional reaction to single decisions affecting any of the three areas.

Leadership tasks

Drawing upon the discipline of social psychology, Adair developed and applied to training the functional view of leadership. The essence of this he distilled into the three interrelated but distinctive requirements of a leader. These are to define and achieve the job or task, to build up and co-ordinate a team to do this, and to develop and satisfy the individuals within the team (Figure 17.3).

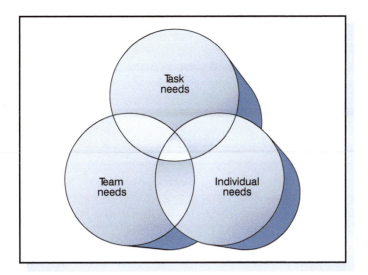

Figure 17.2
Adair's model of
action centred
leadership

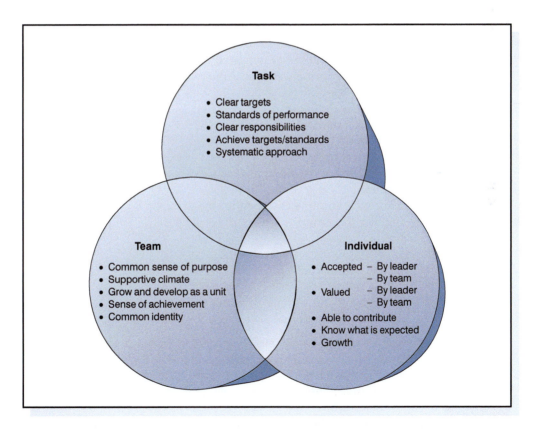

Figure 17.3
The leadership needs

Culture change through teamwork

1. *Task needs.* The difference between a team and a random crowd is that a team has some common purpose, goal or objective, e.g. a football team. If a work team does not achieve the required results or meaningful results, it will become frustrated. Organizations have to make a profit, to provide a service, or even to survive. So anyone who manages others has to achieve results: in operations/production, marketing, selling or whatever. Achieving objectives is a major criterion of success.

2. *Team needs.* To achieve these objectives, the group needs to be held together. People need to be working in a co-ordinated fashion in the same direction. Teamwork will ensure that the team's contribution is greater than the sum of its parts. Conflict within the team must be used effectively; arguments can lead to ideas or to tension and lack of co-operation.

3. *Individual needs.* Within working groups, individuals also have their own set of needs. They need to know what their responsibilities are, how they will be needed, how well they are performing. They need an opportunity to show their potential, take on responsibility and receive recognition for good work.

The task, team and individual functions for the leader are as follows:

a) *Task functions*
- Defining the task.
- Making a plan.
- Allocating work and resources.
- Controlling quality and tempo of work.
- Checking performance against the plan.
- Adjusting the plan.

b) *Team functions*
- Setting standards.
- Maintaining discipline.
- Building team spirit.
- Encouraging, motivating, giving a sense of purpose.
- Appointing sub-leaders.
- Ensuring communication within the group.
- Training the group.

c) *Individual functions*
- Attending to personal problems.
- Praising individuals.
- Giving status.
- Recognising and using individual abilities.
- Training the individual.

The team leader's or facilitator's task is to concentrate on the small central area where all three circles overlap. In a business that is introducing TQM this is the 'action to change' area, where the leaders are attempting to manage the change from *business as usual* to *TQM equals business as usual,* using the cross-functional quality improvement teams at the interface.

In the action area the facilitator's or leader's task is to try to satisfy all three areas of need by achieving the task, building the team and satisfying individual needs. If a leader concentrates on the task, e.g. in going all out for production schedules, while neglecting the training, encouragement and motivation of the team and individuals, (s)he may do very well in the short term. Eventually, however, the team members will give less effort than they are capable of. Similarly, a leader who concentrates only on creating team spirit, while neglecting the task and the individuals, will not receive maximum contribution from the people. They may enjoy working in the team but they will lack the real sense of achievement that comes from accomplishing a task to the utmost of the collective ability.

So the leader/facilitator must try to achieve a balance by acting in all three areas of overlapping need. It is always wise to work out a list of required functions within the context of any given situation, based on a general agreement on the essentials:

- *Planning,* e.g. seeking all available information.
    - Defining group task, purpose or goal.
    - Making a workable plan (in right decision-making framework).
    - *Initiating,* e.g. briefing group on the aims and the plan.
    - Explaining why aim or plan is necessary.
    - Allocating tasks to group members.
    - *Controlling,* e.g. maintaining group standard.
    - Influencing tempo.
    - Ensuring all actions are taken towards objectives.
    - Keeping discussions relevant.
    - Prodding group to action/decision
- *Supporting,* e.g. expressing acceptance of people and their contribution.
    - Encouraging group/individuals.
    - Disciplining group/individuals.
    - Creating team spirit.
    - Relieving tension with humour.
    - Reconciling disagreements or getting others to explore them.
- *Informing,* e.g. clarifying task and plan.
    - Giving new information to the group, i.e. keeping them 'in the picture'.
    - Receiving information from the group.
    - Summarising suggestions and ideas coherently.
- *Evaluating,* e.g. checking feasibility of an idea.
    - Testing the consequences of a proposed solution.
    - Evaluating group performance.
    - Helping the group to evaluate its own performance against standards.

Situational leadership

In dealing with the task, the team and with any individual in the team, a style of leadership appropriate to the situation must be adopted. The teams and the individuals within them will, to some extent, start 'cold', but they will develop and grow in both strength and experience. The interface with the leader must also change with the change in the team, according to the Tannenbaum and Schmidt model (Figure 17.4).

Initially a very directive approach may be appropriate, giving clear instructions to meet agreed goals. Gradually, as the teams become more experienced and have some success, the facilitating team leader will move through coaching and support to

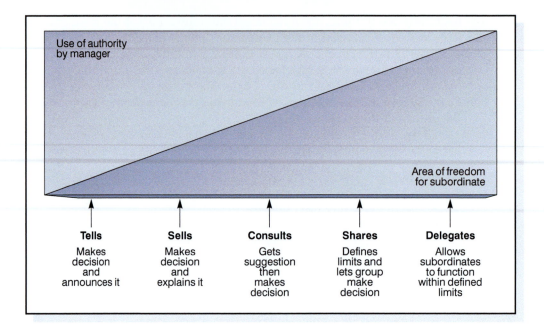

Figure 17.4
Continuum of leadership behaviour

less directing and eventually a less supporting and less directive approach – as the more interdependent style permeates the whole organization.

This equates to the modified Blanchard model in Figure 17.5, where directive behaviour moves from high to low as people develop and are more easily empowered. When this is coupled with the appropriate level of supportive behaviour, a directing style of leadership can move through coaching and supporting to a delegating style. It must be stressed, however, that effective delegation is only possible with developed and capable 'followers', who can be fully empowered.

One of the great mistakes in recent years has been the expectation by management that teams can be put together with virtually no training or development (S1 in Figure 17.5) and that they will perform as a mature team (S4). The Blanchard model emphasizes that there is no quick and easy 'tunnel' from S1 to S4. The only route is the planned climb through S2 and S3.

STAGES OF TEAM DEVELOPMENT

Original work by Tuckman suggested that when teams are put together, there are four main stages of team development, the so-called forming (awareness), storming (conflict), norming (co-operation) and performing (productivity). The characteristics of each stage and some key aspects to look out for in the early stages are given below.

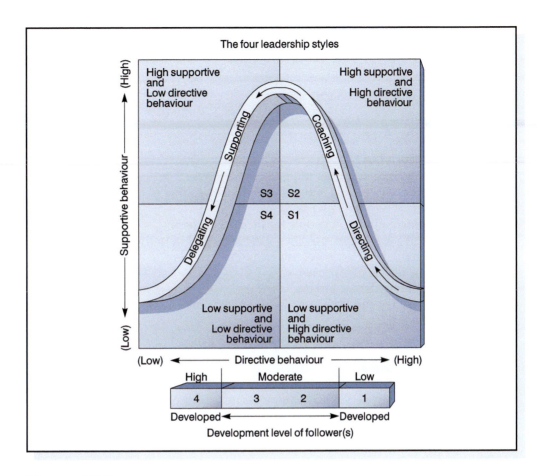

Figure 17.5
Situational leadership – progressive empowerment through TQM

Forming – awareness
Characteristics:
- Feelings, weaknesses and mistakes are covered up.
- People conform to established lines.
- Little care is shown for others' values and views.
- There is no shared understanding of what needs to be done.

  Watch out for:

- Increasing bureaucracy and paperwork.
- People confining themselves to defined jobs.
- The 'boss' is ruling with a firm hand.

Storming – conflict
Characteristics:
- More risky, personal issues are opened up.
- The team becomes more inward-looking.

- There is more concern for the values, views and problems of others in the team.
  Watch out for:
- The team becomes more open, but lacks the capacity to act in a unified, economic, and effective way.

Norming – co-operation

Characteristics:
- Confidence and trust to look at how the team is operating.
- A more systematic and open approach, leading to a clearer and more methodical way of working.
- Greater valuing of people for their differences.
- Clarification of purpose and establishing of objectives.
- Systematic collection of information.
- Considering all options.
- Preparing detailed plans.
- Reviewing progress to make improvements.

Performing – productivity

Characteristics:
- Flexibility.
- Leadership decided by situations, not protocols.
- Everyone's energies utilized.
- Basic principles and social aspects of the organization's decisions considered.

The team stages, the task outcomes and the relationship outcomes are shown together in Figure 17.6. This model which has been modified from Kormanski may be used as a framework for the assessment of team performance. The issues to look for are:

1. How is leadership exercised in the team?
2. How is decision making accomplished?
3. How are team resources utilized?
4. How are new members integrated into the team?

Teams which go through these stages successfully should become effective teams and display the following attributes.

Attributes of successful teams

Clear objectives and agreed goals.
No group of people can be effective unless they know what they want to achieve, but it is more than knowing what the objectives are. People are only likely to be committed to them if they can identify with and have ownership of them – in other words, objectives and goals are agreed by team members.

Often this agreement is difficult to achieve but experience shows that it is an essential prerequisite for the effective group.

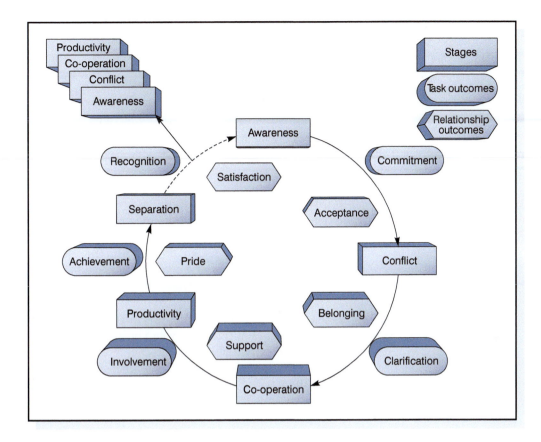

Figure 17.6
Team stages and outcomes

Openness and confrontation

If a team is to be effective, then the members of it need to be able to state their views, their differences of opinion, interests and problems, without fear of ridicule or retaliation. No teams work effectively if there is a cut-throat atmosphere, where members become less willing or able to express themselves openly; then much energy, effort and creativity are lost.

Support and trust

Support naturally implies trust among team members. Where individual group members do not feel they have to protect their territory or job, and feel able to talk straight to other members, about both 'nice' and 'nasty' things, then there is an opportunity for trust to be shown. Based on this trust, people can talk freely about their fears and problems and receive from others help they need to be more effective.

Co-operation and conflict

When there is an atmosphere of trust, members are more ready to participate and are committed. Information is shared rather than hidden. Individuals listen to the ideas of others and build on them. People find ways of being more helpful to each other and the group generally. Co-operation causes high morale – individuals accept each other's strengths and weaknesses and contribute from their pool of knowledge of skill. All abilities, knowledge and experience are fully utilized by the group; individuals have no inhibitions about using other people's abilities to help solve their problems, which are shared.

Allied to this, conflicts are seen as a necessary and useful part of the organizational life. The effective team works through issues of conflict and uses the results to help objectives. Conflict prevents teams from becoming complacent and lazy, and often generates new ideas.

Good decision-making

As mentioned earlier, objectives need to be clearly and completely understood by all members before good decision making can begin. In making decisions effective, teams develop the ability to collect information quickly then discuss the alternatives openly. They become committed to their decisions and ensure quick action.

Appropriate leadership

Effective teams have a leader whose responsibility it is to achieve results through the efforts of a number of people. Power and authority can be applied in many ways, and team members often differ on the style of leadership they prefer. Collectively, teams may come to different views of leadership but, whatever their view, the effective team usually sorts through the alternatives in an open and honest way.

Review of the team processes

Effective teams understand not only the group's character and its role in the organization, but how it makes decisions, deals with conflicts, etc. The team process allows the team to learn from experience and consciously to improve teamwork. There are numerous ways of looking at team processes – use of an observer, by a team member giving feedback, or by the whole group discussing members' performance.

Sound inter-group relationships

No human being or group is an island; they need the help of others. An organization will not achieve maximum benefit from a collection of quality improvement teams that are effective within themselves but fight among each other.

Individual development opportunities

Effective teams seek to pool the skills of individuals, and it necessarily follows that they pay attention to development of individual skills and try to provide opportunities for individuals to grow and learn, and of course have fun.

Once again, these ideas are not new but are very applicable and useful in the management of teams for quality improvements, just as Newton's theories on gravity still apply!

PERSONALITY TYPES AND THE MBTI

No one person has a monopoly of 'good characteristics'. Attempts to list the qualities of the ideal manager, for example, demonstrate why that paragon cannot exist. This is because many of the qualities are mutually exclusive, for example.

Highly intelligent	*v*	Not *too* clever
Forceful and driving	*v*	Sensitive to people's feelings
Dynamic	*v*	Patient
Fluent communicator	*v*	Good listener
Decisive	*v*	Reflective

Although no individual can possess all these and more desirable qualities, a team often does.

A powerful aid to team development is the use of the Myers-Briggs Type Indicator (MBTI). This is based on an individual's preferences on four scales for:

- Giving and receiving 'energy'
- Gathering information
- Making decisions
- Handling the outer world.

Its aim is to help individuals understand and value themselves and others, in terms of their differences as well as their similarities. It is well researched and non-threatening when used appropriately.

The four MBTI preference scales, which are based on Jung's theories of psychological types, represent two opposite preferences:

- *Extroversion – Introversion* – how we prefer to give/receive energy or focus our attention.
- *Sensing – intuition* – how we prefer to gather information.
- *Thinking – Feeling* – how we prefer to make decisions.
- *Judgement – Perception* – how we prefer to handle the outer world.

To understand what is meant by preferences, the analogy of left- and right-handedness is useful. Most people have a preference to write with either their left or their right hand. When using the preferred hand, they tend not to think about it, it is done naturally. When writing with the other hand, however, it takes longer, needs careful concentration, seems more difficult, but with practice would no doubt become easier. Most people *can* write with and use both hands, but tend to prefer one over the other. This is similar to the MBTI psychological preferences: most people are able to use both preferences at different times, but will indicate a preference on each of the scales.

In all, there are eight possible preferences – E or I, S or N, T or F, J or P, i.e. two opposites for each of the four scales. An individual's *type* is the combination and interaction of the four preferences. It can be assessed initially by completion of a simple questionnaire. Hence, if each preference is represented by its letter, a person's type may be shown by a four letter code – there are sixteen in all. For example, ESTJ represents an *extrovert* (E) who prefers to gather information with *sensing* (S), prefers to make decisions by *thinking* (T) and has a *judging* (J) attitude towards the world, i.e. prefers to make decisions rather than continue to collect information. The person with opposite preferences on all four scales would be an INFP, an introvert who prefers intuition for perceiving, feelings or values for making decisions, and likes to maintain a perceiving attitude towards the outer world.

The questionnaire, its analysis and feedback must be administered by a qualified MBTI practitioner, who may also act as external facilitator to the team in its forming and storming stages.

Type and teamwork

With regard to teamwork, the preference types and their interpretation are extremely powerful. The *extrovert* prefers action and the outer world, whilst the *introvert* prefers ideas and the inner world.

Sensing-thinking types are interested in facts, analyse facts impersonally and use a step-by-step process from cause to effect, premise to conclusion. The *sensing-feeling* combinations, however, are interested in facts, analyse facts personally and are concerned about how things matter to themselves and others.

Intuition-thinking types are interested in possibilities, analyse possibilities impersonally and have theoretical, technical or executive abilities. On the other hand, the *intuition-feeling* combinations are interested in possibilities, analyse possibilities personally, and prefer new projects, new truths, things not yet apparent.

Judging types are decisive and planful, they live in orderly fashion, and like to regulate and control. *Perceivers,* on the other hand are flexible, live spontaneously, and understand and adapt readily.

As we have seen, an individual's type is the combination of four preferences on each of the scales. There are sixteen combinations of the preference scales and these may be displayed on a *type table* (Figure 17.7). If the individuals within a team are prepared to share with each other their MBTI preferences, this can dramatically increase understanding and frequently is of great assistance in team development and good team working. The similarities and differences in behaviour and personality can be identified. The assistance of a qualified MBTI practitioner is absolutely essential in the initial stages of this work.

INTERPERSONAL RELATIONS – FIRO-B AND THE ELEMENTS

The FIRO-B (Fundamental Interpersonal Relations Orientation-Behaviour) is a powerful psychological instrument which can be used to give valuable insights into the needs individuals bring to their relationships with other people. The instrument assesses needs for **inclusion, control** and **openness** and therefore offers a framework for understanding the dynamics of interpersonal relationships.

Use of the FIRO instrument helps individuals to be more aware of how they relate to others and to become more flexible in this behaviour. Consequently it enables people to build more productive teams through better working relationships.

Since its creation by William Schutz in the 1950s to predict how military personnel would work together in groups, the FIRO-B instrument has been used throughout the

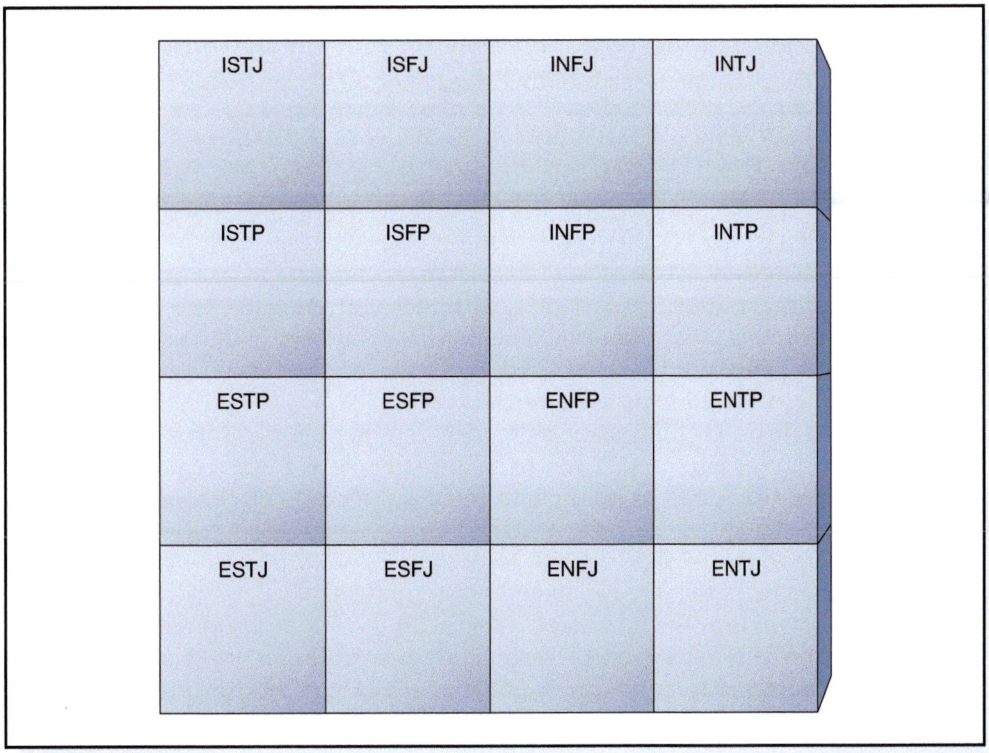

ISTJ	ISFJ	INFJ	INTJ
ISTP	ISFP	INFP	INTP
ESTP	ESFP	ENFP	ENTP
ESTJ	ESFJ	ENFJ	ENTJ

Figure 17.7
MBTI type table form

world by managers and professionals to look at management and decision making styles. Through its ability to predict areas of probable tension and compatibility between individuals, the FIRO-B is a highly effective team building tool which can aid in the creation of the positive environment in which people thrive and achieve improvements in performance.

The theory underlying the FIRO-B incorporates ideas from the work of Adomo, Fromm and Bion and it was first fully described in Schutz's book, *FIRO: A Three Dimensional Theory of Personal Behaviour* (1958). In his more recent book *The Human Element* (1994), Schutz developed the instrument into a series of 'elements', B, F, S, etc. and offers strategies for heightening our awareness of ourselves and others.

The FIRO-B takes the form of a simple-to-complete questionnaire, the analysis of which provides scores that estimate the levels of behaviour with which the individual is comfortable, with regard to his/her needs for inclusion, control and openness. Schutz described these three dimensions in the form of the decisions we make in our relationships regarding whether we want to be:

- 'in' or 'out' – inclusion
- 'up' or 'down' – control
- 'close' or 'distant' – openness

Culture change through teamwork

The FIRO-B estimates our unique level of needs for each of these dimensions of interpersonal interaction.

The instrument further divides each of these dimensions into:

i) the behaviour we feel most comfortable **exhibiting towards** other people – **expressed** behaviours, and

ii) the behaviour we **want from** others – **wanted** behaviours.

Hence, the FIRO-B 'measures', on a scale of 0–9, each of the three interpersonal dimensions in two aspects (Table 17.1).

The **expressed** aspect of each dimension indicates the level of behaviour the individual is most comfortable with towards others, so high scores for the expressed dimensions would be associated with:

High scored expressed behaviours

Inclusion – Makes efforts to include other people in his/her activities – tries to belong to or join groups and to be with people as much as possible.

Control – tries to exert control and influence over people and tell them what to do.

Openness – Makes efforts to become close to people – expresses friendly open feelings, tries to be personal and even intimate.

Low scores would be associated with the opposite expressed behaviour.

The **wanted** aspect of each dimension indicates the behaviour the individual prefers others to adopt towards him/her, so high scores for the wanted dimensions would be associated with:

High scored wanted behaviours

Inclusion – Wants other people to include him/her in their activities – to be invited to belong to or join groups (even if no effort is made by the individual to be included).

Control – Wants others to control and influence him/her and be told what to do.

Openness – Wants others to become close to him/her and express friendly, open, even affectionate feelings.

Low scores would be associated with the opposite wanted behaviours.

It is interesting to look at typical manager FIRO-B profiles, based on their scores for the six dimensions/aspects in Table 17.1. Figure 17.8 shows the average of a sample of 700 middle and senior managers in the UK with boundaries at one sigma, plotted on expressed/wanted scales for the three dimensions.

On average the managers show a higher level of expressed inclusion – including people in his/her activities – than wanted inclusion. Similarly, and not surprisingly perhaps, expressed control – trying to exert influence and control over others – is higher in managers than wanted control. When it comes to openness, the managers tend to want others to be open, rather than be open themselves.

Table 17.1 The FIRO-B interpersonal dimensions and aspects

	Inclusion	Control	Openness
Expressed behaviour	Expressed inclusion	Expressed control	Expressed openness
Wanted behaviour	Wanted inclusion	Wanted control	Wanted openness

Modified from: W. Schutz (1978) *FIRO Awareness Scales Manual*, Palo Alto, CA, Consulting Psychologists Press.

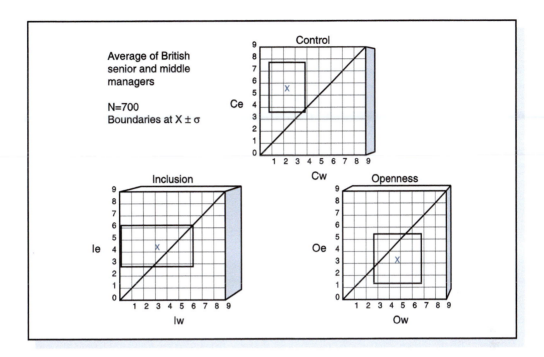

Figure 17.8
Typical manager profiles (FIRO-B)

It is even more interesting to contemplate these results when one considers the demands of total quality management, employee involvement and self-directed teams. These tend to require from managers certain behaviours, for example, lower levels of expressed control and higher levels of wanted control, so that the people feel empowered. Similarly, managers are encouraged to be more open. These, however, are **opposite** to the apparent behaviours of the sample of managers shown graphically in Figure 17.8. It is not surprising then that people believe 'TQM has failed' in some organizations, where managers were being asked to empower employees and be more open – and who can argue against that – yet their basic underlying needs caused them to behave in the opposite way.

Understanding what drives these behaviours is outside the scope of this book but other FIRO and Element instruments can help individuals to further develop understanding of themselves and others. FIRO and Schutz's Elements instruments for measuring **feelings** (F) **self-concept** (S) can deepen the awareness of what lies behind our behaviours with respect to inclusion, control and openness. The reader is advised to undertake further reading and seek guidance from a trained administrator of these instruments, but the overall relationship between the B and F instruments is given below.

Behaviours related to: | **Feelings related to:**
Inclusion | Significance
Control | Competence
Openness | Likeability

Culture change through teamwork

Issues around control behaviour then may arise because of underlying feelings about competence. Similarly, underlying feelings concerning significance may lead to certain inclusion behaviours.

FIRO-B in the workplace

The inclusion, control and openness dimensions form a cycle (Figure 17.9), which can help groups of people to understand how their individual and joint behaviour develops as teams are formed. Given in Table 17.2 are the considerations, questions and outcomes under each dimension. If inclusion issues are resolved first it is possible to progress to dealing with the control issues, which in turn must be resolved if the openness issues are to be dealt with successfully. As a team develops, it travels around the inclusion, control and openness cycle time and time again. If the issues are not resolved in each dimension, further progress in the next dimension will be hindered – it is difficult to deal with issues of control if unresolved inclusion issues are still around and people do not know whether they are 'in' or 'out' of the group. Similarly it is difficult to be open if it is not clear where the power base is in the group.

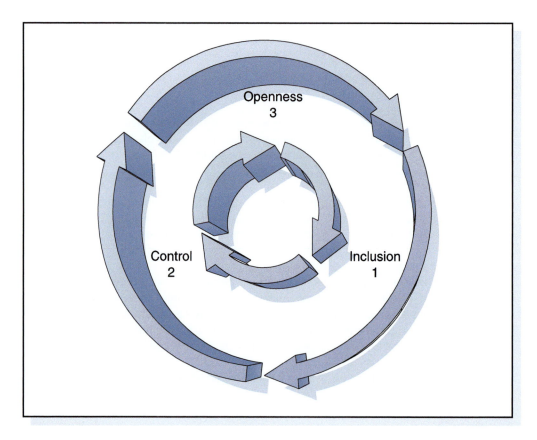

Figure 17.9
The inclusion, control and openness cycle

Table 17.2 Considerations, questions and outcomes for the FIRO-B dimensions

Dimension	Considerations	Some typical questions	If resolved we get:	If not resolved we get:
Inclusion	Involvement – how much you want to include other people in your life and how much attention and recognition you want.	Do I care about this? Do I want to be involved? Does this fit with my values? Do I matter to this group? Can I be committed? . . . leading to . . . . Am I 'in' or 'out'?	A feeling of belonging A sense of being recognized and valued Willingness to become committed	A feeling of alienation A sense of personal insignificance No desire for commitment or involvement
Control	Authority, responsibility, decision making, influence.	Who is in charge here? Do I have power to make decisions? What is the plan? When do we start? What support do I have? What resources do I have? . . . leading to . . . . Am I 'up' or 'down'?	Confidence in self and others Comfort with level of responsibility Willingness to belong	Lack of confidence in leadership Discomfort with level of responsibility – fear of too much – frustration with too little 'Griping' between individuals
Openness	How much are we prepared to express our true thoughts and feelings with other individuals.	Does she like me? Will my work be recognized? Is he being honest with me? How should I show appreciation? Do I appear aloof? . . . leading to. . . Am I 'open' or 'closed'?	Lively and relaxed atmosphere Good-humored interactions Open and trusting relationships	Tense and suspicious atmosphere Flippant or malicious humor Individuals isolated

This I-C-O cycle has lead to the development by the author and his colleagues of an 'openness model' which is in three parts. Part 1 based on the premise that to participate productively in a team individuals must firstly be involved and then committed. Figure 17.10 shows some of the questions which need to be answered and the outcomes from this stage. Part 2 deals with the control aspects of empowerment and management and Figure 17.11 summarizes the questions and outcomes. Finally Part 3, summarized in Figure 17.12, ensures openness through acknowledgement and trust. The full openness cycle (Figure 17.13) operates in a clockwise direction so that trust leads to more involvement, further commitment, increased empowerment, etc. Of course, if progress is not made round the cycle and trust is replaced by fear, it is possible to send the whole process into reverse – a negative cycle of suspicion fault-finding, abdication and confusion (Figure 17.14). Unfortunately this will be recognized as the culture in some organizations where the focus of enquiry is 'what has gone wrong' leading to 'whose fault was it?'

Fortunately, individuals and organizations seem keen to learn ways to change these negative communications that sour relationships, dampen personal satisfaction

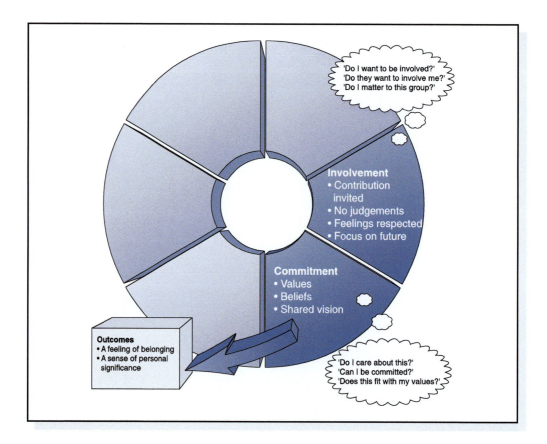

Figure 17.10
The openness model, Part 1 Inclusion: involvement, inviting contribution, responding

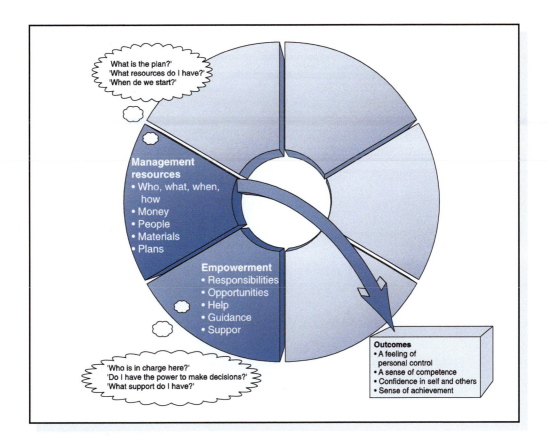

Figure 17.11
The openness model, Part 2 Control: choice, influence, power

and reduce productivity. The inclusion, control openness cycle is a useful framework for helping teams to pass successfully through the forming and storming stages of team development. As teams are disbanded for whatever reason, the process reverses and the first thing which goes is the openness.

At the organizational level, Figure 17.15 shows how the inclusion, control and openness cycle highlights how individual and corporate behavior can interact to stifle improvement. Simple cultural and organizational factors can combine to create conditions whereby the power of the broader workforce to accelerate change and improvement is negated. Increasing the extent to which the workforce is included in the goals and objectives of the change, and empowering them to make change happen, results in greater levels of contribution to improvement. By valuing and recognizing the increased contribution, leaders increase levels of inclusion and reinforce the cycle and the negative spiral is reversed.

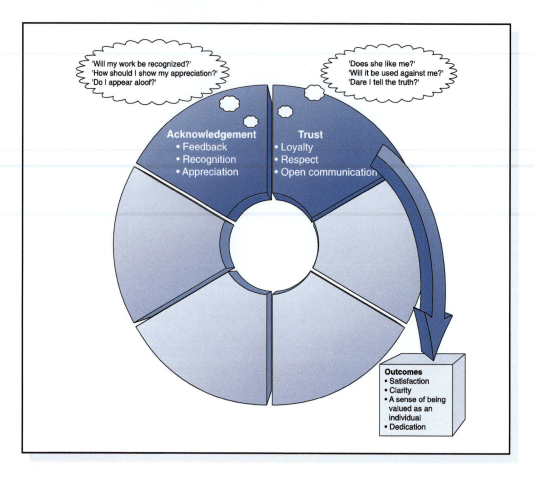

Figure 17.12
The openness model, Part 3 Openness: expression of true thoughts and feelings with respect for self and others

The five 'A' stages for teamwork

The awareness provided by the use of the MBTI and FIRO-B instruments helps people to appreciate their own uniqueness and the uniqueness of others – the foundation of mutual respect and for building positive, productive and high performing teams.

For any of these models or theories to benefit a team, however, the individuals within it need to become **aware** of the theory, e.g., the MBTI or FIRO-B. They then need to **accept** the principles as valid, **adopt** them for themselves in order to **adapt** their behaviour accordingly. This will lead to individual and team **action** (Figure 17.16).

In the early stages of team development particularly, the assistance of a skilled facilitator to aid progress through these stages is necessary. This is often neglected, causing failure in so many team initiatives. In such cases the net output turns out to be lots of nice warm feelings about 'how good that team workshop was a year ago', but the nagging reality that no action came out and nothing has really changed.

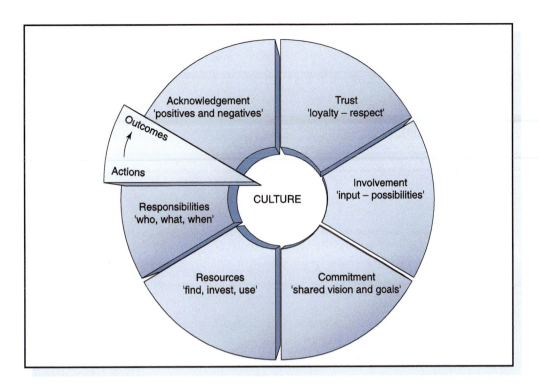

Figure 17.13
The full openness model

Figure 17.14
The negative cycle

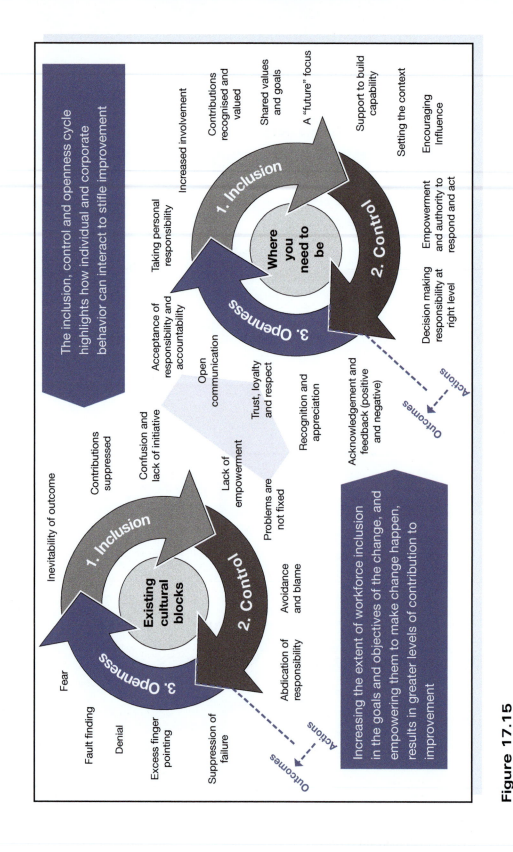

Figure 17.15
The extent to which corporate culture can block improvement

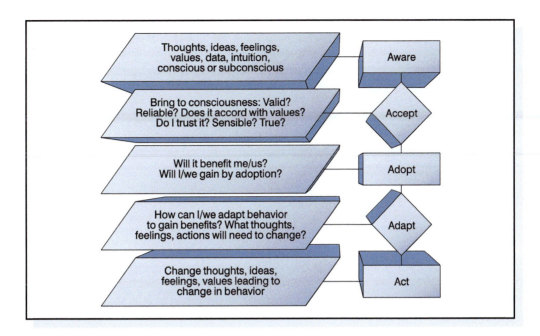

Figure 17.16
The five 'A' stages for teamwork

BIBLIOGRAPHY

Adair, J., *Effective Teambuilding* (2nd edn), Pan Books, London, 1987.

Atkinson, P., *Creating Culture Change: The Key to Successful Total Quality Management*, IFS, Bedford, 1990.

Katzenbach, J.R. and Smith, D.K., *The Wisdom of Teams – Creating the High Performance Organization*, McGraw-Hill/Harvard Business School, Boston, MA, 1994.

Kormanski, C., 'A situational leadership approach to groups using the Tuckman Model of Group Development', *The 1985 Annual: Developing Human Resources*, University Associates, San Diego, CA, 1985.

Kormanski, C. and Mozenter, A., 'A new model of team building: a technology for today and tomorrow', *The 1987 Annual: Developing Human Resources*, University Associates, San Diego, CA, 1987.

Krebs Hirsh, S., *MBTI Team Building Program, Team Member's Guide*, Consulting Psychologists Press, Palo Alto, CA, 1992.

Krebs Hirsh, S. and Kummerow, J.M., *Introduction to Type in Organizational Settings*, Consulting Psychologists Press, Palo Alto, CA, 1987.

McCaulley, M.H., 'How individual differences affect health care teams', *Health Team News*, 1 (8), pp.1–4, 1975.

Myers-Briggs, I., *Introduction to Type: a description of the theory and applications of the Myers-Briggs Type Indicator*, Consulting Psychologists Press, Palo Alto, 1987.

Scholtes, P.R., *The Team Handbook, Joiner Associates*, Madison, NY, 1990.

Culture change through teamwork

Schutz, W., *FIRO: A Three-Dimensional Theory of Interpersonal Behaviour*, Mill Valley WSA, CA, 1958.

Schutz, W., *FIRO: Awareness Scales Manual*, Consulting Psychologists Press, Palo Alto, CA, 1978.

Schutz, W., *The Human Element – Productivity, Self-esteem and the Bottom Line*, Jossey-Bass, San Francisco, CA, 1994.

Tannenbaum, R. and Schmidt, W.H., 'How to choose a leadership pattern', *Harvard Business Review*, May–June, 1973.

Tuckman, B.W. and Jensen, M.A., 'States of small group development revisited', *Group and Organizational Studies*, 2 (4), pp. 419–427, New York, 1977.

Wilkinson, A. and Willmott, H (eds), *Making Quality Critical – New Perspectives on Organizational Change*, Routledge, London, 1995.

CHAPTER HIGHLIGHTS

The need for teamwork

- The only efficient way to tackle process improvement or complex problems is through teamwork. The team approach allows individuals and organizations to grow.
- Within fragmented supply chains there is often a need for effective teams that cross organizational boundaries.
- Employees will not engage continual improvement without commitment from the top, a quality 'climate' and an effective mechanism for capturing individual contributions.
- Teamwork for quality improvement is driven by a strategy, needs a structure and must be implemented thoughtfully and effectively.

Running process management and improvement teams

- Process management and improvement teams are groups brought together by management to improve a process or tackle a particular problem on a project basis. The running of these teams involves several factors: selection and leadership, objectives, meetings, assignments, dynamics, results and reviews.
- The need for training in the basic skills of team leadership should not be underestimated if successful outcomes are sought.

Teamwork and action-centred leadership

- Early work in the field of human relations by McGregor, Maslow and Hertzberg was useful to John Adair in the development of his model for teamwork and action-centred leadership.
- Adair's model addresses the needs of the task, the team and the individuals in the team, in the form of three overlapping circles. There are specific task, team and individual functions for the leader, but (s)he must concentrate on the small central overlap area of the three circles.
- The team process has inputs and outputs. Good teams have three main attributes: high task fulfilment, high teams maintenance and low self-orientation.
- In dealing with the task, the team and its individuals, a situational style of leadership must be adopted. This may follow the Tannenbaum and Schmidt, and Blanchard models through directing, coaching and supporting to delegating.

Stages of team development

- When teams are put together, they pass through Tuckman's forming (awareness), storming (conflict), norming (co-operation) and performing (productivity) stages of development.
- Teams that go through these stages successfully become effective and display clear objectives and agreed goals, openness and confrontation, support and trust, co-operation and conflict, good decision-making, appropriate leadership, review of the team processes, sound relationships and individual development opportunities.

Personality types and the MBTI

- A powerful aid to team development is provided by the Myers-Briggs Type Indicator (MBTI).
- The MBTI is based on individuals' preferences on four scales for giving and receiving 'energy' (extroversion-E or introversion-I), gathering information (sensing-S or intuition-N), making decisions (thinking-T or feeling-F) and handling the outer world (judging-J or perceiving-P).
- An individual's type is the combination and interaction of the four scales and can be assessed initially by completion of a simple questionnaire. There are sixteen types in all, which may be displayed for a team on a type table.

Interpersonal relations – FIRO-B and the elements

- The FIRO-B (Fundamental Interpersonal Relations Orientation – Behaviour) instrument gives insights into the needs individuals bring to their relationships with other people.
- The FIRO-B questionnaire assesses needs for inclusion, control and openness, in terms of expressed and wanted behaviour.
- Typical manager FIRO-B profiles conflict with some of the demands of TQM and can, therefore, indicate where particular attention is needed to achieve successful TQM implementation.
- The inclusion, control, and openness dimensions form an 'openness' cycle which can help groups to understand how to develop their individual and joint behaviours as the team is formed. An alternative negative cycle may develop if the understanding of some of these behaviours is absent.
- Increasing the extent to which the workforce is included in the goals & objectives of change and empowering them to make change happen results in greater levels of contribution to improvement; by valuing and recognizing the increased contribution, leaders increase levels of inclusion and reinforce the cycle and the negative spiral is reversed.
- The five As: for any of the teamwork models and theories, the individuals must become aware, need to accept, adopt and adapt, in order to act. A skilled facilitator is always necessary.

Chapter 18

Communications, innovation and learning

COMMUNICATING THE QUALITY STRATEGY

People's attitudes and behaviour clearly can be influenced by communications; one only has to look at the media or advertising to understand this. The essence of changing attitudes is to gain acceptance for the need to change, and for this to happen it is essential to provide relevant information, convey good practices and generate interest, ideas and awareness through excellent communication processes. This is possibly the most neglected part of many organizations' operations, yet failure to communicate effectively creates unnecessary problems, resulting in confusion, loss of interest and eventually in declining quality through apparent lack of guidance and stimulus.

Total quality management will significantly change the way many organizations operate and 'do business'. This change will require direct and clear communication from the top management to all staff and employees, to explain the need to focus on processes. Everyone will need to know their roles in understanding processes and improving their performance.

Whether a strategy is developed by top management for the direction of the business/organization as a whole, or specifically for the introduction of TQM, that is only half the battle. An early implementation step must be the clear widespread communication of the strategy.

An excellent way to accomplish this first step is to issue a total quality message that clearly states top management's commitment to quality and outlines the role everyone must play. This can be in the form of a quality policy (see Chapters 3 and 4) or a specific statement about the organization's intention to integrate quality into the business operations. Such a statement might read:

> The Board of Directors (or appropriate title) believe that the successful implementation of total quality management is critical to achieving and

maintaining our business goals of leadership in quality, delivery and price competitiveness.

We wish to convey to everyone our enthusiasm and personal commitment to the total quality approach, and how much we need your support in our mission of business improvement. We hope that you will become as convinced as we are that business and process improvement is critical for our survival and continued success.

We can become a total quality organization only with your commitment and dedication to improving the processes in which you work. We will help you by putting in place a programme of education, training and teamwork development, based on business and process improvement, to ensure that we move forward together to achieve our business goals.

The quality director or 'TQM co-ordinator' should then assist the senior management team to prepare a directive. This must be signed by all business unit, division or process leaders, and distributed to everyone in the organization. The directive should include the following:

- Need for improvement.
- Concept for total quality.
- Importance of understanding business processes.
- Approach that will be taken and people's roles.
- Individual and process group responsibilities.
- Principles of process measurement.

The systems for disseminating the message should include all the conventional communication methods of seminars, departmental meetings, posters, newsletters, intranet, etc. First line supervision will need to review the directive with all the staff, and a set of suitable questions and answers may be prepared in support.

Once people understand the strategy, the management must establish the infrastructure (see Chapter 13). The required level of individual commitment is likely to be achieved, however, only if everyone understands the aims and benefits of TQM, the role they must play and how they can implement process improvements. For this understanding a constant flow of information is necessary, including:

1. When and how individuals will be involved.
2. What the process requires.
3. The successes and benefits achieved.

The most effective means of developing the personnel commitment required is to ensure people know what is going on. Otherwise they will feel left out and begin to believe that TQM is not for them, which will lead to resentment and undermining of the whole process. The first line of supervision again has an important part to play in ensuring key messages are communicated and in building teams by demonstrating everyone's participation and commitment.

In the Larkins' excellent book *Communicating Change* (1994), the authors referred to three 'facts' regarding the best ways to communicate change to employees.

1. Communicate directly to supervisors (first-line).
2. Use face-to-face communication.
3. Communicate relative performance of the local work area.

The language used at the 'coal face' will need attention in many organizations. Reducing the complexity and jargon in written and spoken communications will facilitate comprehension. When written business communications cannot be read or understood easily, they receive only cursory glances, rather than the detailed study they require. *Simplify and shorten* must be the guiding principles. The communication model illustrated in Figure 18.1 indicates the potential for problems through environmental distractions, mis-matches between sender and receiver (or more correctly, decoder) in terms of attitudes – towards the information and each other – vocabulary, time pressures, etc.

All levels of management should introduce and stress 'open' methods of communication, by maintaining open offices, being accessible to staff/employees and taking part in day-to-day interactions and the detailed processes. This will lay the foundation for improved interactions *between* staff and employees, which is essential for information flow and process improvement. Opening these lines of communication may lead to confrontation with many barriers and much resistance. Training and the behaviour of supervisors/managements should be geared to helping people accept responsibility for their own behaviour, which often creates the barriers, and for breaking the barriers down by concentrating on the process rather than 'departmental' needs.

Resistance to change will always occur and is to be expected. Again first line management should be trained to help people deal with it. This requires an understanding of the dynamics of change and the support necessary – not an obsession with forcing people to change. Opening up lines of communication through a previously closed system, and publicising people's efforts to change and their results, will aid the process. Change can be – even should be – exciting if employees start to share their development, growth, suggestions and questions. Management needs to encourage and participate in this by creating the most appropriate communication systems.

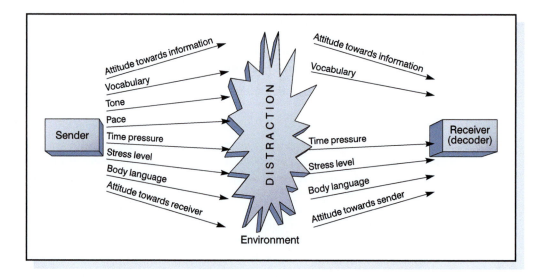

Figure 18.1
Communication model

The people in most organizations fall into one of four 'audience' groups, each with particular general attitudes towards TQM:

- *Senior managers,* who should see TQM as an opportunity, both for the organization and themselves.
- *Middle managers,* who may see TQM as another burden without any benefits, and may perceive a vested interest in the status quo.
- *Supervisors* (first-line or junior managers), who may see TQM as another 'flavour of the period' or campaign, and who may respond by trying to keep heads down so that it will pass over.
- *Other employees,* who may not care, so long as they still have jobs and get paid, though these people must be the custodians of the delivery of quality to the customer and own that responsibility.

Senior management needs to ensure that each group sees TQM as being beneficial to them. Total quality training material and support (whether internal from a quality director and team or from external consultants) will be of real value only if the employees are motivated to respond positively to them. The implementation strategy then must be based on two mutually supporting aspects:

1. 'Marketing' any TQM initiatives.
2. A positive, logical process of communication designed to motivate.

There are of course a wide variety of approaches to, and methods of, TQM and business improvement. Any individual organization's quality strategy must be designed to meet the needs of its own structure and business, and the state of commitment to continuous improvement activities. These days very few organizations are starting from a green-field site. The key is that groups of people must feel able to 'join' the quality process at the most appropriate point for them. For middle managers to be convinced that they must participate, TQM must be presented as the key to help them turn the people who work for them into 'total quality employees'.

The noisy, showy, hype-type activity is not appropriate to any aspect of TQM. TQM 'events' should of course be fun, because this is often the best way to persuade and motivate, but the value of any event should be judged by its ability to contribute to understanding and the change to TQM. Key words in successful exercises include 'discovery', affirmation, participation and team-based learning. In the difficult area of dealing with middle and junior managers, who can and will prevent change with ease and invisibility, the recognition that progress must change from being a threat to a promise will help. In any workshops designed for them, managers and supervisors should be made to feel recognized, not victimized and the programmes should be delivered by specially trained people. The environment and conduct of the workshops must also demonstrate the organization's concern for quality.

The key medium for motivating the employees and gaining their commitment to quality is face-to-face communication and *visible* management commitment. Much is written and spoken about leadership, but it is mainly about communication. If people are good leaders, they are invariably good communicators. Leadership is a human interaction depending on the communications between the leaders and the followers. It calls for many skills that can be *learned* from education and training, but must be *acquired* through practice.

COMMUNICATION, LEARNING, EDUCATION AND TRAINING

It may be useful to consider why people learn. They do so for several reasons, some of which include:

a) Self-betterment
b) Self-preservation
c) Need for or to take responsibility
d) To become properly accountable
e) Saving time or effort
f) Sense of achievement
g) Pride of work
h) Curiosity.

So communication and training can be a powerful stimulus to personal development at the workplace, as well as achieving improvements for the organization. This may be useful in the selection of the appropriate method(s) of communication, the principal ones being:

- *Verbal communication* either between individuals or groups, using direct or indirect methods, such as public address and other broadcasting systems and recordings.
- *Written communication* in the form of notices, bulletins, information sheets, reports, e-mail and recommendations.
- *Visual communication* such as posters, films, video, internet/intranet, exhibitions, demonstrations, displays and other promotional features. Some of these also call for verbal and written communication.
- *Example,* through the way people conduct themselves and adhere to established working codes and procedures, through their effectiveness as communicators and ability to 'sell' good practices.

The characteristics of each of these methods should be carefully examined before they are used in helping people to learn.

It is the author's belief that education and training is the single most important factor in actually improving quality and business performance, once there has been commitment to do so. For education and training to be the effective, however, it must be planned in a systematic and objective manner to provide the right sort of learning experience. Education and training must be continuous to meet not only changes in technology but also changes in the environment in which an organization operates, its structure and perhaps most important of all the people who work there.

Education and training cycle of improvement

Education and training activities can be considered in the form of a cycle of improvement (Figure 18.2), the elements of which are the following.

Ensure education and training is part of the policy

Every organization should define its policy in relation to education and training The policy should contain principles and goals to provide a framework within which learning experiences may be planned and operated. This policy should be communicated to all levels.

Establish objectives and responsibilities for education and training

When attempting to set education and training objectives three essential requirements must be met:

1. Senior management must ensure that learning outcomes are clarified and priorities set.
2. The defined education and training objectives must be realizable and attainable.

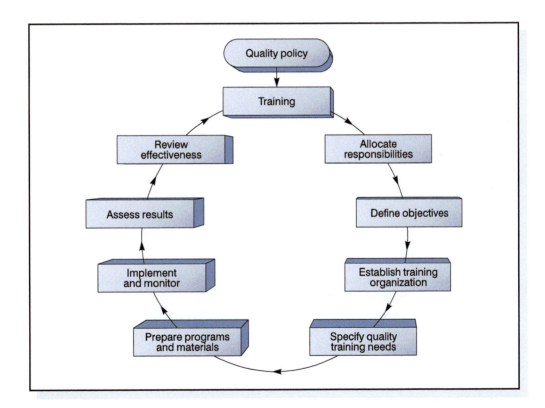

Figure 18.2
The quality training cycle

3. The main objectives should be 'translated' for all areas in the organization. Large organizations may find it necessary to promote a phased plan to identify these.

The following questions are useful first steps when identifying education and training objectives:

- How are the customer requirements transmitted through the organization?
- Which areas need improved performance?
- What changes are planned for the future?
- What are the implications for the process framework?

Education and training must be the responsibility of line management, but there are also important roles for the individuals concerned.

Establish the platform for a learning organization

The overall responsibility for seeing that education and training is properly organized must be assumed by one or more designated senior executives. All managers have a responsibility for ensuring that personnel reporting to them are properly trained and competent in their jobs. This responsibility should be written into every manager's job description. The question of whether line management requires specialized help should be answered when objectives have been identified. It is often necessary to use specialists, who may be internal or external to the organization.

Specify education and training needs.

The next step in the cycle is to assess and clarify specific education and training needs. The following questions need to be answered:

a) Who needs to be educated/trained?
b) What competencies are required?
c) How long will the education/training take?
d) What are the expected benefits?
e) Is the training need urgent?
f) How many people are to be educated/trained?
g) Who will undertake the actual education/training?
h) What resources are needed, e.g. money, people, equipment, accommodation, outside resources?

Prepare education/training programmes and materials.

Senior management should participate in the creation of overall programmes, although line managers should retain the final responsibility for what is implemented, and they will often need to create the training programmes themselves.

Training programmes should include:

- The training objectives expressed in terms of the desired behaviour.
- The actual training content.
- The methods to be adopted.
- Who is responsible for the various sections of the programme.

Implement and monitor education and training

The effective implementation of education and training programmes demands considerable commitment and adjustment by the trainers and trainees alike. Training is a progressive process, which must take into account any learning problems of the trainees.

Assess the results

In order to determine whether further education or training is required, line management should themselves review performance when training is completed. However good the training may be, if it is not valued and built upon by managers and supervisors, its effect can be severely reduced.

Review effectiveness of education and training

Senior management will require a system whereby decisions are taken at regular fixed intervals on:

- The policy.
- The education and training objectives.
- The education/training organization.
- The progress towards a learning organization.

Even if the policy remains constant, there is a continuing need to ensure that new education and training objectives are set either to promote work changes or to raise the standards already achieved.

The purpose of management system audits and reviews is to assess the effectiveness of the management effort. Clearly, adequate and refresher training in these methods is essential if such checks are to be realistic and effective. Audits and reviews can provide useful information for the identification of changing quality-training needs.

The education/training organization should be reviewed in the light of the new objectives, and here again it is essential to aim at continuous improvement. Training must never be allowed to become static, and the effectiveness of the organization's education and training programmes and methods must be assessed systematically.

A SYSTEMATIC APPROACH TO EDUCATION AND TRAINING FOR QUALITY

Education and training for quality should have, as its first objective, an appreciation of the personal responsibility for meeting the 'customer' requirements by everyone from the most senior executive to the newest and most junior employee. Responsibility for the training of employees in quality rests with management at all levels and, in particular, the person nominated for the co-ordination of the organization's quality effort. Education and training will not be fully effective, however, unless responsibility for the deployment of the policy rests clearly with the Chief Executive. One objective of this policy should be to develop a *climate* in which everyone is quality conscious and acts with the needs of the customer in mind at all times.

The main elements of effective and systematic quality training may be considered under four broad headings

- Error/defect/problem prevention.
- Error/defect/problem reporting and analysis.
- Error/defect/problem investigation.
- Review.

The emphasis should obviously be on error, defect or problem prevention, and hopefully what is said under the other headings maintains this objective.

Error/defect/problem prevention

The following contribute to effective and systematic training for prevention of problems in the organization:

1. An issued quality policy.
2. A written quality management system.
3. Job specifications that include quality requirements.
4. Effective steering committees, including representatives of both management and employees.
5. Efficient 'housekeeping' standards.
6. Preparation and display of maps, flow diagrams and charts for all processes.

Error/defect/problem reporting and analysis

It will be necessary for management to arrange the necessary reporting procedures, and ensure that those concerned are adequately trained in these procedures. All errors, rejects, defect, defectives, problems, waste, etc. should be recorded and analysed in a way that is meaningful for each organization, bearing in mind the corrective action programmes that should be initiated at appropriate times.

Error/defect/problem investigation

The investigation of errors, defects and problems can provide valuable information that can be used in their prevention. Participating in investigations offers an opportunity for training. The following information is useful for the investigation:

a) Nature of problem.
b) Date, time and place.
c) Product/service with problem.
d) Description of problem.
e) Causes and reasons behind causes.
f) Action advised.
g) Action taken to prevent recurrence.

Review of quality training

Review of the effectiveness of quality training programmes should be a continuous process. However, the measurement of effectiveness is a complex problem. One way of reviewing the content and assimilation of a training course or programme is to monitor behaviour during quality audits. This review can be taken a stage further by comparing employees' behaviour with the objectives of the quality-training programme. Other measures of the training processes should be found to establish the benefits derived.

Education and training records

All organizations should establish and maintain procedures for the identification of education and training needs and the provision of the actual training itself. These procedures should be designed (and documented) to include all personnel. In many situations it is necessary to employ professionally qualified people to carry out specific tasks, e.g. accountants, lawyers, engineers, chemists, etc., but it must be recognized that all other employees, including managers, must have or receive from the company the appropriate education, training and/or experience to perform their jobs. This leads to the establishment of education and training records.

Once an organization has identified the special skills required for each task, and developed suitable education and training programmes to provide competence for the tasks to be undertaken, it should prescribe how the competence is to be demonstrated. This can be by some form of examination, test or certification, which may be carried out in-house or by a recognized external body. In every case, records of personnel qualifications, education, training and experience should be developed and maintained. National vocational qualifications may have an important role to play here.

At the simplest level this may be a record of tasks and a date placed against each employee's name as he/she acquires the appropriate skill through education and training. Details of attendance on external short courses, in-house induction or training schemes complete such records. What must be clear and easily retrievable is the status of training and development of any single individual, related to the tasks that he/she is likely to encounter. Clearly, as the complexity of jobs increases and managerial activity replaces direct manual skill, it becomes more difficult to make decisions on the basis of such records alone. Nevertheless, they should document the basic competency requirements and assist the selection procedure.

STARTING WHERE AND FOR WHOM

Education and training needs occur at four levels of an organization:

- *Very senior management* (strategic decisions-makers).
- *Middle management* (tactical decision-makers or implementers of policy).
- *First level supervision and quality team leaders* (on-the-spot decision-makers).
- *All other employees* (the doers).

Neglect of education/training in any of these areas will, at best, delay the implementation of TQM and the improvements in performance. The provision of training for each group will be considered in turn, but it is important to realize that an integrated programme is required, one that includes follow-up learning-based activities and encourages exchange of ideas and experience.

Very senior management

The chief executive and his team of strategic policy makers are of primary importance, and the role of education and training here is to provide awareness and instil commitment to quality. The importance of developing real commitment must be established and often this can only be done by a free and frank exchange of views between trainers and trainees. This has implications for the choice of the trainers themselves, and the fresh-faced graduate, sent by the 'package consultancy' operator into the lion's den of a boardroom, will not make much impression with the theoretical approach that he or she is obliged to bring to bear. The author recalls thumping many a boardroom table, and using all his experience and whatever presentation skills he could muster, to convince senior managers that without a TQM-based approach they would achieve disappointing outcomes. It is a sobering fact that the pressure from competition and customers has a much greater record of success than enlightenment, although dragging a team of senior managers down to the shop floor to show them the results of poor management was successful on more than one occasion.

Executives responsible for marketing, sales, finance, design, operations, purchasing, personnel, distribution, etc. all need to understand quality. They must be shown how to define the policy and objectives, how to establish the appropriate organization, how to clarify authority and generally how to create the atmosphere in which total quality will thrive. This is the only group of people in the organization that can ensure that adequate resources are provided and directed at:

1. Meeting customer requirements – internally and externally.
2. Setting standards to be achieved – zero failure.
3. Monitoring of quality performance – quality related costs.
4. Introducing a good quality management system – prevention.
5. Implementing process control and improvement methods – SPC, Six Sigma, Lean.
6. Spreading the idea of quality throughout the whole workforce – TQM.

Middle management

The basic objectives of management quality training should be to make managers conscious and anxious to secure the benefits of the total quality effort. One particular 'staff' manager will require special training – the quality manager, who will carry the responsibility for management of the quality management system, including its design, operation and review.

The middle managers should be provided with the technical skills required to design, implement, review and change the parts of the quality management system that will be under their direct operational control. It will be useful throughout the training programmes to ensure that the responsibilities for the various activities in each of the functional areas are clarified. The presence of a highly qualified and experienced quality manager should not allow abdication of these responsibilities, for the internal 'consultant' can easily create not-invented-here feelings by writing out procedures without adequate consultation of those charged with implementation.

Middle management should receive comprehensive training on the philosophy and concepts of teamwork, and the techniques and applications of SPC, Six Sigma, Lean and whatever other approaches are to be used. Without the teams and tools, the quality management system will lie dormant and lifeless. It will relapse into a paper generating system, fulfilling the needs of only those who thrive on bureaucracy. They need to learn how to put this lot together in a planning–process–people–performance value chain that is sustainable for the future.

First-level supervision

There is a layer of personnel in many organizations which plays a vital role in their inadequate performance – 'foremen' and first line supervisors – the forgotten men and women of industry and commerce. Frequently promoted from the 'shop floor' (or recruited as graduates in a flush of conscience and wealth!) these people occupy one of the most crucial managerial roles, often with no idea of what they are supposed to be doing, without an identity and without training. If this behaviour pattern is familiar and is continued, then TQM is doomed.

The first level of supervision is where the implementation of total quality is actually 'managed'. Supervisors' training should include an explanation of the principles of TQM, a convincing exposition on the commitment to quality of the senior management and an explanation of what the quality policy means for them. The remainder of their training needs to be devoted to explaining their role in the operation of the quality management system, teamwork, SPC, etc. and to gaining *their* commitment to the concepts and techniques of total quality.

It is often desirable to involve the middle managers in the training of first line supervision in order to:

- Ensure that the message they wish to convey through their tactical manoeuvres is not distorted.
- Indicate to the first line supervision that the organization's whole management structure is serious about quality, and intends that everyone is suitably trained and concerned about it too. One display of arrogance towards the training of supervisors and the workforce can destroy such careful planning, and will certainly undermine the educational effort.

All other employees

Awareness and commitment at the point of production or service delivery is just as vital as at the very senior level. If it is absent from the latter, the TQM programme will not begin; if it is absent from the shop floor, total quality will not be implemented. The training here should include the basics of quality and particular care should be given to using easy reference points for the explanation of the terms and concepts. Most people can relate to quality and how it should be managed, if they can think about the applications in their own lives and at home. Quality is really such common sense that, with sensitivity and regard to various levels of intellect and experience, little resistance should be experienced.

All employees should receive detailed training on the processes and procedures relevant to their own work. Obviously they must have appropriate technical or 'job' training, but they must also understand the requirements of their customers. This is frequently a difficult concept to introduce, particularly in the non-manufacturing areas, and time and follow-up assistance needs to be given if TQM is to take hold. It is always bad management to ask people to follow instructions without understanding why and where they fit into their own scheme of things.

TURNING EDUCATION AND TRAINING INTO LEARNING

For successful learning, training must be followed up. This can take many forms, but the managers need to provide the lead through the design of improvement projects, coaching and 'surgery' workshops.

In introducing statistical methods of process control, for example, the most satisfactory strategy is to start small and build up a bank of knowledge and experience. Sometimes it is necessary to introduce SPC techniques alongside existing methods of control (if they exist), thus allowing comparisons to be made between the new and old methods. When confidence has been established from these comparisons, the SPC methods will almost take over the control of the processes themselves. Improvements in one or two areas of the organization's operations, by means of this approach, will quickly establish the techniques as reliable methods of controlling quality, and people will learn how to use them effectively.

The author and his colleagues have found that a successful formula is the in-company training course(s) plus coaching and follow-up workshops. A properly constructed and managed coaching approach is needed to guide staff through projects and ensure the techniques learnt in the 'classroom' are directly applied to 'live' improvement situations. Usually a workshop or seminar is followed within a few weeks by 'surgery' workshop at which participants on the initial training course(s) present the results of their efforts to improve processes, and use the various methods. The presentations and specific implementation problems are discussed. A series of such workshops will add continually to the follow-up, and can be used to initiate process or quality improvement teams. Wider organizational presence and activities are then encouraged by the follow-up activities.

Information and knowledge

Information and knowledge are two words used very frequently in organizations, often together in the context of 'Knowledge Management' and 'Information Technology', but how well are they managed and what is their role in supporting TQM?

Recent researchers and writers on knowledge management have drawn attention to the distinction between explicit knowledge – that which we can express to others – and tacit knowledge – the rest of our knowledge, which we cannot easily communicate in words or symbols.

If much of our knowledge is tacit, perhaps we do not fully know what we know and it can be very difficult to explain or communicate what we know. Explicit knowledge can be put into a form that we can communicate to others – the words, figures and models in this book are an example of that. In many organizations, how-ever, especially the service sector, much of people's valuable and useful knowledge is tacit rather than explicit.

The creation and expression of knowledge takes place through social interaction between tacit and explicit knowledge and the matrix in Figure 18.3 shows this as four modes of knowledge conversion.

Socialization allows the conversion of tacit knowledge in one individual into tacit knowledge in other people, primarily through sharing experiences. The conversion of tacit knowledge to explicit knowledge is *externalization*, which in the process of making it readily communicable. *Internalization* converts explicit knowledge to tacit knowledge by translating it into personal knowledge – this could be called learning. The conversion of forms of explicit knowledge, such as creating frameworks, is *combination*.

Explicit knowledge as information

When knowledge is made explicit by putting it into words, diagrams or other representations, it can then be typed, copied, stored and communicated electronically – it becomes *information*. Perhaps then a useful definition for information is something that is or can be made explicit. Information, which represents captured knowledge, has value as an input to human decision making and capabilities. Tacit knowledge remains intrinsic to individuals and only they have the capacity to act effectively in its use.

These ideas about information and knowledge enable us to substitute information for explicit knowledge, and simply *knowledge* – in the business sense of capacity to act effectively – as tacit knowledge. This clarification of the distinction between

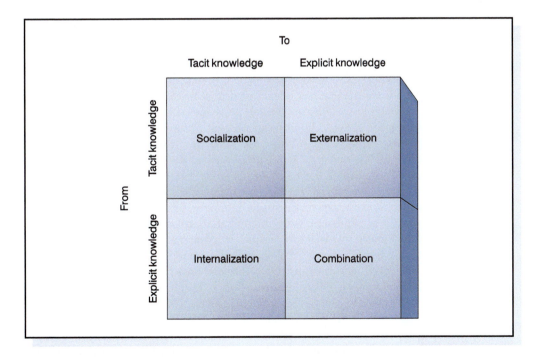

Figure 18.3
Modes of knowledge conversion

information and knowledge makes the knowledge conversion framework more directly applicable to inter-organizational interaction.

In the same way, externalization is capturing people's knowledge – their capacity to act in their business roles – by making it explicit and turning it into information, as in the form of written documentation or structured business processes. This remains information until other people internalize it to become part of their own knowledge – or capacity to act effectively. Having a document on a server or bookshelf does not make individuals knowledgeable, nor does reading it. Knowledge comes from understanding the document by integrating the ideas into existing experience and knowledge, and thus providing the capacity to act usefully in new ways. In the case of written documents, language and diagrams are the media by which the knowledge is transferred. The information presented must be actively interpreted and internalized, however, before it becomes new knowledge to the reader.

The process of internalization is essentially that of knowledge acquisition, which is central to the whole idea of learning, knowledge management and knowledge transfer. Understanding the nature of this process is extremely valuable in implementing effective business improvements and in adding greater value to customers.

Socialization refers to the transfer of one person's knowledge to another person, without an intermediary of captured information in documents. It is a most powerful form of knowledge transfer. As we know from childhood people learn from other people far more effectively than they learn from books and documents, in both obvious and subtle ways. Despite technological advances that allow people to telecommute and work in different locations, organizations function effectively mainly because people who work closely together have the opportunity for rich interaction and learning on an ongoing and often informal basis. This presents challenges, of course, in today's 'Virtual Organizations'.

The learning-knowledge management cycle

One way of thinking about learning and knowledge management is as a dynamic cycle from tacit knowledge to explicit knowledge and back to tacit knowledge. In other words, people's knowledge is externalized into information, which to be useful must then be internalized by others (learned) to become part of their knowledge, as illustrated in Figure 18.4. This flow from knowledge to information and back to knowledge constitutes the heart of organizational learning and knowledge management. Direct sharing of knowledge through socialization is also vital. In large organizations, however, capturing whatever is possible in the form of documents and other digitized representations means that information can be stored, duplicated, shared and made available to people on whatever scale desired.

The fields of learning and knowledge management encompass all the human issues of effective externalization, internalization and socialization of knowledge. As subsets of those fields, information management and document management address the middle part of the cycle, in which information is stored, disseminated and made easily available on demand. It is a misnomer to refer to information sharing technology, however advanced, as knowledge management. Effective implementation of those systems must address how people interact with technology in an organizational context, which only then is beginning to address the real issues of knowledge (see Dawson, 2000).

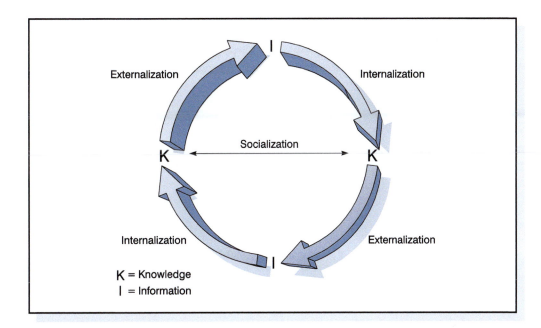

Figure 18.4
The knowledge management cycle

THE PRACTICALITIES OF SHARING KNOWLEDGE AND LEARNING

In world-class organizations there is clear evidence that knowledge is shared to maximize performance, with learning, innovation and improvement encouraged. In such establishments information (explicit knowledge) is collected, structured and managed in alignment with and support of the organization's policies and strategies. These days this is often achieved in large organizations through 'intranet' and/or common network mounted file servers. These can provide common access to online reports, training material and performance figures versus targets. The key is to provide appropriate access for both internal and external users to relevant information, with assurance required of its validity, integrity and security.

Cultivating, developing and protecting intellectual property can be the key in many sectors to maximising customer value, for example in professional services, such as legal and consultancy. Firms in this sector can survive only if they are constantly seeking to acquire, increase, use and transfer knowledge effectively. This in turn has strong links with learning and innovation. The proper use and management of relevant information and knowledge resources leads to the generation of creative thinking and innovative solutions. The aim must be to make information available as widely as possible on the most appropriate basis to improve general understanding and increase efficiency. If this is to be linked to questions in employee surveys, internal valuation that knowledge is being captured and used effectively can take place.

In the EFQM Excellence Model there are now clear feedback loops of 'Innovation and Learning', both from the results – on people, customers, society and key performance – to the enablers and within the enablers themselves. For example, key performance outcomes, such as profit, market share, cash flow, can and should be used to identify problems, areas for improvements or even strengths in aspects of the organization's leadership, its strategies and its processes. Equally the way people are managed and partnerships established can be improved, innovatively based on perception measures from customers. The results from the society in which we operate can lead to innovations in resource management.

In some companies and industries information on each customer is vital to identify specific needs and to aid communication, planning and monitoring progress. The whole area of customer's relationship management (CRM) relies on some sort of personal records for each customer being kept. This presents modern methods of knowledge management and information technology with the opportunity to demonstrate their powers.

There may also be learning and innovation loops and opportunities within the enabler's criteria. For example, key performance indicators related to the management of processes, such as cycle time or conversion of bids to businesses won, can help to generate innovations in strategies, particularly if there is a good alignment between the two. Information on staff, on customers and other sources of information can be readily kept on databases and their effective and efficient use can mean the difference between success and failure in many industries and sectors.

BIBLIOGRAPHY

Adair, J., *The Challenge of Innovation*, Talbot Adair Press, Guildford, 1990.
Dawson, R., *Developing Knowledge-based Client Relationships*, Butterworth-Heinemann, Oxford, 2000.
Ettlie, J.E., *Managing Technological Innovation*, Wiley, London, 2000.
Francis, D., *Unblocking the Organizational Communication*, Gower, Aldershot, 1990.
Larkin, T.J. and Larkin, S., *Communicating Change – Winning Employee Support for New Business Goals*, McGraw-Hill, New York, 1994.
Purdie, M., *Communicating for Total Quality*, British Gas, London, 1994.
Tidd, J., Bessant, J. and Pavitt, K., *Managing Innovation*, Wiley, London, 2001.
Zairi, M., *Best Practice Process Innovation Management*, Butterworth-Heinemann, Oxford, 1999.

CHAPTER HIGHLIGHTS

Communicating the total quality strategy

- People's attitudes and behaviour can be influenced by communication, and the essence of changing attitudes is to gain acceptance through excellent communication processes.
- The strategy and changes to be brought about through TQM should be clearly and directly communicated from top management to all staff/employees. The first step is to issue a 'total quality message'. This should be followed by a signed TQM directive.

- People must know when and how they will be brought into the TQM process, what the process is, and the successes and benefits achieved. First-line supervision has an important role in communicating the key messages and overcoming resistance to change.
- The complexity and jargon in the language used between functional groups needs to be reduced in many organizations. Simplify and shorten are the guiding principles.
- 'Open' methods of communication and participation should be used at all levels. Barriers may need to be broken down by concentrating on process rather than 'departmental' issues.
- There are four audience groups in most organizations – senior managers, middle managers, supervisors and employees – each with different general attitudes towards TQM. The senior management must ensure that each group sees TQM as being beneficial.
- Good leadership is mostly about good communications, the skills of which can be learned through training but must be acquired through practice.

Communication, learning, education and training

- There are four principal types of communication: verbal (direct and indirect), written, visual and by example. Each has its own requirements, strengths and weaknesses.
- Education and training is the single most important factor in improving quality and performance, once commitment is present. This must be objectively, systematically and continuously performed.
- All education and training should occur in an improvement cycle of ensuring it is part of policy, establishing objectives and responsibilities, establishing a platform for a learning organization, specifying needs, preparing programmes and materials, implementing and monitoring, assessing results and reviewing effectiveness.

A systematic approach to education and training for quality

- Responsibility for education and training of employees rests with management at all levels. The main elements should include error/defect/problem prevention, reporting and analysis, investigation and review.
- Education and training procedures and records should be established to show how job competence is demonstrated.

Starting where and for whom?

- Education and training needs occur at four levels of the organization: very senior management, middle management, first level supervision and quality team leaders, and all other employees.

Turning education and training into learning

- For successful learning all quality training should be followed up with improvement projects and 'surgery' workshops.

- It is useful to draw the distinction between explicit knowledge (that which we can express to others) and tacit knowledge (the rest of our knowledge which cannot be communicated in words or symbols).
- The creation and expression of knowledge takes place through social interaction between tacit and explicit knowledge, which takes the form of socialization, externalization, internalization and combination.
- When knowledge is made explicit it becomes 'information', which in turn has value as an input to human decision-making and capability. Tacit knowledge (simply 'knowledge') remains intrinsic to individuals who have the capacity to act effectively in its use.
- One way of thinking about learning and knowledge management is as a dynamic cycle from tacit knowledge to explicit knowledge (information) and back to tacit knowledge.

The practicalities of sharing knowledge and learning

- In world-class organizations there is clear evidence that knowledge is shared to maximize performance, with learning, innovation and improvement encouraged. This is often achieved through an 'intranet' or common network mounted file servers, providing common on-line access to information.
- Managing intellectual property is key to success in many sectors and this has strong links with learning and innovation. Where information must be made available as widely as possible, internal performance of this aspect can be valuable.
- The clear feedback loops of 'innovation and learning' in the EFQM Excellence Model drive increased understanding of the linkages between the results and the enablers, and between the enabler criteria themselves.

Part V Discussion questions

1. The so-called process approach has certain implications for organizational structures. Discuss the main organizational issues influencing the involvement of people in process improvement.
2. Various TQM teamwork structures are advocated by many writers. Describe the role of the various 'quality teams' in the continuous improvement process. How can an organization ensure that the outcome of teamwork is consistent with its mission?
3. Describe the various types of quality teams which should be part of introducing a total quality approach. Explain the organizational requirements associated with these and give some indication of how the teams operate.
4. A large insurance company has decided that teamwork is to be the initial focus of its TQM programme. Describe the role of a Quality Council or Steering Group and Process Quality Teams in managing teamwork initiatives in quality improvement.
5. Explain the difference between Quality Improvement Teams and Quality Circles. What is their role in quality improvement activities?
6. Discuss some of the factors that may inhibit teamwork activities in a TQM programme.
7. Suggest an organization for teamwork in a quality improvement programme and discuss how the important aspects must be managed, in order to achieve the best results from the use of teams. Describe briefly how the teams would proceed, including the tools they would use in their work.
8. Describe in full the various types of quality teams which are necessary in a total quality programme. Give some indication of how the teams operate at each level and, using the 'DRIVER' model, discuss the problem-solving approach that may be adopted.
9. Discuss the various models for teamwork within a total quality approach to business performance improvement. Explain through these models the role of the individual in TQM, and what work can be carried out in this area to help teams through the 'storming' stage of their development.
10. Teamwork is one of the key 'necessities' for TQM. John Adair's 'Action Centred Leadership' model is useful to explain the areas which require attention for successful teamwork. Explain the model in detail showing your understanding of each of the areas of 'needs'. Pay particular attention to the needs of the individual, showing how a psychometric instrument, such as the Myers Briggs Type Indicator (MBTI) or FIRO-B, may be useful here.

Part VI

Implementation

Joy's soul lies
in the doing.

William Shakespeare, 1564–1616,
from 'Troilus and Cressida'

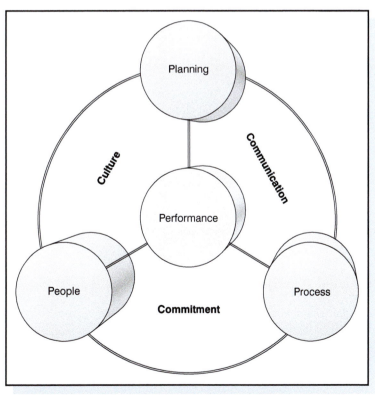

Chapter 19

Implementing TQM

TQM AND THE MANAGEMENT OF CHANGE

The author recalls the managing director of a large support services group who decided that a major change was required in the way the company operated if serious competitive challenges were to be met. The Board of Directors went away for a weekend and developed a new vision for the company and its 'culture'. A human resources director was recruited and given the task of managing the change in the people and their 'attitudes'. After several 'programmes' aimed at achieving the required change, including a new structure for the organization, a staff appraisal system linked to pay and seminars to change attitudes, very little change in actual organizational behaviour had occurred.

Clearly something had gone wrong somewhere. But what, who, where? Everything was wrong, including what needed changing, who should lead the changes and, in particular, how the change should be brought about. This type of problem is very common in organizations which desire to change the way they operate to deal with increased competition, a changing market place, new technology, different business rules. In this situation many organizations recognize the need to move away from an autocratic management style, with formal rules and hierarchical procedures, and narrow work demarcations. Some have tried to create teams, to delegate (perhaps for the first time), and to improve communications.

Some of the senior managers in such organizations recognize the need for change to deal with the new realities of competitiveness, but they lack an understanding of how the change should be implemented. They often believe that changing the formal organizational structure, having 'culture change' programmes and new payment systems will, by themselves, make the transformations. In much research work carried out by the European Centre *for* Business Excellence, the research and education division of Oakland Consulting, it has been shown that there is almost an inverse relationship between successful change and having formal organization-wide 'change programmes'. This is particularly true if one functional group, such as HR (human resources) or OD (organizational development), 'owns' the programme.

In several large organizations in which total quality management has been used successfully to effect change, the senior management did not focus on formal structures and systems, but set up *process-management* teams to solve real business or organization problems. The key to success in this area is to align the employees of the business, their roles and responsibilities with the organization and its *processes*. When an organization focuses on its key processes, that is the activities and tasks themselves, rather than on abstract issues such as 'culture' and 'participation', then the change process can begin in earnest.

An approach to change based on process alignment, and starting with the vision and mission statements, analysing the critical success factors *and* moving on to the core processes, is the most effective way to engage the staff in an enduring change process (see Chapter 4). Many change programmes do not work because they begin trying to change the knowledge, attitudes and beliefs of individuals. The theory is that changes in these areas will lead to changes in behaviour throughout the organization. It relies on a form of religion spreading through the people in the business.

What is often required, however, is virtually the opposite process, based on the recognition that people's behaviour is determined largely by the roles they have to take up. If we create for them new responsibilities, team roles and a process driven environment, a new situation will develop, one that will force their attention and work on the processes. This will change the culture. *Teamwork* is an especially important part of the TQM model in terms of bringing about change. If changes are to be made in quality, costs, market, product or service development, close co-ordination among the marketing, design, production/operations and distribution groups is essential. This can be brought about effectively by multifunctional teams working on the processes and understanding their interrelationships. *Commitment* is a key element of support for the high levels of co-operation, initiative and effort that will be required to understand and work on the labyrinth of processes existing in most organizations. In addition to the knowledge of the business as a whole, which will be brought about by an understanding of the mission, CSF, process breakdown links, certain *tools, techniques* and *interpersonal skills* will be required for good *communication* around the processes. These are essential if people are to identify and solve problems as teams.

If any of these elements are missing the total quality underpinned change process will collapse. The difficulties experienced by many organizations' formal change processes are that they tackle only one or two of these necessities. Many organizations trying to create a new philosophy based on teamwork fail to recognize that the employees do not know which teams need to be formed round their process – which they begin to understand together, perhaps for the first time – and further recognition that they then need to be helped as individuals through the forming-storming-norming-performing sequence, will generate the interpersonal skills and attitude changes necessary to make the new 'structure' work.

Organizations will avoid the problems of 'change programmes' then by concentrating on 'process alignment' – recognizing that people's roles and responsibilities must be related to the processes in which they work. Senior managers may begin the task of process alignment by developing a self-reinforcing cycle of *commitment, communication* and *culture* change. In the introduction of total quality for managing change, timing can be critical and an appropriate starting point can be a broad review of the organization and the changes required by the top management team. By gaining this shared diagnosis of what changes are required, what the 'business' problems are and/or what must be improved, the most senior executive mobilizes the initial commitment that is vital to begin the change process. An important element here is to get the top team working as a team, and techniques such as MBTI and/or FIRO-B will play an important part (see Chapter 17). See also Chapter 9 for the figure of 8 change management framework derived from research carried out by the ECforBE.

PLANNING THE IMPLEMENTATION OF TQM

The task of implementing TQM can be daunting and the chief executive faced with this may draw little comfort from the 'quality gurus'. The first decision is where to begin and this can be so difficult that many organizations never get started. This has been called TQP – total quality paralysis!

The chapters of this book have been arranged in an order which should help senior management bring total quality into existence. The preliminary stages of understanding and commitment are vital first steps which also form the foundation of the whole TQM structure. Too many organizations skip these phases, believing that they have the right attitude and awareness, when in fact there are some fundamental gaps in their 'quality credibility'. These will soon lead to insurmountable difficulties and collapse of the edifice.

While an intellectual understanding of quality provides a basis for TQM, it is clearly only the planting of the seed. The understanding must be translated into commitment, policies, plans and actions for TQM to germinate. Making this happen requires not only commitment, but a competence in leadership and in making changes. Without a strategy to implement TQM through process management, capability and control, the expended effort will lead to frustration. Poor quality management can become like poor gardening – a few weed leaves are pulled off only for others to appear in their place days later, plus additional weeds elsewhere. Problem solving is very much like weeding, tackling the root causes, often by digging deep, is essential for better control.

Individuals working on their own, even with a plan, will never generate optimum results. The individual effort is required in improvement but it must be co-ordinated and become involved with the efforts of others to be truly effective. The implementation begins with the drawing up of a quality policy statement, and the establishment of the appropriate organizational structure, both for managing and encouraging involvement in quality through teamwork. Collecting information on how the organization operates, including the costs of quality, helps to identify the prime areas in which improvements will have the largest impact on performance. Planning improvement involves all managers but a crucial early stage involves putting quality management systems in place to drive the improvement process and make sure that problems remain solved forever, using structured corrective action procedures.

Once the plans and systems have been put into place, the need for continued education, training and communication becomes paramount. Organizations which try to change the culture, operate systems, procedures or control methods without effective, honest two-way communication will experience the frustration of being a 'cloned' type of organization which can function but inspires no confidence in being able to survive the changing environment in which it lives.

An organization may, of course, have already taken several steps on the roads to TQM. If good understanding of quality and how it should be managed already exists, there is top management commitment, a written quality policy and a satisfactory organizational structure, then the planning stage may begin straight away. When implementation is contemplated, priorities amongst the various projects must be identified. For example, a quality management system which conforms to the requirements of ISO 9000 may already exist and this step will not be a major task, but introducing a quality costing system may well be. It is important to remember, however, that a review of the current performance in all the areas, even when well established, should be part of normal operations to ensure continuous improvement.

These major steps may be used as an overall planning aid for the introduction of TQM, and they should appear on a planning or Gantt chart. Major projects should be time-phased to suit individual organizations' requirement, but this may be influenced by outside factors, such as pressure from a customer to introduce statistical process control (SPC) or to operate a quality system which meets the requirements of a standard. The main projects may need to be split into smaller sub-projects, and this is certainly true of management system work, the introduction of SPC, Six Sigma, Lean and improvement teams.

The education and training part will be continuous and draw together the requirements of all the steps into a cohesive programme of introduction. The timing of the training inputs, follow-up sessions and advisory work should be co-ordinated and reviewed, in terms of their effectiveness, on a regular basis. It may be useful at various stages of the implementation to develop checks to establish the true progress. For example, before moving from understanding to trying to obtain top management commitment, objective evidence should be obtained to show that the next stage is justified.

Following commitment being demonstrated by the publication of a signed quality policy, there may be the formation of a council, and/or steering committee(s). Delay here will prevent real progress being made towards TQM through teamwork activities.

The launch of process improvement requires a balanced approach and the three major components must be 'fired' in the right order to lift the campaign off the ground. If teams are started before the establishment of a good system of management, there will be nothing to which they can adhere. Equally, if SPC is introduced without a good system of data recording and standard operating procedures, the techniques will simply measure how bad things are. A quality management system on its own will give only a weak thrust which must have the boost of improvement teams and SPC to make it come 'alive'.

An effective co-ordination of these three components will result in quality improvement through increased capability. This should turn to consistently satisfied customers and, where appropriate, increase preservation of market share.

Total quality management may be integrated into the strategy of any organization through an understanding of the core business processes and involvement of the people. The recommended framework is shown in Figure 19.1 and it all starts with the vision, goals, strategies and mission which should be fully thought through, agreed and shared in the business. What follows determines whether these are achieved. The factors which are critical to success, the CSFs – the building blocks of the mission – are then identified. The key performance indicators (KPIs), the measures associated

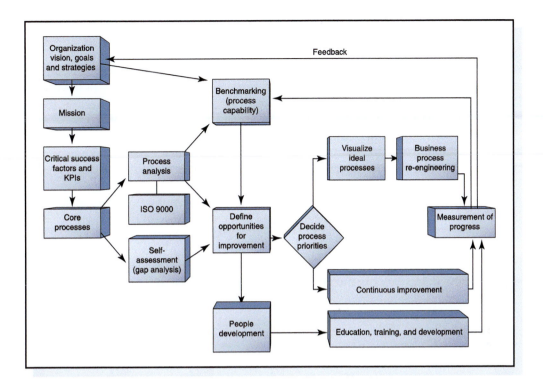

Figure 19.1
The framework for implementation of TQM

with the CSFs, tell us whether we are moving towards or away from the mission or just standing still.

Having identified the CSFs and KPIs, the organization should know what are its *core processes*. This is an area of potential bottleneck for many organizations because, if the core processes are not understood, the rest of the framework is difficult to implement. If the processes are known, we can carry out process analysis, sometimes called mapping and flowcharting, to fully understand our business and identify opportunities for improvement. By the way, ISO 9000 standard based systems should drop out at this stage, rather than needing a separate and huge effort and expense.

Self-assessment to the European (EFQM) Excellence Model or Baldrige Quality model, and benchmarking, will identify further improvement opportunities. This could create a very long list of things to attend to, many of which require people development, training and education. What is clearly needed next is prioritization – to identify those processes which are run pretty well – they may be advertising/ promoting the business or recruitment/selection processes, and subject them to a continuous improvement regime. For those processes which we identify as being poorly carried out, perhaps marketing, forecasting, training or even financial management, we may subject them to a complete re-visioning and redesign activity, perhaps even revision of the business itself. That is where BPR comes in.

Performance-based measurement of all processes and people development activities is necessary to determine progress and feedback to the benchmarking and strategic planning activities, so that the vision, goals, mission and critical success factors may be examined and reconstituted, if necessary, to meet new requirements for the organization and its customers, internal and external.

World-class organizations, of which there need to be more in most countries, are doing *all* of these things. They have implemented their version of the framework and are achieving world-class performance and results. What this requires first, of course, is world-class leadership and commitment.

In many successful companies TQM is not the very narrow set of tools and techniques often associated with failed 'programmes' in organizations in various parts of the world. It is part of a broad-based approach used by world-class companies, such as those presented in the case studies section of this book, to achieve organizational excellence, based on customer results, the highest weighted category of all the quality and excellence awards. TQM embraces *all* of these areas. If used properly, and fully integrated into the business, it will help any organization deliver its goals, strategy and targets, including those in the public sector. This is because it is about people and their identifying, understanding, managing and improving processes – the things any organization has to do particularly well to achieve its objectives.

CHANGE CURVES AND STAGES

A useful device in thinking about any implementation programme is the change curve (Figure 19.2). This represents a journey on which employees need to be taken if change is to lead to actions and be successful. The stages that can be expected on such a well-managed journey are:

- *Unaware*: employees have a sense that something needs to be done but not what, how or why.
- *Awareness*: employees starting to become aware of what changes are needed, where the business wants to be, and how to get there.
- *Comprehension*: employees understand the desired environment and have a more detailed understanding about how the organization intends to reach its goals; they will be concerned about how the change is likely to affect them.
- *Conviction*: employees understand the change and are reaching their own conclusions with respect to the necessity and feasibility of the change and the effect upon them.
- *Action*: employees will be ready to take part in actions and plans that will deliver the change.

The change curve is useful to plan the various changes of involvement and engagement. For example, Figure 19.3 shows three stages (there may be more, depending on the organization size and complexity) in which the change curve repeats, in this case through different layers of management and staff. It shows how the impetus for change develops to a point where it becomes 'irresistible:'

Stage 1 – **CE Exec**; passion and conviction to lead the transformation:

- Clear and compelling driver for change
- Sense of urgency for the change

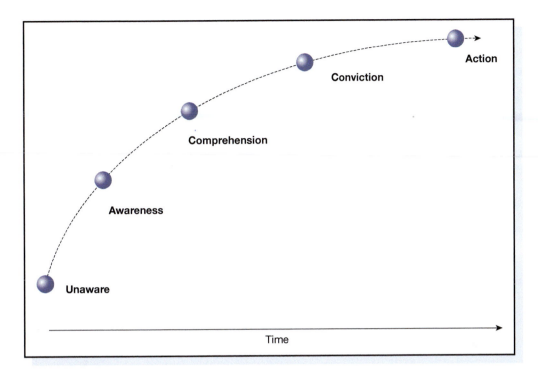

Figure 19.2
The change curve

- Clear vision and goals
- Power to make the change happen.

Stage 2 – **Middle Managers**; enabled to lead their teams through the transformation:

- Understand what and why change is happening
- Support the need to change
- Have confidence that the chief executive can successfully lead the transformation
- Understand how the change will affect their team and those with whom they interact
- Understand the steps to be taken to effect the change
- Have confidence that the 'system' is being considered
- Have sufficient information to guide their teams
- Have appropriate 'tools' to help guide change
- Feel that they can rely on support from their leaders
- Have a clear idea of their contribution to the change and clarity over the effect on them.

Stage 3 – **Employees**; willing and able to change in line with the desired state:

- Understand what and why change is happening
- Support the need to change

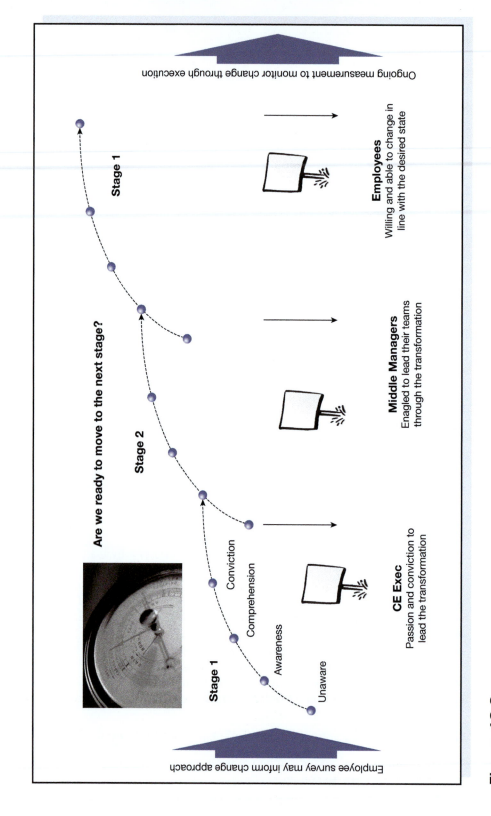

Figure 19.3
The stages of change

- Have confidence that their line managers can help them through the change
- Understand how the change will affect them and those with whom they work
- Understand the steps to be taken
- Have access to the information they need to help them understand/monitor change
- Have confidence that they will be equipped to deal with changes that affect them
- Have a clear idea of their contribution to the change.

Overcoming resistance to change

There can be many reasons why people resist change and, as far as possible, they need to be identified and thought about if that resistance is to be overcome. Common reasons include:

- *Perceived negative outcomes* – change can unleash a multitude of fears, of the unknown, loss of status or position, loss of freedom, loss of responsibility and/or authority, and loss of good working conditions and, of course, money.
- *Fear of more work* and less opportunity for rewards.
- *Habits must be broken* – dozens of inter-related habits lead to a style of management.
- *Lack of communication* – the organization does not effectively communicate the what, why and how of change and does not clearly spell out expectations for future performance.
- *Failure to align with the organization as a whole* – the organization's structure, business systems, technology, core competencies, employee knowledge and skills and culture are not aligned and/or integrated with the change effort.
- *Employee rebellion* – there is a view that people do not resist the intrusion of something new into their lives as much as they resist the resulting loss of control; this can be a huge yet 'hidden' factor.

A complementary change curve can be helpful in overcoming resistance to change (Figure 19.4). The stages likely to be encountered include:

- *Pre-awareness*: a sense exists that something needs to be done but not what, how or why.
- *Awareness*: thoughts about what changes are needed, where we want to be and how to get there are coming into focus but are not yet defined.
- *Self concern*: the desired environment and possibly some elements of the projects are now in detail, whereupon the concern 'How will this affect me?' becomes primary. At some point between self-concern and mental tryout the transition to acceptance begins.
- *Mental tryout*: changes are beginning to be viewed as inevitable, attitudes shift to 'How do I make this work for me?'
- *Hands-on*: simulation of the new environment in the form of pilot projects, prototypes or training is formalized. The point of no return is reached

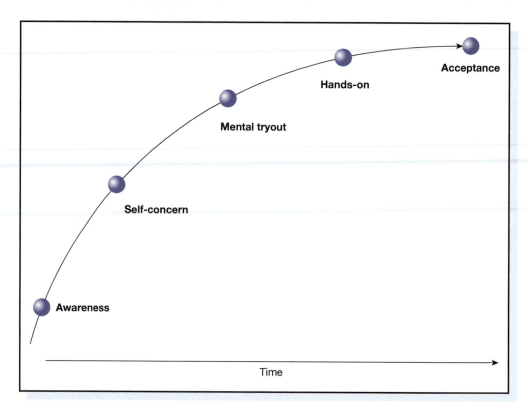

Figure 19.4
Overcoming resistance to change

somewhere between hands-on and acceptance when the momentum for the change and near-acceptance of change have increased to the point that turning back is unlikely.

- *Acceptance*: the changed order of things is achieved and the new environment becomes status quo.

Managing change effectively

In Chapter 9 on Benchmarking and Change Management, the figure of 8 framework for the effective management of change was introduced (Figure 19.5). The essence of achieving successful change during the implementation of approaches such as TQM, Lean Six Sigma, Continuous Improvement involves:

- *Establish a need to change*; if you want people to change don't give them a choice.
- *Create a clear and compelling vision* that shows people how their lives will be better; without this, any transformation effort can easily degrade into a long list of incompatible and time-consuming, often confusing, projects that can take people in the wrong direction.

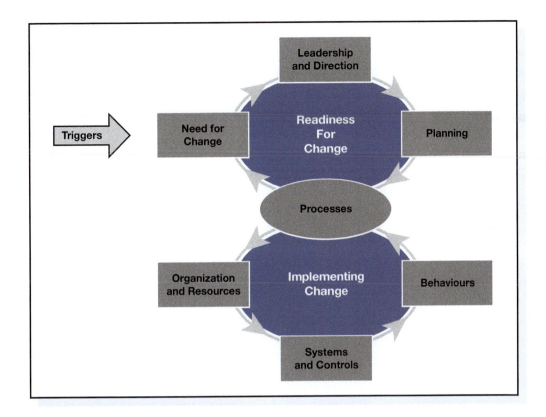

Figure 19.5
The organizational change framework

- *Go for true performance results and create early wins*; successful change programmes begin with results – clear, tangible, bottom line results and the earlier they occur the better.
- *Communicate, communicate, communicate*; poor and/or inadequate communication is one of the chief reasons for failed change efforts.
- *Build a strong, committed, guiding coalition*; that includes top management
- *Redesign the processes, systems and structures*; if we fail to look at the processes that are needed for the 'new order of things,' together with the structures and systems around them, a lot of old behaviour can just be reinforced – the new desired behaviours need to be rewarded.
- AND FINALLY – *People do not resist their own ideas*; people who participate in deciding what and how things will change not only are more likely to support the change but are actually changed themselves by the act of participation.

Using consultants to support change and implementation

Sometimes viewed as a costly necessity by business owners, the role of a consultant can be perceived as intrusive and disruptive. Used wisely, however, consultants can provide specialist skills, offer sound advice, suggest practical solutions and inject new life into a business without costing the earth or upsetting staff.

However well a business is managed, problems exist in any organization, caused either by external or internal events. Good leaders recognize this fact and tackle problems head on, plan changes to address them and implement solutions. In some problem-solving or change management situations, the best strategy is to combine internal and external resources in client/consultant teams. This can enhance the skills and competence of internal people, significantly reduce the costs of solving the problem or making the changes effectively, and avoid the internal resistance that can be associated with consultants.

Performance problems typically manifest themselves in gaps between expected and actual performance, whether this is the quality of a product in the eyes of a customer, failure to deliver something on time, excessive costs or poor profitability. So the first step in identifying problems is to provide accurate performance data. Sometimes performance gaps require specialist skills which do not exist within an organization, and it is not feasible to develop the in-house capability. If the problems are non-recurring and there will be no ongoing use for the skills, the use of a consultant is appropriate. They can also fill a competence gap until management can obtain the necessary skills or find time to close the gaps in performance.

If the decision is made to employ a consultant, users have to be clear about who should be involved during the assignment and ensure that staff have the time required to do the work that the organization has to undertake. They should manage each stage of the consultancy and ensure that all stakeholders are signed up to the diagnosis, the recommendations, the planning and to implementation.

The brief for potential consultancies should always be carefully defined. It should include a short description of the organization and of the problem or situation which has led to requiring a consultant. An outline of what the consultancy is expected to achieve, the starting date and expected duration of the work, some idea of how the consultancy is expected to proceed and who might be involved should also be in the initial briefing. It will be helpful to think through whether the consultant will be asked to:

- perform a specific task, e.g. draw up a quality plan, develop a strategy
- help people think and talk through what needs to be done regarding a particular task
- help improve the way people communicate with and relate to each other
- help resolve a conflict
- conduct training/coaching to help people learn specific skills or increase awareness.

How to choose a consultant

The professional and ethical standards that guide the consultant chosen should complement the organization's values and philosophy. It will then be necessary to determine how specifically the consultant's knowledge, competence and past experience relates to the client's own particular issues and needs. It will be important to find someone who seems to offer an effective mix of thought leadership, expertise and experience. They should also be seen to work closely with clients to address their needs, both in the short and long term.

The ethos driving the consultant's approach should be to ensure sustained improvement as they transfer their knowledge to the client organization. They should also provide customer feedback that confirms an understanding of the real issues, and should be adept at applying a set of basic consulting competences such as diagnosis, strategic planning, change management and execution, in addition to specific technical skills, such as Lean Six Sigma.

During the consultancy

Consultancy can be a sensitive matter, so it is vital that the consultant has the appropriate interpersonal competences, including problem root cause analysis and confrontation skills, risk management, collaboration, conflict management and relationship building. They should also be recognized by a trade body, such as the Management Consultancies Association (www.mca.com). Consultants are there to assist, so clients should not be defensive – they need to describe and explain all the issues surrounding the project. Clients should not be afraid of providing too much

Getting the best from an on-going consultancy project

- closely liaise with the consultancy team

- define, with the consultant, what information will be available and when

- identify and communicate to the consultancy team any problems so remedial action can be taken promptly

- obtain routine progress reports on the project from the consultant, with interim results

- hold regular progress reviews with actions agreed in writing

- where necessary and where agreed, provide staff, facilities and information promptly

- be open to innovative approaches and methodologies proposed by the consultancy team, but seek supporting evidence for these

- establish an agreed protocol for dealing with proposed changes in the client/consultant team

- ensure client staff and the consultancy team are briefed on the confidentiality state of information

- undertake a joint post-project completion review with the consultancy team within six months of completion to build on the experience.

information; consultants much prefer this to a lack of detail. Make sure you have identified the potential 'life of project' (the time the project will take, including implementation), the expected benefits as well as the initial investment required.

A consultancy project usually commences with a diagnostic of the situation, hopefully with an internal team, and employs a variety of techniques to induce high levels of internal commitment in the change process, building support for actions necessary for its success. A final and very important step at the end of the project or programme of work is to undertake a joint post-programme/project completion review with the consultancy team to see what has been gained and learned from the project or programme, and to give feedback to the consultant-client team. This should take place within six months of completion.

SUSTAINED IMPROVEMENT

Never-ending or continuous improvement is probably the most powerful concept to guide management. It is a term not well understood in many organizations, although that must change if those organizations are to survive. To maintain a wave of sustained improvement, it is necessary to develop generations of managers who not only understand but are dedicated to the pursuit of never-ending improvement in meeting external and internal customer needs.

The concept requires a systematic approach that has the following components:

- *Planning* the processes and their inputs.
- *Providing* the inputs.
- *Operating* the processes.
- *Evaluating* the outputs.
- *Examining* the performance of the processes.
- *Modifying* the processes and their inputs.

This system must be firmly tied to a continuous assessment of customer needs, and depends on a flow of ideas on how to make improvements, reduce variation and generate greater customer satisfaction. It also requires a high level of commitment and a sense of personal responsibility in those operating the processes. Another useful rubric for process improvement that is often associated with IT/software implementation is: standardize; simplify; shrink; share; satisfy.

The never-ending improvement cycle ensures that the organization learns from results, standardizes what it does well in a documented quality management system and improves operations and outputs from what it learns. But the emphasis must be that this is done in a planned, systematic and conscientious way to create a climate – a way of life – that permeates the whole organization.

There are three basic principles of sustained improvement:

- Focusing on the *customer*.
- Understanding the *process*.
- All *employees* committed to quality.

1. Focusing on the customer

An organization must recognize, throughout its ranks, that the purpose of all work and all efforts to make improvements is to serve the customers better. This means that it must always know how well its outputs are performing, in the eyes of the customer, through measurement and feedback. The most important customers are the external ones, but the quality chains can break down at any point in the flows of work. Internal customers therefore must also be well served if the external ones are to be satisfied and remain loyal.

2. Understanding the process

In the successful operation of any process it is essential to understand what determines its performance and outputs. This means intense focus on the design and control of the inputs, working closely with suppliers, and understanding process flows to eliminate bottlenecks and reduce waste. If there is one difference between management/supervision in the Far East and the West, it is that in the former management is closer to and more involved in the processes. It is not possible to stand aside and manage in never-ending improvement. TQM in an organization means that everyone has the determination to use their detailed knowledge of the processes and make improvements, and use appropriate tools and techniques to analyse and create action plans.

3. All employees committed to quality

Everyone in the organization, from top to bottom, from offices to technical service, from headquarters to local sites, must play their part. People are the source of ideas and innovation, and their expertise, experience, knowledge and co-operation have to be harnessed to get those ideas implemented.

When people are treated like machines, work becomes uninteresting and unsatisfying. Under such conditions it is not possible to expect quality services and reliable products. The rates of absenteeism and of staff turnover are measures that can be used in determining the strengths and weaknesses of management style and people's morale, in any organization.

The first step is to convince everyone of their own role in total quality. Employers and managers must, of course, take the lead and the most senior executive has a personal responsibility. The degree of management's enthusiasm and drive will determine the ease with which the whole workforce is motivated.

Most of the work in any organization is done away from the immediate view of management and supervision, and often with individual discretion. If the co-operation of some or all of the people is absent, there is no way that managers will be able to cope with the chaos that will result. This principle is extremely important at the points where the processes 'touch' the outside customer. Every phase of these operations must be subject to continuous improvement and for that everyone's co-operation is required.

Never-ending improvement is the process by which greater customer satisfaction is achieved. Its adoption recognizes that quality is a moving target but its operation actually results in quality.

A model for total quality management

The concept of total quality management is basically very simple. Each part of an organization has customers, whether within or without, and the need to identify what the customer requirements are, and then set about meeting them, forms the core of a total quality approach. Good *performance* requires the three hard management necessities: *planning*, including the right policies and strategies, *processes* and supporting management systems and improvement tools, such as statistical process control (SPC), and *people* with the right knowledge, skills and training (Figure 19.6). These are complementary in many ways, and they share the same requirement for an uncompromising top level commitment, the right culture and good communications. This must start with the most senior management and flow down through the organization. Having said that, teamwork, SPC, or a quality management system, may be used as a spearhead to drive TQM through an organization. The attention to many aspects of a company's operations – from purchasing through to distribution, from data recording to control chart plotting – which are required for the successful introduction of a good quality management system, or the implementation of SPC,

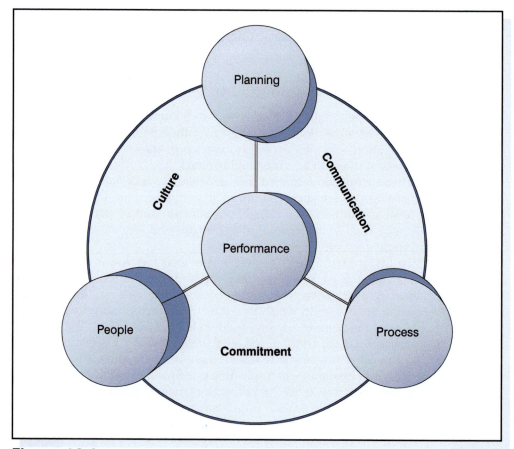

Figure 19.6
A model for total quality management (TQM)

will have a 'Hawthorne effect', concentrating everyone's attention on the customer-supplier interface, both inside and outside the organization.

Total quality management calls for consideration of processes in all the major areas: marketing, design, procurement, operations, distribution, etc. Clearly, these each require considerable expansion and thought, but if attention is given to all areas, using the concepts of TQM, then very little will be left to chance. Much of industry commerce and the public sector would benefit from the continuous improvement cycle approach represented in Figure 19.7, which also shows the 'danger gaps' to be avoided. This approach will ensure the implementation of the management commitment represented in the quality policy, and provide the environment and information base on which teamwork thrives.

By implementing the concepts of total quality and the approach of managing, controlling and improving business process, many organizations have improved their day-to-day operations. Established quality management systems, statistical process control methods and an enhanced quality consciousness have allowed these organizations to provide customers with quality products and services that match with customer requirements and organization policies and that have been properly planned, developed, designed, produced and deployed. As a result, there have been reductions in development and design problems, defects, errors, installation problems, service failures and market claims and complaints. Excellent product and service quality has been achieved to generate satisfied and loyal customers. Results also include improved reliability, safety and environmental outcomes that meet society's needs. Together with positive economic outcomes these improvements have enabled many organizations to acquire a world-class reputation.

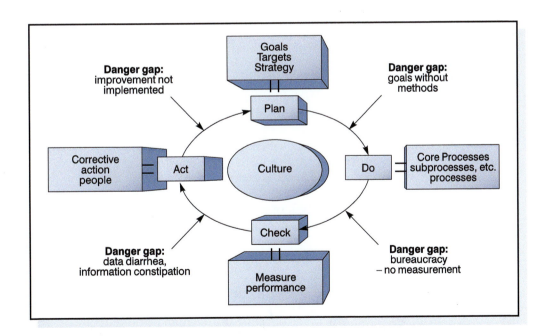

Figure 19.7
TQM implementation – all done with the Deming continuous improvement cycle

BIBLIOGRAPHY

Albin, J.M., *Quality Improvement in Employment and other Human Services – Managing for Quality through Change*, Paul Brookes Pub., 1992.

Antony, J. and Preece, D., *Understanding, Managing and Implementing Quality*, Routledge, London, 2002.

Ciampa, D., *Total Quality – A User's Guide for Implementation*, Addison-Wesley, Reading, Mass., 1992.

Cook, S., *Customer Care – Implementing Total Quality in Today's Service Driven Organization*, Kogan Page, London, 1992.

Economist Intelligence Unit, *Making Quality Work: Lessons from Europe's Leading Companies*, EIU, London, 1992.

Fox, R., *Six Steps to Total Quality Management*, McGraw-Hill, NSW, 1994.

Hiam, A., *Closing the Quality Gap – Lessons from America's Leading Companies*, Prentice-Hall, Englewood Cliffs, NJ, 1992.

Kanji, G.P. and Asher, M., *100 Methods for Total Quality Management*, Sage, London, 1996.

Morgan, C. and Murgatroyd, S., *Total Quality Management in the Public Sector*, Open University Press, Milton Keynes, 1994.

Munro-Faure, L. and Munro-Faure, M., *Implementing Total Quality Management*, Pitman, London, 1992.

Oakland, J.S., *Total Organizational Excellence*, Butterworth-Heinemann, Oxford, 2006.

Oakland, J.S., *Oakland on Quality Management*, Butterworth-Heinemann, Oxford, 2006.

Senge, P.M., Kleiner, A., Roberts, C., Ross, R., Roth, G. and Smith, B., *A Fifth Discipline Resource – the dance of change*, Nicholas Brearley, London, 2000.

Strickland, F., *The Dynamics of Change*, Routledge, London, 1998.

CHAPTER HIGHLIGHTS

TQM and the management of change

- Senior managers in some organizations recognize the need for change to deal with increasing competitiveness, but lack an understanding of how to implement the changes.
- Successful change is effected not by focusing on formal structures and systems, but by aligning process management teams. This starts with writing the mission statement, analysis of the critical success factors (CSFs) and understanding the critical or key processes.
- Senior managers may begin the task of process alignment through a self-reinforcing cycle of commitment, communication and culture change.

Planning the implementation of TQM

- Making quality happen requires not only commitment but competence in the mechanics of TQM. Crucial early stages will comprise establishment of the appropriate organization structure; collecting information, including quality costs; teamwork; quality systems; and training.

- The launch of quality improvement requires a balanced approach, through systems, teams and tools.
- A new implementation framework allows the integration of TQM into the strategy of an organization through an understanding of the core business processes and involvement of people. This leads through process analysis, self-assessment and benchmarking to identifying opportunities for improvement, including people development.
- The process opportunities should be prioritized into continuous improvement and re-engineering/redesign. Performance-based measurement determines progress, and feeds back to the strategic framework.

Change curves and stages

- A useful device in thinking about any implementation programme is the change curve which represents a journey on which employees need to be taken if change is to lead to actions and be successful. The stages that can be expected on such a well-managed journey are: unaware, awareness, comprehension, conviction and action. Use of the curve leads to the planning of stages of involvement and engagement for executives, middle management and employees.
- There can be many reasons why people resist change and these need to be identified and thought about if that resistance is to be overcome. A complementary change curve can be helpful in overcoming resistance to change.
- The essence of achieving successful change during the implementation of approaches such as TQM, Lean Six Sigma, Continuous Improvement involves: establishing a need to change; creating a clear and compelling vision; going for true performance results and creating early wins; communicating well; building a strong, committed, guiding coalition; redesigning the processes, systems and structures.

Use of consultants to support change and implementation

- Used wisely consultants can provide specialist skills, offer sound advice, suggest practical solutions and inject new life into a business without costing the earth or upsetting staff – the best strategy is to combine internal and external resources in client/consultant teams.
- The brief for potential consultancies should always be carefully defined and include a short description of the organization, the problem or situation, an outline of what the consultancy is expected to achieve, the starting date and expected duration of the work, some idea of how the consultancy is expected to proceed and who might be involved.
- The professional and ethical standards that guide the consultant chosen should complement the client organization's values and philosophy and the ethos driving the consultant's approach should be to ensure sustained improvement as they transfer their knowledge to the client.

Sustained improvement

- Managers need to understand and pursue never-ending improvement. This should cover planning and operating processes, providing inputs, evaluating outputs, examining performance, and modifying processes and their inputs.
- There are three basic principles of continuous improvement: focusing on the customer, understanding the process and seeing that all employees are committed to quality.
- In the model for TQM the customer-supplier chains form the core, which is surrounded by the hard management necessities of planning, processes and people. These are complementary and share the same needs – for top level commitment, the right culture and good communications.

Part VI Discussion questions

1. You have just joined a company as the Quality Executive. The method of quality control is based on the use of inspectors who return about 15 per cent of all goods inspected for modification, rework or repair. The monthly cost accounts suggest that the scrap rate of raw materials is equivalent to about 10 per cent of the company's turnover and that the total cost of employing the inspectors is equal to about 15 per cent of the direct labour costs. Outline your plan of action to address the situation over the first 12 months.

2. You have recently been appointed as Transport Manager of the haulage division of an expanding company and have been alarmed to find that maintenance costs seem to be higher than you would have expected in an efficient organization. Outline some of the measures that you would take to bring the situation under control.

3. TQM has been referred to as 'a rain dance to make people feel good without impacting on bottom line results'. It was also described as 'flawed logic that confuses ends with means, processes with outcomes'. The arguments on whether to focus on budget control through financial management or quality improvement through process management clearly will continue in the future. Discuss the problems associated with taking a financial management approach which has been the traditional method used by many organizations.

4. You are a management consultant who has been invited to make a presentation on Total Quality Management to the board of directors of a company employing around 200 people. They are manufacturers of injection moulded polypropylene components for the automotive and electronics industries, and they also produce some lower technology products, such as beverage bottle crates. As they hope to supply the Ford Motor Company with a new product line and achieving their supplier registration status is vital, the board have asked you to stress the role of quality systems and statistical process control (SPC) in TQM. Prepare your presentation, including references to appropriate models.

5. Describe the key stages in integrating total quality management into the strategy of an organization. Illustrate your answer by reference to one of the following types of organization: a large national automotive manufacturer, an international petro-chemical company, a national military service or a large bank.

6. What are the critical elements of integrating total quality management or business improvement into the strategy of an organization? Illustrate your approach with reference to an organization with which you are familiar, or which you have heard about and studied.

7. You are the new Quality Director of part of a large electrical component manufacturing assembly and service company. Some members of the top management team have had some brief exposure to Six Sigma and Lean, and you have been appointed to consider plans for implementation. Set down your arguments and plans for the process which you might initiate to deal with this situation. Your plans should include reference to any training needs, outside help and additional internal appointments required, with timescales.

8. You have been appointed as an external personal advisor to the Chief Executive of ONE of the following:

National Outminster Bank or
Portstown Royal Infirmary or
University of Leedford.

The members of the top management have had some brief exposure to 'Business Excellence' and you have been appointed to help the Chief Executive lay down plans for its implementation.

Choose any of the above organizations and set down plans for the process which you would initiate to help the Chief Executive achieve this. Your plans should be as fully developed as possible and include reference to any training needs, further outside help and any internal appointments required, with a realistic timescale.

Case studies

READING, USING AND ANALYSING THE CASES

The cases in this book provide a description of what occurred in ten different organizations, regarding various aspects of their quality and operational performance improvement efforts. They may each be used as a learning vehicle as well as providing information and description which demonstrate the application of the concepts and techniques of TQM and business excellence.

The objective of writing the cases has been to offer a resource through which the student of TQM (including the practising manager) understands how organizations which adopt the approach operate. It is hoped that the cases provide a further useful and distinct contribution to TQM education and training.

The case material is suitable for practising managers, students on undergraduate and postgraduate courses, and all teachers of the various aspects of business management and TQM. The cases have been written so that they may be used in three ways:

1. As orthodox cases for student preparation and discussion
2. As illustrations, which teachers may also use as support for their other methods of training and education
3. As supporting/background reading on TQM.

If used in the orthodox way, it is recommended that first the case is read to gain an understanding of the issues and to raise questions which may lead to a collective and more complete understanding of the organization, TQM and the issues in the particular case. Second, case discussion or presentations in groups will give practice in putting forward thoughts and ideas persuasively.

The greater the effort put into case study preparation, analysis and discussions in groups, the greater will be the individual benefit. There are, of course, no perfect or tidy cases in any subject area. What the directors and managers of an organization actually did is not necessarily the best way forward. One object of the cases is to make the reader think about the situation, the problems and the progress made, and what improvements or developments are possible.

Each case emphasizes particular challenges or issues which were apparent for the organization. This may have obscured other more important ones. Imagination, innovation and intuition should be as much a part of the study of a case as observation and analysis of the facts and any data available.

TQM cases, by their nature, can be very complicated and, to render the cases in this book useful for study, some of the complexity has been omitted. This simplification is accompanied by the danger of making the implementation seem clear-cut and obvious, but that is never the case with TQM!

The main objective of each description is to enable the reader to understand the situation and its implications, and to learn from the particular experiences. The cases are not, in the main, offering specific problems to be solved. In using the cases, the reader/student should try to:

- *Recognize or imagine* the situation in the organization
- *Understand* the context and objectives of the approaches adopted
- *Analyse* the different parts of the case (including any data) and their interrelationships
- *Determine* the overall structure of the situation/problem/case
- *Consider* the different options facing the organization
- *Evaluate* the options and the course of action chosen, using any results stated
- *Consider any recommendations* which should be made to the organization for further work, action or implementation.

The set of cases has been chosen to provide a good coverage across different types of industry and organization, including those in the service, manufacturing and public sectors. The value of illustrative cases in an area such as TQM is that they inject reality into the conceptual frameworks developed by authors on the subject. The cases are all based on real situations and are designed to illustrate certain aspects of managing change in organizations, rather than to reflect good or poor management practice. The cases may be used for analysis, discussion, evaluation and even decision-making within groups without risk to the individuals, groups or organization(s) involved. In this way, students of TQM and business excellence may become "involved" in many different organizations and their approaches to TQM implementation, over a short period and in an efficient and effective way.

The organizations described here have faced translating TQM theory into practice, and the description of their experiences should provide an opportunity for the reader of TQM literature to test his/her preconceptions and understanding of this topic. All the cases describe real TQM processes in real organizations and we are grateful to the people involved for their contribution to this book.

John Oakland

FURTHER READING

Easton, G., *Learning from Case Studies* (2nd edition), Prentice-Hall, UK, 1992.

CASE STUDY ORDER

Each of the case studies in this section may be used as illustrations and the basis of discussions on a number of parts of the text. The cases are listed in the order given below.

C1 **Nissan** – TQM objectives management process
C2 **Shell Services** – Sustainable business improvement in a global corporation
C3 **Lloyd's Register** – Improvement programme – group business assurance
C4 **STMicroelectronics** – TQM implementation and policy deployment
C5 **TNT Express** – Business process management
C6 **Fujitsu UK & Ireland** – Process management and improvement at the heart of the BMS
C7 **Car Care Plan** – Simplifying business processes to secure competitive advantage
C8 **ABB** – Building quality and operational excellence
C9 **EADS (Airbus Group)** – Lean Six Sigma approach to performance improvement
C10 **NHS** – Establishing a capability for continuous quality improvement

Case study 1 TQM objectives management process in Nissan

INTRODUCTION

Nissan is a Japanese multinational automotive manufacturer headquartered in Japan. It was a core member of the Nissan Group but has become more independent following mergers and restructuring. Along with its normal range of models Nissan also produces a range of luxury models branded as Infiniti.

Nissan is approximately the sixth largest automaker in the world behind General Motors, Volkswagen, Toyota, Hyundai and Ford. The company's global headquarters is located in Nishi-ku, Yokohama. In 1999, Nissan entered a two-way alliance with Renault of France. In 2010 Daimler became involved in a triple alliance to allow for the increased sharing of technology and development

costs, encouraging global cooperation and mutual development. In 2010, Nissan announced that its hybrid technology was no longer based on Toyota's. Nissan have a UK factory in Washington, Tyne and Wear.

TQM OBJECTIVES MANAGEMENT PROCESS IN NISSAN

The aim of the TQM approach in Nissan is in line with the corporate objectives and drives business success through total customer satisfaction (Figure C1.1). The quality approach encompasses quality of design, quality of production and quality of sales and services.

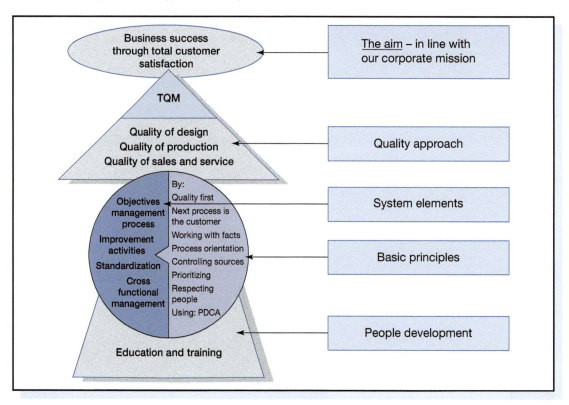

Figure C1.1
Nissan TQM approach

434

The company believes it can continually improve the quality of its:

- products and processes
- work to make it more effective and efficient
- people through education, training, motivation and clarity of direction.

The basic Nissan TQM principles are:

- Quality first
- Next process is the customer
- Working with facts and data
- Process orientation
- Controlling sources
- Prioritization
- Respecting people
- Using PDCA (Plan, Do, Check, Act).

The TQM promotion structure is shown in Figure C1.2. This drives the relationship between TQM and the annual objectives/mid-term plan (Figure C1.3).

Nissan's quality leadership begins with the classic vision and mission statements:

> **Nissan Corporate Vision:** Enriching people's lives
> **Nissan Corporate Mission:** Nissan provides unique and innovative automotive products

and services that deliver superior measurable value to all stakeholders

The foundation supporting these is trust and consistently beating the industry average on quality, safety and environmental performance. Significantly in Nissan quality is not only about the end product, it involves every aspect of the business from research and development right through to the quality of customer service. This is captured in the quotation from CEO Carlos Ghosn:

> For Nissan, Quality is about Trust. Building Trust requires putting our Customer at the centre of our company. Putting our Customer at the centre of our company means putting Quality at the centre of our company. Quality Leadership is a Corporate Commitment for Nissan.

TQM OBJECTIVES MANAGEMENT PROCESS – HOSHIN KANRI OR POLICY DEPLOYMENT

In Nissan, the TQM OMP is the way they set and cascade the policies and objectives which in turn are designed to clarify what they are trying to

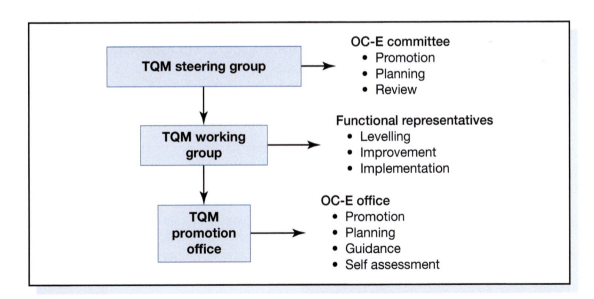

Figure C1.2
TQM promotion structure

Case 1 TQM objectives management process in Nissan 435

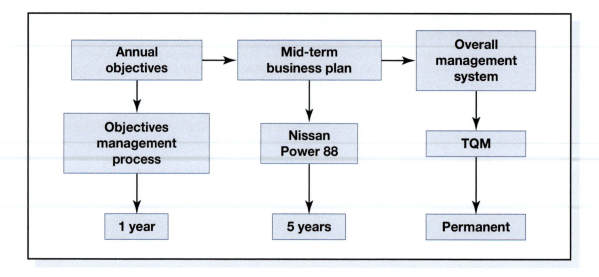

Figure C1.3
Relationship between annual objectives/MTP & TQM

achieve as a company, the role of everyone in achieving it and how they intend to go about it.

The key principles of the Nissan objectives management process are:

- Focus on the process as well as the results
- Prioritize the 'Vital Few' objectives
- Top down/bottom up
- Speak with 'Facts and Data'
- Define the 'How to Achieve'
- Teamwork
- 'Catchball'.

This is all done through a classical cascading Hoshin Kanri objectives management process, shown in Figure C1.4, which leads to a deployment schedule and a mid-term objectives plan. Key definitions used in this approach include:

- **Policy statement:** statement of intent
- **Objectives:** details of what must be achieved
- **Targets:** define the level to be achieved and by when
- **Strategies:** how to achieve the objective
- **Control items:** allow progress to be monitored
- **Responsibilities:** lead and support roles
- **Prioritization:** policy or routine management.

Nissan distinguish between policy and routine management objectives – policy management objectives are breakthrough objectives needing

significant additional resources (4Ms – man; materials; method; machines) together with cross functional support; whilst routine management objectives can be achieved with current resources and support systems, even though they are still vital to achieve business success.

The TQM objectives management process in Nissan has six key steps:

1. Select breakthrough issues   PLAN
2. Create annual objectives plans   "
3. Implementation Planning   "
4. Implementation   DO
5. Diagnosis   CHECK
6. Actions   ACT

SELECTED BREAKTHROUGH ISSUES

Identify sources

- Company Corporate Vision and Mission
- NML/NISA Mid-Term Plans
- Business Plans
- Review of the current year's Annual Objectives Plans (Diagnosis)
- Internal and External influences:
  - **SWOT** (Strengths; Weaknesses; Opportunities; Threats)
  - **PESTEL** (Political; Economic; Social; Technological; Environmental; Legal)

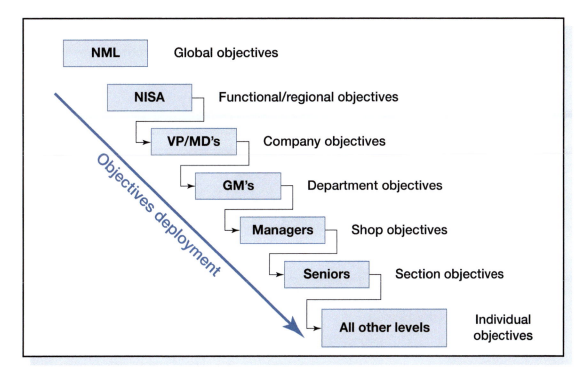

Figure C1.4
Objectives deployment

- **Political**
- Tax Policy; Labour Law; Trade Restrictions; Tariffs and Political Stability
- **Economic**
- Economic Growth; Interest Rates; Exchange Rates; Inflation Rate
- **Social**
- Health Consciousness; Population Growth; Age Distribution; Career Attitudes; Emphasis on Safety
- **Technological**
- R&D Activity; Automation; Technology Incentives; Rate of Technology Change
- **Environmental**
- Climate; Climate Change
- **Legal**
- Discrimination Law; Consumer Law; Antitrust Law; Employment Law; Health and Safety Law.

Create objective and strategies

- Objective/Strategy Proposal (OSP) Sheets (e.g. Figure C1.5) are raised for each generic issue,

i.e. Quality, Cost, Delivery, People and New Model Preparation and others as appropriate
- Using 'Facts and Data'
- 'Catchball' required
- Lead department to coordinate each generic issue.

The benefits of using the OSP sheet approach are that it clarifies the objectives and targets to be achieved, identifies the 'gap between where we are today and where we want to be', uses an analytical approach to review the current condition, identifies the key issues that need to be addressed, establishes the strategies required to be put in place, communicates relevant information in a visual format and the gap analysis determines whether the objective is breakthrough or routine based on the need for extra resources (4Ms).

CREATE ANNUAL OBJECTIVES PLANS

This starts with development of the policy statement which indicates the overall intent and puts objectives into context. What makes a good objective/

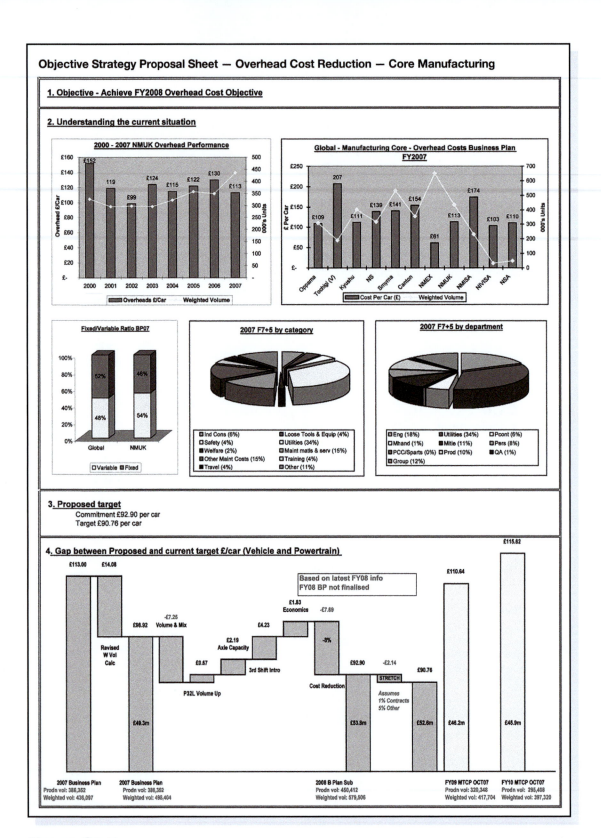

Figure C1.5
OSP sheet example

Objective Strategy Proposal Sheet — Overhead Cost Reduction — Core Manufacturing *(continued)*

4. Gap between Proposed and current target £/car - X12C Submission Jan 08 (Vehicle only)

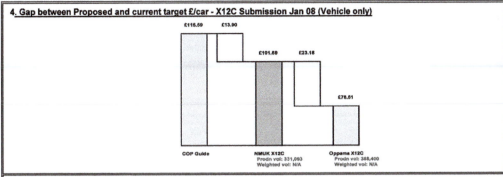

5. Analysing the factors

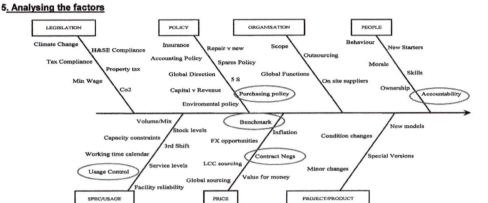

6. Key factors

1) Benchmarking - Achievable targets, vehicle and/or PT, Best Practice, visibility, inconsistencies, lack of transparency, capital vs. revenue, fixed vs. variable, allocations.

2) Accountability - Relatively low amount so low priority, disempowered by GE, actions not followed through, L&O Meeting does not bite, no direction.

3) Purchasing Policy - GE inefficient and ineffective, GE little technical knowledge, GE no ownership/accountability, no FS&S resource in purchasing, buying not purchasing, outsourced too far.

4) Usage Control - spend up to budget, no spend up to budget, false economies e.g. conveyor oiling, 3-shift impact, environmental pressures.

5) Contract Negotiations - Electricity to March 09 so must address in BP08, RNPO dictate two year contract.
       - Mitie spec changes for 3-shift / parts warehouse / NDS / on site suppliers, 5S service shortfall. 34 of 35 contractors resisting.

6) Global Direction - Vendor tooling and FOREX.

7. Strategies

Objective	Target	Strategy	Control item	Timing	Responsibility
Achieve Overhead cost Target	£90.76/Car @ 579k W/Vol	1a) Review cost allocation i.e. G&A, Kyushu & NDS (recharges) 1b) Benchmark activity with Oppama - identify opportunity areas - deep dive the opportunities		Qtr 1 Qtr 1 Qtr 2	Fin Eng/Fin/Others All Depts
		2a) Restructure L&O Meeting - Finance led, action points minuted, agree attendees. 2b) Allocate budget responsibility to Managers 2c) Managers to get OSP VCS objectives.		Qtr 1 Qtr 1 Qtr 1	Fin Fin HR
		3a) Document processes and review; NMUK, GE (NTCE, NDE). 3b) Optimise organisation - NP21, K2, NMUK vs. GE resource, P-Card, E-Auction.		Qtr 2 Qtr 2	Purch Purch/Fin
		4a) Review metering - what to meter; meters being monitored. 4b) Revive maintenance and consumable stores activities and engage consultants like Rockwell where appropriate 4c) Build statistical database to monitor usage e.g. gloves		Qtr 1 Qtr 1 Qtr 1	Eng Fin/Eng/PC Purch
		5a) Make financial case to show policy costing money 5b) CO^2 reduction OSP will support this 5c) Review the specification of all major contracts 5d) Review the cost allocation and recharging procedure		Qtr 1 Qtr 1 Qtr 1 Qtr 1	Purch/Eng/Fin Fin/Eng/PC/SCM/NDS Eng/Pur Fin/Eng/PC/SCM/NDS
		6a) Review vendor tooling - piece price vs. capital costs 6b) Review FOREX - Purchasing vs. Treasury		Qtr 2 Qtr 2	Purch/Fin Purch/Fin

Figure C1.5
continued

strategy is that it is clearly derived from the policy statement and it describes 'What' must be achieved – indicates the intended result. Of course, it must be supported by top level management and must be a challenge to the company or department. The strategies answer the 'How' to the 'What' of objectives.

To aid the process, Nissan use the SMART objectives/strategies approach:

- **Specific:** should be brief statements and within appropriate boundaries
- **Measurable:** quantifiable, you need to be able to measure whether or not you have achieved your objective (use a benchmark where possible)
- **Achievable:** get agreement from management and your team on 'What you want to do' before any work begins on 'How to do it'
- **Realistic:** can you achieve them within the timeframe, budget and business context?
- **Timely:** must be time limited.

Targets and control items are vital here and Nissan try to use measures that indicate results achieved or the activity undertaken, e.g. results of meetings rather than number of meetings. On average, one or two control items are deployed per strategy and numerical measures are used for both objectives and strategies.

In order to cascade the plan it is broken down, where possible using a 'Z-Pattern,' i.e. each strategy becomes an objective on a lower level Annual Objectives Plan. The control item from the higher level strategy becomes the lower-level target and the lower-level Annual Objectives Plan is then built as normal.

At this stage it is important to include a link checking/review stage because annual objectives plans are difficult to get right first time. The alternative is to wait for the first diagnosis to find out how good it is but the cost of getting it wrong is much higher than the cost of link checking/review.

IMPLEMENTATION PLANNING Ë IMPLEMENTATION

This stage begins with breaking the strategy down into specific tasks, using clear and unambiguous task descriptions. It is then necessary to allocate Ownership and Estimate Timing, matching tasks to skills and responsibilities of staff. Nissan use Master Schedules to monitor progress of the activities. Tips from Nissan at this point include concentrating on specific tasks, not strategic approaches, reviewing plans monthly and making tasks one month or less so it is possible to see progress and catch slippage, and getting commitment and realistic estimates by involving the people in the estimation of times to complete tasks. Figure C1.6 shows typical Master Schedules that help deliver this.

In implementation planning Nissan believe it is important to establish methods of communication which:

- set the context for the objective, making it obvious why they are required, especially 'Breakthrough Items'
- explain why targets, strategies and control items were chosen
- is understood by all of the target audience (unambiguous).

Figures C1.7 and C1.8 show examples of visual management communication methods employed in Nissan.

DIAGNOSIS

Objectives diagnosis is done by a formal review of objectives, strategies, targets and control items using the Nissan Review Form RF1 document. This is top management led (EVP/SVP/VP/GM) and part of a monthly company review at OC-E (EVP/SVP/VP), a monthly functional review (SVP/VP/GM) and a monthly review at the VP/MD to GM level.

Figure C1.9 shows a typical Review Form 1 Scorecard in standard A3 format, the purpose of which is to identify progress of Objectives (Results) and Strategies (Process) and identify good points/bad points and future actions. Three consecutive red signals on the RF1 requires an RF2 document to be produced.

ACTION

Diagnosis findings are used to identify objectives and strategies that are failing, request an analytical approach to 'Control the Source' of failing objectives and strategies, highlight activities needing remedial action, re-direct resources to where they are more needed and to stop reoccurrence of problems. The 'Catchback Plan' is set down in the RF2 form which:

ID	Task Name	J	A	S	O	N	D	J	F	M
1	**NMUK manufacturing cost planning:**									
2	Finance attend #11 GMCM FY '05 manufacturing budget guidelines meeting			12/09 □ 14/09						
3	Validate #11 GMCM FY '05 manufacturing budget guidelines			19/09 □ 10/11						
4	IPEC #17 (official release of FY '06 manufacturing budget guide)			05/10 ◆ 06/10 ◆ 10/11						
5	NMUK MC approval of cost targets by category by department				11/11 ◆ 17/11					
6	Imazu-san visit (NML SVP manufacturing) for budget approval				21/11 □ 02/12					
7	Breakdown of company cost targets to shop/manager level									
8	**Objective/strategy development:**									
9	Review of NE manufacturing value-up mid term plan (2005–2007)		15/07							
10	Development of NMUK value-up mid term plan (2005–2007)	07	26/08							
11	MD/TQM office to issue list of '06 OSP sheets to be developed			26/09 ◆						
12	Preparation of '06 OSP sheets by departments (include 'best of the best' benchmark data)			26/09 □ 28/10						
13	Departments to issue OSP sheets to TQM office				31/10 ◆					
14	Director's lodging event (proposed Slaley Hall)				04/11 ◆ 05/11					
15	Produce draft '06 NMUK annual objectives plan (AOP)				07/11 ◆					
16	MD office communication of company objectives to managers				14/11 ◆					
17	**Objective/strategy deployment:**									
18	Development of department OSP sheets				31/10 □ 25/11					
19	Developmental lodging events				11/11 ◆ 02/12					
20	Development of departmental annual objectives plans				21/11 ◆ 02/12					
21	Departments to issue AOPs to TQM office				05/12 ◆					
22	TQM office to 'link check' company and departmental objectives				05/12 ◆ 09/12					
23	Formal issue of '06 NMUK annual objectives plan (AOP)				12/12 ◆					
24	Preparation of departmental TQM boards (directors)				21/11 ◆ 16/12					
25	Completion of TQM boards (directors)				16/12 ◆					
26	Preparation of shop/manager level annual objectives plans (AOP)				14/11 ◆ 09/12					
27	Preparation of shop/manager TQM boards (managers)				28/11 ◆ 09/12					
28	Completion of TQM boards (managers)				16/12 ◆					
29	Development of manufacturing cross shop job share plans (production level)				14/11 □ 16/12					
30	Manufacturing cross shop job share meetings (production lead)				09/01: ◆ 03/02					
31	Preparation of senior level Annual objectives plans (AOP)				21/11 ◆ 16/12					
32	Preparation of supervisor/engineer/controller/AO/TO/grade annual objectives sheet (AOS)				05/12 □ 03/02					
33	Departmental communication of objectives to staff				09/01 ◆ 03/02					
34	TQM office to issue policy and objectives posters and cards				16/01 ◆ 03/02					

Figure C1.6
Master schedules

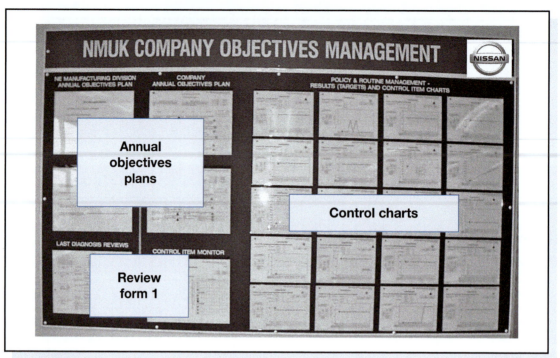

Figure C1.7
Communication – visual management

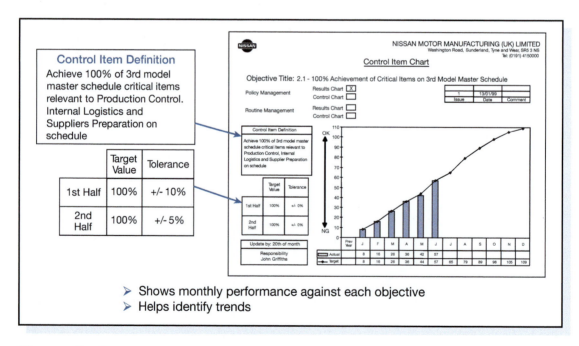

➤ Shows monthly performance against each objective
➤ Helps identify trends

Figure C1.8
Visual management (control charts)

OC-E MTOP RF1 Scorecard (Results)

NISSAN POWER 88 [OC-E Mid-Term Plan Policy Statement]

"Power" derives its significance from the strengths and efforts we will apply to our Brands and Sales. We will renew our focus on the overall customer experience. "88" refers to the measurable rewards from achieving our plan. We aim to achieve a Global Market Share of 8% and we will increase our Corporate Operating Profit to a Sustainable 8%.

Legend:
- Achievement of Target
- Between Target & Commitment
- Below Commitment

Trend: ↑ / → / ↓

Issue: I — Issued after April OC-E Meeting; N — Issue N (After March OC-E Meeting); Date; Comment

Results & Judgement

Objective or Strategy	O or S	EVP/SVP/CVP	VP/GM	Contr.	FYXX Target Commit.	FYXX Contr.	FYYY Target Commit.	FYZZ Target Commit.	Actual Results/Target (September) Results	Judgement A	M	J	J	A	S	Trend	Forecast (Based on EOY Target)	Forecast (Green/Amber/Red)	Good Points/Bad Points/Future Actions
1.0 Reinforce Brand & Sales Power	O														●				Annual Survey
1.1 Improve Brand Power	S									N/A	N/A	N/A			●	↑		●	
1.2 Improve Overall Opinion	S									N/A	N/A	N/A	N/A		N/A	→			TBA
1.3 Improve Purchase Consideration	S									N/A	N/A	N/A	N/A		N/A	↑		●	
1.4 Increase Sales Quality	S									●	●	●	●	●	N/A	←		●	
1.5 Increase Service Quality	S									●	●	N/A	N/A		N/A	←		●	
1.6 Increase Loyalty Ratio	S									N/A	N/A	N/A	N/A		N/A	→		●	
1.7 Increase Network Power	S									N/A	N/A	N/A	N/A		N/A	←		●	
1.8 Improve Cost of Ownership	S									N/A	N/A	N/A	N/A		N/A	N/A		N/A	
1.9 Improve Product Quality	S									●	●	●	●	●	●	↑		●	
1.10 Achieve "Top 20" European	S									N/A	N/A	N/A	N/A		N/A	→		●	
1.11 Achieve Zero Emission Leadership	S									N/A	N/A	N/A	N/A		N/A	↑		●	
2.0 Increase Regional % Market Share	O									●	●	●	●	●	●	↑		●	
2.1 Business Expansion (Passenger Vehicles)	S									●	●	●	●	●	N/A	↑		●	
2.2 Business Expansion (Fleet)	S									●	●	●	●	●	N/A	←		●	
2.3 Business Expansion (Russia)	S									●	●	●	●	●	●	←		●	
2.4 Business Expansion (LCV)	S									●	●	●	●	●	●	→		●	
2.5 Business Expansion (Infiniti)	S									N/A	N/A	N/A	N/A	N/A	N/A	↑		●	
3.0 Deliver Regional % COP Growth	O									●	●	●	●	●	●	N/A		●	
3.1 Deliver Nissan Passenger Vehicle Profit	S									●	●	●	●	●	●	←		●	
3.2 Improve Nissan After Sales Profit	S									●	●	●	●	●	●	N/A		●	
3.3 Deliver Nissan LCV Profit	S									●	●	●	●	●	●	N/A		●	
3.4 Deliver Infiniti Vehicle Profit	S									●	●	●	●	●	●	N/A		●	
3.5 Achieve G&A Cost Budget	S									●	●	●	●	●	●	N/A		●	
3.6 Reduce Currency Risk	S									●	●	●	●	●	●	↓		●	
3.7 Control Investment Costs/Budget	S									●	●	●	●	●	●	←		●	
3.8 Reduce Total Inventory (Value & Stock Level)	S									●	●	●	●	●	N/A	↑		●	
3.9 Deliver Cost Leadership (TdC by Model)	S									N/A	●	●	●	●	N/A	↑		●	

Figure C1.9

RF1 scorecard results

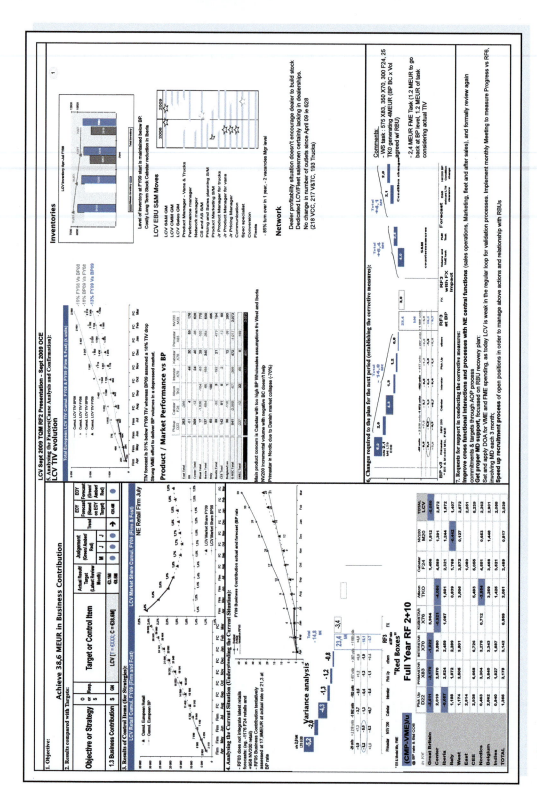

Figure C1.10
Review form 2 (RF2)

- applies a systematic approach to problem solving using facts and data not 'gut feel'
- analyses the current condition
- identifies the root cause of a deviation from the planned achievement
- develops countermeasures for the root cause
- considers how to standardize countermeasures
- uses a standard pro-forma document to improve consistency across all users
- simplifies feedback onto a single sheet of A3.

Figure C1.10 gives an example of a completed RF2 form.

THE TQM OBJECTIVES MANAGEMENT PROCESS IN NISSAN IN SUMMARY

Figure C1.11 provides a simple summary of how the TQM OMP works in Nissan, including the supporting documentation. This helps the senior management in the business to give clear direction, address critical breakthrough issues, get strong support and break habits in order to create new breakthroughs.

From benchmarking studies Nissan have found that too many organizations try to implement an improvement methodology/programme selected by the senior management team, which quite often is implemented without understanding the rationale or the real issues, and yet people are expected to deliver measurable improvement. Often results are linked to individual remuneration, there are lots of initiatives without any real drive, and this can result in lack of sustainability. In Nissan they identify issues, look at priority, urgency, cost benefit, resources and select appropriate tools for the job, aligned with the needs of the business.

Brand image is a very important part of managing quality in Nissan, building on the relationships with customers to create an exemplar company, recognized for excellence in everything they do. The TQM OMP is key to ensuring everyone has 'business acumen', with all levels understanding what the customer wants and their responsibilities for delivering it. Having a pipeline of professional, highly skilled and capable people, with the right training and development so they understand the tools techniques, ensures effective deployment of policies and strategies. This is coupled with good change management – the ability of the people in the organization to understand the need to change and align with what they need to achieve.

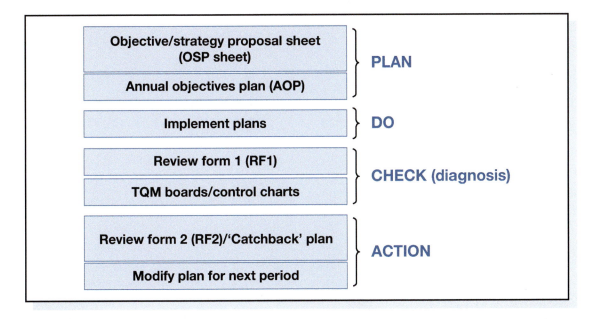

Figure C1.11
TQM objectives management process structure (PDCA based)

ACKNOWLEDGEMENT

The author is grateful for the help and information provided by Kieran McGonagle, TdC Coordination, John Martin and Chris Benardis, Nissan Motor Manufacturing, in the preparation of this case study.

DISCUSSION QUESTIONS

1. Evaluate the Hoshin Kanri based approach of objectives management adopted by Nissan, in relation to the nature of their business in the automotive sector; could this type of approach be adapted for an organization operating in the health or education sector?
2. Discuss the leadership, commitment and strategic requirements of the type of approach used in Nissan.
3. What role could benchmarking play in the development of the OMP methods used?

Case study 2   *Sustainable business improvement in a global corporation – Shell Services*

BACKGROUND

Setting up a new global organization is a challenge in itself. To do this by harmonising existing, but different business operations across the world into a single, global organization adds another level of complexity. This case study describes how Shell Services enabled such a transformation by developing and putting in place a set of tools, processes and systems that became known as the Shell Services Quality Framework, or SQF. To put the organization into context, Shell Services comprised several companies across the globe employing some 6,500 staff with a turnover in excess of £1 billion.

With a clear focus on becoming a customer-centric organization, there was a need to look at the core processes required to sustain improved business performance as perceived by customers. Many of these processes were broken. At the same time, it was recognized that without helping the people in the organization to embrace the values, behaviours and competencies necessary to become customer-centric, the vision could not be achieved. Finally, both people and process improvements had to be underpinned by a quality framework that could be used to define standards, targets and metrics as well as tracking performance improvements over time (Figure C2.1).

With such a diverse and complex organization, no one existing quality model was seen as offering a suitable basis for harmonisation and inclusivity. Although some proprietary models were favoured locally, there was seen to be benefit in seeking to bring together the best of these into a Shell specific product. Criteria such as simplicity with completeness, inclusion of best practice, availability of supporting tools and suitability for self-assessment were chosen and several well-known quality improvement approaches were researched to arrive at the SQF (Figure C2.2). Each model contributed attributes and strengths, but no single model offered the power, simplicity and completeness of the SQF.

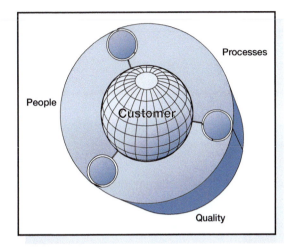

Figure C2.1
Components of a customer-centric strategy

STRUCTURE OF THE SQF

At the top level, the SQF is a simple but powerful construct consisting of five key chevrons. Four of these are enablers – namely Purpose, People, Resources and Process. The fifth is the Results chevron, which focuses on tracking performance improvement as a result of implementing the framework (Figure C2.3).

Although this may seem a simple construct it has proved tremendously valuable, even at the top level, to ask a simple question about each of the five chevrons. A satisfactory answer is somewhat more difficult to provide than a business leader might expect.

Each of the chevrons is broken down into level 2 and level 3 components in order to define key descriptors, for which tiers of practice, including best practice, can be defined at level 4. This is best illustrated in Figure C2.4, where the Process chevron is taken down to level four of the framework.

447

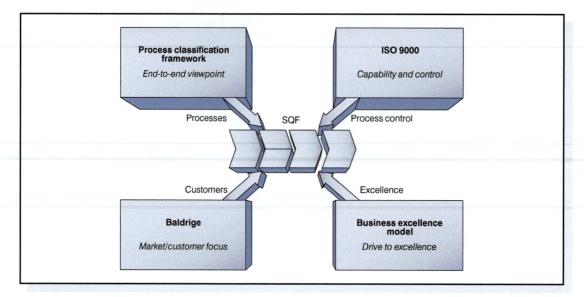

Figure C2.2
SQF heritage

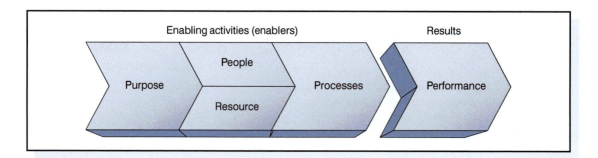

Figure C2.3
The SQF – a simple but powerful construct

GETTING STARTED WITH THE SQF

As part of the validation process for the SQF a baseline assessment was carried out across the organization to determine the starting point for performance improvement and to 'prove' the SQF in practice. This yielded valuable data which served both objectives. Some 80 managers and leaders were interviewed and asked to assess where they thought their part of the organization was in comparison to the tiers of practice in level 4 of the SQF. A fundamental finding was

confirmation of virtually no performance measurement in many areas – indeed the organization did not rate at all against this component. Other key findings were: good articulation of aspirations and purpose, but limited cascade and execution through the line; very few common processes implemented across the organization; supplier relationships not effectively managed; lots of initiatives activity around knowledge management and virtual team working but little collection of institutional knowledge and intellectual capital; many valuable initiatives in place to improve

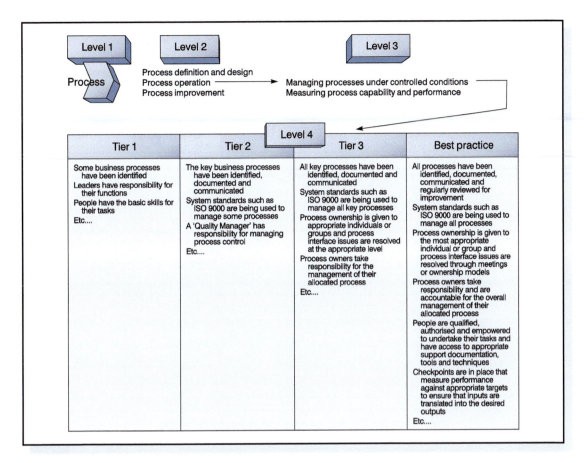

Figure C2.4
Cascading the SQF down to best practice

overall performance but signs of initiative overload and limited capacity to follow-through.

Although a sobering exercise it proved invaluable in demonstrating the need for a systematic approach to improving the business, and in all areas. Figures C2.5 and C2.6 illustrate some of the findings from the baseline activity.

SQF TOOLS AND TECHNIQUES

In order to achieve a sustained and consistent approach in using the SQF it was recognized that a set of supporting tools and techniques were required, which together would allow the full benefits of the SQF to be realized in achieving improved business performance. This had to be in place in order to begin to address complex, outdated and broken processes across the organization. This was the genesis of the Business Improvement System shown in Figure C2.7.

IMPROVEMENT TOOLS AND TECHNIQUES

In order to operationalize this system a structured approach to process improvement was developed around the DRIVER methodology. This provided the tools and techniques to analyse business problems in detail. It also provided a systematic way of identifying and implementing solutions and supported the use of tools such as affinity diagrams, cause and effect analysis, force field analysis, metaplanning and Pareto analysis. The six steps within DRIVER are outlined in Table C2.1 (see also Chapter 13).

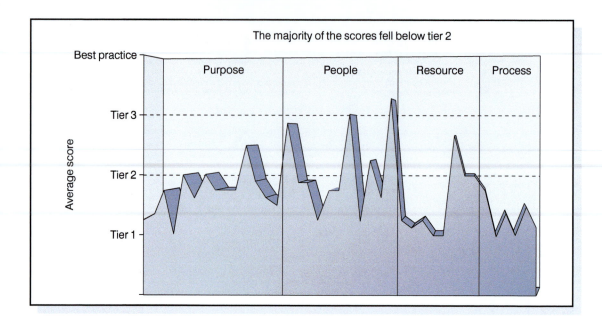

Figure C2.5
Average baseline findings

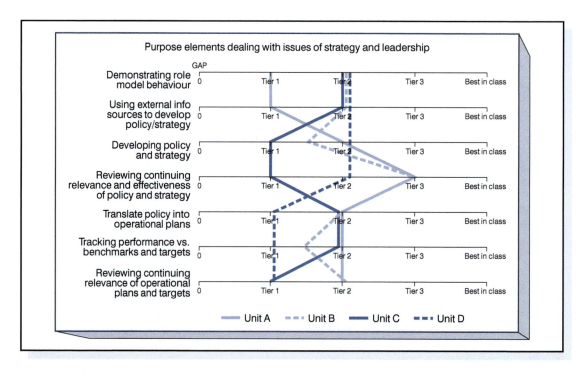

Figure C2.6
Purpose elements dealing with issues of strategy and leadership

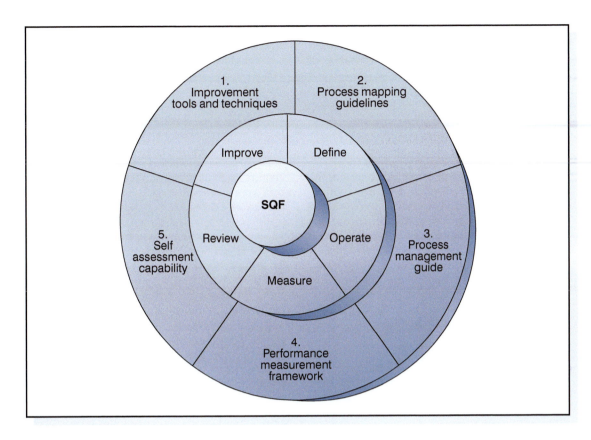

Figure C2.7
Business improvement system (BIS)

Table C2.1 The six steps in DRIVER

Stage	Output
D Define scope of process improvement	Purpose, scope and success criteria
R Review 'as is' process	Defined process, problems and performance
I Identify 'should be' process	Defined improvement (process, problem prevention and performance)
V Verify improvements	Prioritized and verified improvements to deliver benefit
E Execute improvement delivery	Deployed improvements
R Reinforce implementation	Measure of improvement benefit

PROCESS MAPPING GUIDELINES

This provided the ability to understand a process by graphical analysis. This was seen as necessary to build a common understanding of issues affecting the process in question, as well as a way of dealing with purpose, inputs, ouputs, resources, controls and interfaces. Such maps were invaluable to understand the 'as is' position when process re-design is appropriate.

PROCESS MANAGEMENT GUIDE

This is where the SQF really came into its own. At level 4 (see Figure C2.4) it provided a clear definition of process operation reflecting best practice, and essentially provided the bedrock on which new processes were defined. This forced issues to be considered such as the operation of processes under controlled conditions, the concept of process ownership, and starting to think about performance measurement.

PERFORMANCE MEASUREMENT FRAMEWORK

This was probably the most difficult aspect of the whole Business Improvement System to implement. A set of project templates were developed that allowed the selection of outcome metrics for any standard business process. Properly assembled, this provided, for the first time, a regime of measurement and tracking for key processes, aimed at achieving the 'perfect transaction' in spirit if not in practice. Considering the results of the original baseline study referred to above, this was the area most in need of improvement.

SELF-ASSESSMENT CAPABILITY

The intention of this component was two-fold: first, to provide an on-going self-assessment capability where templates, questionnaires, scoring sheets and action plans were provided based on the SQF;

second, to allow those parts of the organization for whom accreditation and recognition was important to use the SQF and the Business Improvement System to test their readiness for achieving their goals.

Once these tools and techniques had been developed and integrated into the Business Improvement System, two pilots were undertaken to validate and fine-tune the approach. The first was improving the accounting processes around joint ventures in Exploration and Production. The result here was a simplified and shortened end-to end process, resources released for more value added tasks and higher customer satisfaction levels. The second pilot addressed problem management in the IS service delivery organization. Again, by looking afresh at the process, particularly at performance tracking, the end result was a global, stable process with defined service levels and performance tracking – and happier customers.

SQF AND BUSINESS IMPROVEMENT

Once the SQF had been developed, proven in pilots and supported by a Business Improvement System, the task then became one of deploying the SQF such that priority areas were addressed,

Table C2.2 Progress in Business Improvement areas

Business Focus	Applied to	Purpose	Examples to date
Alignment	Plans, Projects, Options	To ensure alignment with business strategy and priorities	Communication Strategy, Leadership Team priorities, Strategy Renewal, HSE Strategy
Choice	Initiatives	To help prioritize by considering purpose, impact, resources, capacity, etc.	Business Improvement Programme, HR Agenda, Stress Management Strategy, Cost Reduction Process, F.I.R.S.T.
Improvement	Process	To improve or redesign core processes, with a focus on best practice	Contract to Billing Process, Account Management Process, Property Sales, Joint Venture Accounting, Service Delivery
Assessment	Organization	To self assess for continuous improvement	Customer Service Teams, Customer Satisfaction with HR
Recognition	Business	To achieve accreditation/recognition against best in class standards	Later

improvements could be sustained and the whole approach could begin to permeate the organization. There is no doubt the SQF and supporting tools can be used wherever business problems exist and several areas were targeted. Perhaps one of the most powerful was in helping the whole organization come together around key management thrusts for the year – **FIRST**: **F**ocus on customers, **I**mprove billing, **R**educe cost, **S**ervice excellence, **T**alent development. Work was carried out with the executive team to arrive at these top five thrusts by using the SQF to distil the really important strategies from a much longer 'wish list', again forcing clarity around real purpose, business impact and performance measures. Using the SQF to help with choices in this way it was possible to reinforce important aspects of organization and culture, so critical when aligning and mobilising support from across the whole organization.

As a summary, progress in the areas of Business Improvement is highlighted in Table C2.2.

It is worth mentioning here another area of SQF deployment in a little more detail: discretionary expenditure – The 'Business Improvement

Programme' in Table C2.2. Every organization spends time and money on undertaking projects, hopefully to improve key processes, market standing, profitability, reputation, etc. Faced with a wide array of some 35 IT related projects, all seemingly justified in their own right and amounting to some $80 million, the question was 'how do we ensure we are working on those projects that will provide best value for money and clear alignment with our business objectives?' Furthermore, 'how do we exercise some degree of control over such a disparate set of projects ranging from billing improvement, through knowledge management to hardware renewal?' And finally, 'how do we ensure a first class strategy ends up in a first class implementation?' The answers lay in using the SQF as a 'filter' to address what may appear to be 20 simple questions, but ones that were to prove to be worth their weight in gold (Figure C2.8).

Taking just one question as an example from Figure C2.8 – the second one under People – 'What is our capacity to implement the degree of change required for successful implementation of this project?' This was a powerful question often

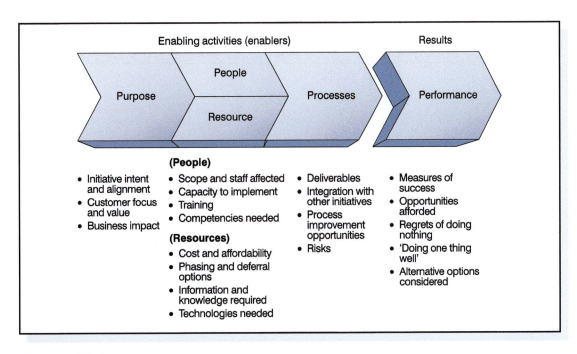

Figure C2.8
SQF twenty questions to ensure a structured approach to any improvement opportunity

Case 2 Business improvement: Shell Services

overlooked in the rush to deliver improvements across the organization. Every organization has a limited capacity for major changes at any one time. Is it feasible to ask people to implement a new billing system while a global help desk is being implemented *and* a major office move is under way?

Through this process of using the SQF to ensure discretionary expenditure was being properly considered, the leadership team found they were able to make better decisions, generate 100 per cent support for priority projects and provide an environment for the best possible chance of successful implementation. Even more than this, some parts of the organization started to ask the question at every management team meeting – 'has this idea been SQF'd?'

On an even more simple scale, value was derived by asking just five key 'acid test' questions about any initiative under way in the organization. Sometimes these proved quite difficult to answer for even the most well-understood projects! These are shown in Figure C2.9.

MY SQF

Once the SQF had been introduced for organizational improvements and it was accepted as a tool for improved decision making, there was another, perhaps more important aspect yet to be developed – using the SQF as a tool to help people in their own personal work planning, and maybe even to start to address the issues of work/life balance.

Putting the SQF into a small leaflet format that could be distributed to each member of staff proved immensely valuable. It helped in discussions between staff and supervisors, where real issues around their workload, priorities and challenges could be discussed in a way that took the heat and emotion out of the debate. For an individual reviewing work priorities, it was important to be absolutely clear about the purpose of a task or project, as well as having a good understanding of what was required to achieve success in implementation. It also meant having the confidence and a fair basis for saying **No!** to some activities.

NEXT STEPS

During the two years of development , testing and implementation of the SQF, it was never found wanting in terms of an area of the business where it could not be applied. One of the most difficult aspects was getting people to accept the simplicity

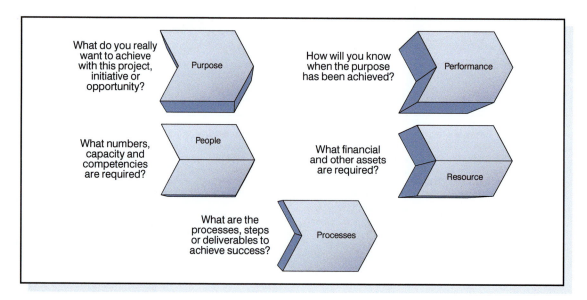

Figure C2.9
Five Easy Questions!

of the construct, and not to look for complexity – as if it were perhaps a test of intellectual rigour. The SQF was deliberately predicated around the fact that it would outlast business fads and theories – keeping it down to earth meant that it was bound to be successful over time and not thrown away when the next idea came along. Managers will keep using the SQF to provide an anchor to business improvement – it costs little, is difficult to argue against and helps the most important asset in the organization – the people!

ACKNOWLEDGEMENT

The author is grateful for the contribution made by Roger Wotton in the preparation of this case study.

DISCUSSION QUESTIONS

1. List and discuss the advantages and disadvantages of developing an organization specific quality framework, such as the SQF, in comparison with adopting an existing framework such as the Baldrige or the EFQM Excellence Model.
2. Critique the high level SQF against the EFQM Excellence Model showing the overlaps and 'underlaps', and providing advice to the management of Shell Services on how their framework could be developed and improved.
3. Public sector organizations could adopt the SQF. Compare and contrast the SQF with other initiatives and standards that have been adopted recently in government agencies and discuss the merits of each.

Case study 3   Lloyd's Register improvement programme – group business assurance

BACKGROUND

The Lloyd's Register Group (LRG) provides independent assurance to companies operating high-risk, capital-intensive assets in the energy and transportation sectors, to enhance the safety of life, property and the environment. This helps its clients to create safe, responsible and sustainable supply chains. Safety has been at the heart of LRG's work since 1760 and the company invests time, money and resources to fulfil Lloyd's Register's mission: to protect life and property and advance transportation and engineering education and research. Through impartial advice, broad knowledge, deep experience and close relationships LRG help ensure a safer world.

LRG is also one of the world leaders in assessing business processes and products to internationally recognized standards. The standards are either those of major independent bodies or ones that LRG have developed themselves. From design and new build to in-service operations and decommissioning, LRG aim to deliver complete lifecycle and risk management solutions to help ensure the safety, integrity and operational performance of assets and systems. Clients are typically managing large-scale, high-value assets where the cost of mistakes can be very high, both financially and in terms of the impact on local communities and the environment. In such a setting, organizations need advice and support they can trust.

LRG employ 8,000 people who work in 186 countries, turn over in excess of £900 million with an operating surplus nearing £100 million.

To quote the Chief Executive,

> Business Assurance is not something the business does. It defines the business. It is the control that ensures our people processes are linked to our clients at a performance standard set by the company. It is the only corporate activity that we want to 'disappear' by growing stronger as the quality system becomes integrated into our normal activity.

OVERVIEW OF THE LLOYD'S REGISTER IMPROVEMENT PROGRAMME

Lloyd's Register is run by following a 'closed-loop top management system', following best practice, which develops and drives the group strategy management system and links the improvement and investment portfolio to it (Figure C3.1).

The company describes a before and after situation as follows.

Before: no group strategy – individual strategies and tactics; no system for exec performance reviews other than financial performance. Prior to the programme there was little or no connection between strategy and operational performance. The two activities were run separately with strategy being mainly about setting a general course. The monthly operating report did not make any reference to strategic objectives or review actual performance against specific non-financial KPIs. Quarterly group performance reviews were entirely based on financial performance.

After: significance of non-financial KPIs – risk, we do it now. After a couple of iterations they now have a system that is used to run Lloyd's Register by the entire executive team. It is led by the CEO and underpins the way LRG work. At a global leadership event (c.100 top people from across LRG) the strategy map, closed loop system and dashboard were the first and last slides the CEO used on the day. Amongst the more significant changes are the inclusion of key account metrics and a corporate risk management framework into the system. The latter is used by the CEO to update the Board to which he reports.

Any improvement activity is driven by gap analysis from the performance metrics, therefore targeting strategic and operational priorities. That way the impact of the improvement activity can be measured in terms of the process output performance.

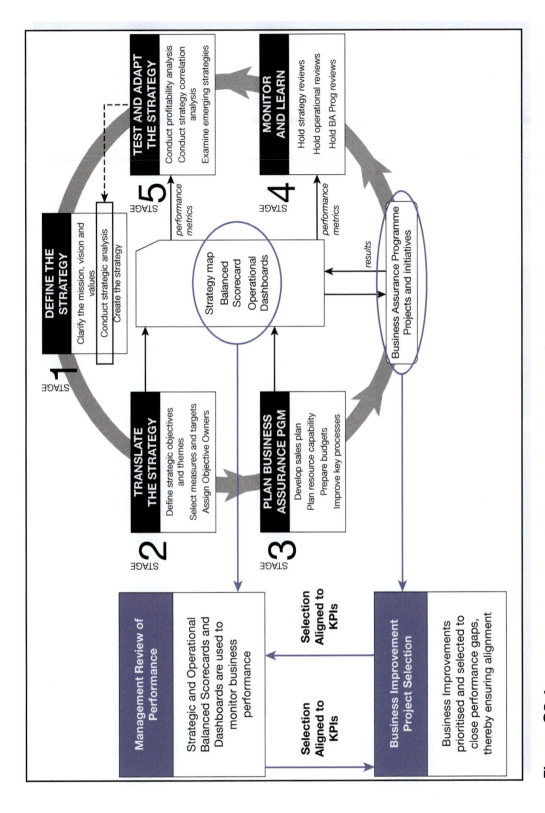

Figure C3.1
Closed loop improvement cycle

The improvement strategy in LRG has been developed from a centralized to a decentralized approach:

- *Driving strategy using closed loop management*: Improvement is at its most powerful when linked to strategy. Use of the Kaplan and Norton closed loop model (Balanced Scorecard) demonstrates where gaps exist and identification of projects that will close these gaps has enabled LRG to truly claim this as a mode of operation.
- *Large scale visible projects driven by an effective PMO*: LRG have positively influenced the leadership team through executing large scale centrally led projects that have been aimed at areas that are visible in the business, with a fully functioning PMO (using the P30 model) and an Executive portfolio board.
- *Decentralized, locally owned, improvement:* Having gained credibility LRG have decentralized through training of people within the business and carrying out a large variety of smaller scale six sigma, lean and consultancy-style interventions, using a light training but heavy coaching approach.
- *Using key input process variables effectively:* LRG believe that an effective programme is determined by a number of KPIs. They have identified and worked on each element to try to ensure that it is in place, effective and appropriate to the business.

LINKING STRATEGY TO IMPROVEMENT

As Figure C3.2 shows, 'Results Maps' link everyday KPIs to 'Strategic Objectives'; these are visual, repeatable and easy to understand. The 'Results Onion' forms the heart of the crucial process of deciding which KPIs are important to the business. It also enables effortless creation of 'composite KPIs' and it also offers a unique way to communicate the linkage between strategy and individual scorecards. These maps were developed with the key stakeholders and for a suite of 10 maps, giving an audit trail that shows why specific KPIs were chosen and how they interact.

As Figure C3.3 shows, using the PMO the executive team have identified the top **30 critical activities** that are happening and need to happen within the business in order for LRG to achieve the

5 year strategic plan. Business assurance has been pivotal in developing the plan and supporting the resultant activities.

The development of this long-term line of sight for major change activities has enabled LRG to prioritize and schedule major change and local change so that the business does not become overloaded. The top 30 initiatives have been widely communicated in order that people are aware of the change agenda.

MANAGEMENT INFORMATION, REPORTING AND REVIEW

Communicating performance: performance information – both business performance and specifically programme performance – is presented through dashboards and reports which use research tested design principles to improve ease of understanding and analytical comparison. Design of graphics, icons, use of colour and real estate are all carefully considered to draw the user to key issues, allow information to be quickly assimilated and ensure consistency across the reporting and review activities.

Great KPIs and metrics: KPIs are designed based upon the strategic objectives detailed on the strategy map, and tested and approved by the leaders of the organization. Improvement work is specifically targeted against improving measured performance in order to achieve the strategy. LRG aim for crystal clear KPI definition. Each KPI is clearly documented, showing calculations, known issues, information sources, production and performance owners (RACI for each). Each KPI has its own definitions sheet and is logged and maintained in the KPI Index database.

Putting the information in the hands of the right people: LR plan to ensure the right performance information – including programme delivery – is reviewed by the right level of the organization. Programme boards involve the relevant executive members and they are trained in their role as board members and champions. In detail, LRG have formal design and terms of reference for meetings, putting improvement and fact-based decision making at the heart of the meetings process. The focus is on prioritized issues (by impact) and the results of agreed actions.

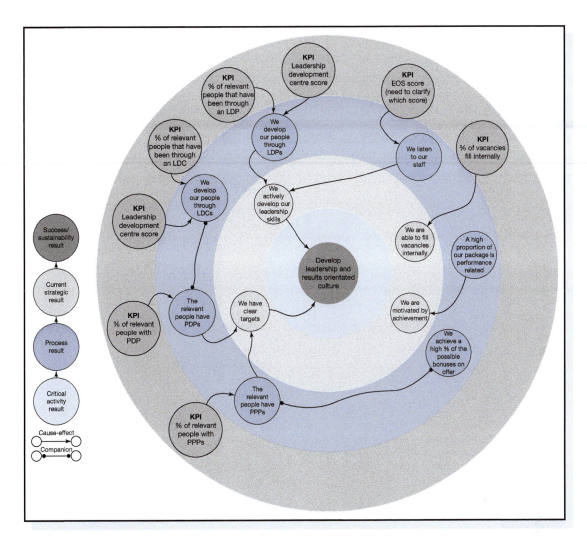

Figure C3.2
Linking strategy to improvement – results maps

MANAGING SELECTION AND DELIVERY

Full governance through OGC P3O: Using the OGC P3O Standard LRG apply full governance to the portfolio of activity within the business. This is managed within PowerSteering software and the processes of identifying, selecting, running, controlling, closing and reporting all activities that have been fully documented within the management system. The process of developing this document was a milestone in maturity as they

were able to examine their own practices and ensure the use of the very best approach available.

A common approach and use of lifecycle gates: LRG are able to manage the full range of activity carried out in this way from large scale Black Belt projects through to minor interventions; all have a common approach and set of lifecycle stage gates.

Common pipeline management: local approach to delivery: This unified approach enables LRG to effectively manage its own resource and the pipeline provides visibility of

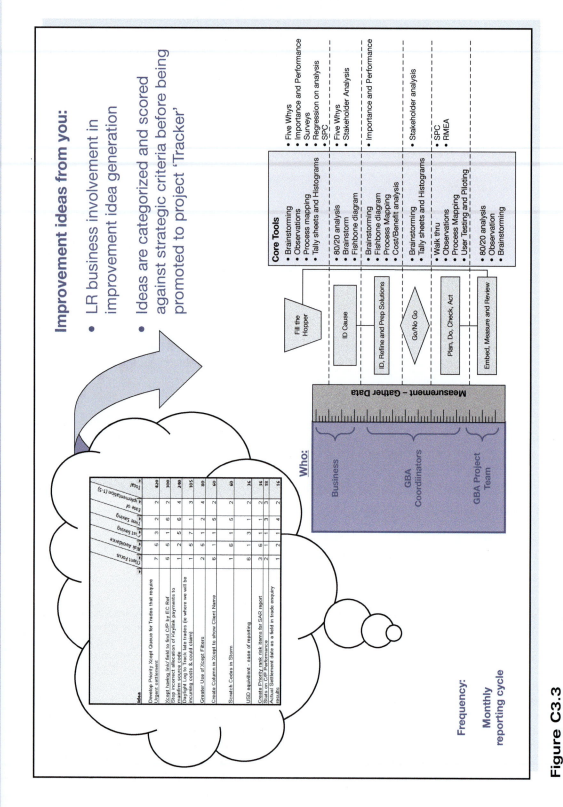

Figure C3.3
Linking strategy to improvement – critical activities

the activity and, importantly, accountability. The PMO typically manages a Pipeline of over 100 opportunities with a current activity of 15 centrally driven projects and over 40 small scale local activities.

DELIVERY – BEING PRACTICAL

Demand-led training – never 'sheep dip': Delivery of identified projects is a key activity within any programme. This builds image, reputation and ultimately credibility. LRG understand this and actively work with this in mind. The engine of the change activity is ensuring the recruitment of talented people and LRG try very hard to recruit the best people, preferably from within the business, and train only those that are interested. LR claim they will NEVER EVER 'sheep dip' people in Six Sigma, firmly believing this leads to corporate cynicism.

Blending Six Sigma, Lean and P3O: LRG have identified that the structured thinking that Six Sigma methodology brings is very well accepted within the business. However they have also worked extensively to blend Lean thinking into the offices through an alliance net work across the business. These two cornerstones, along with the P3O framework for governance, ensure clear documentation, with well-recorded outcomes, ensuring 'learn as we go' and that knowledge is retained.

Active engagement of sponsors: LRG work extensively to ensure that project sponsors are actively engaged with each project's steering committee, each playing a working role in the activities. The steering committee has been identified as a key success factor in projects that work well – they have their roots in this forum. The relationship between this working group and the 'Portfolio Board' is also important.

High satisfaction and success: This approach to delivery works with both very large and very small projects; sponsor satisfaction and project success rate are consistently above 95 per cent.

HOW THE IMPROVEMENT PROGRAMME ENGAGES WITH THE BUSINESS

Management engagement: LRG are active in influencing all levels within the company, from the group director of business assurance through to a network of over 25 dedicated business assurance leaders through each geographical region and business stream. This ensures that business assurance is always on the agenda and well represented.

Associated people engagement: Each project has a dedicated change methodology attached to it. The development of CHAT (Change Acceptance Techniques) ensures that the purpose of each project is well understood by all. An effective communication plan is developed and tools are in place to work through the change challenges that the projects inevitably bring. This is vitally important and recognized as a key success factor.

Whole business engagement: LRG 'Inform, Remind, Reassure' as appropriate through the Intranet with the 'whole business', this approach providing a broad coverage. On a project-specific level LRG use more personal techniques, such as 'Lunch & Learns' and 'Local Briefings', which make good use of the alliance network, printed media, eLearning, WebEx, being sensitive to the fact that the business is widespread with a diverse culture. There is a 'staff suggestion scheme' that enables everybody to be able to enter ideas through feedback handling schemes.

Width of project engagement: LRG are moving from being highly centralized around the UK headquarters to a decentralized approach. Projects are active in diverse locations such as Hong Kong, Rotterdam, Utrecht, Athens, Houston, Canada , Kuala Lumpur , Germany, Korea and Aberdeen. The diversity of problems that these are aimed at is remarkable, ranging from cash management, improved cycle time for surveyor certification, lifting appliance inspection, development of a new management system, ISO 14001 implementation, HR starter/leaver process, creation of a technical centre of excellence, SMART bidding, SMART project management, transfer of class cycle time reduction and value of group support.

DEVELOPING LOCAL SKILLS AND IMPROVEMENT ABILITY

Avoiding 'sheep-dip' cynicism – Honeycomb, a new approach: Early on LRG recognized that 'sheep dip' Green Belt training was not the route they wanted to use, but they also knew that a critical success factor was to get people involved at a local level. To achieve this they developed a twin approach Green Belt as a method of developing wider alliance and 'Honeycomb' as a method of developing light process improvement skills within the business.

Lots of small projects, making a big difference, with effective coaching: Over six waves of Honeycomb training have been completed with over 40 closed projects, but more importantly there are 40 people in the business who have each made a small difference. LRG recognize that coaching is a critical success factor and the lack of effective coaching can be a limiting factor in the speed of deployment.

DEVELOPING LOCAL SKILLS AND IMPROVEMENT ABILITY

The Honeycomb approach works at an operational level within the business to develop local Business Improvement capability. It is aimed at removing local 'pain points' that may not be addressed by larger projects. In this way it improves local process efficiency and effectiveness by empowering local businesses to take ownership for local process improvement and remove any 'learned helplessness'.

This enables the business to self-help, drive continuous improvement at the operational level and create stronger awareness of the customer. The practical details include a two-day training on process fundamentals, incorporating Lean Six Sigma and some common sense! Managers are informed through the alliance of the goals and an engagement workshop is held with local management before the deployment begins.

Projects are scoped in the training and attendees start working on them. Each attendee is given a coach from the team and regular coaching sessions and broader 'clinics' are held to help them through the project work. Regular coaching by the group business assurance team has been a critical success factor in making this work.

EVERYDAY PRINCIPLES: IS IT CATCHING FIRE?

It started in Rotterdam where the local management recognized how this activity was helping to develop talent and leadership. They saw an opportunity in the workshop with the 'honeycombers' and identified a wealth of areas for improvement. People were identified as being suitable Green Belts and were trained.

The Rotterdam Improvement programme was born and the activity has now spread across the other businesses within that geographical location. Project success has been at a regional level with the redesign of the assessor certification process, leaning the system and leading to increased billing opportunities. The Rotterdam programme has developed a life of its own with the GBA role being a very light touch, on an intermittent basis.

EVERYDAY PRINCIPLES: IS IT STICKING?

The improvement programme is now part of the LRG Graduate process and in response to this an academy has been developed which details the 'basic quality and improvement principles.' The syllabus is based around the UK Chartered Quality Institute (CQI) body of knowledge (BoK) for a 'Foundation in Quality.' All Business Assurance staff are encouraged to become Chartered Quality Professionals as this adds to their portfolio of knowledge and encourages CPD (continuous professional development) through the Institute.

Typical external presentations run at eight events per year, and the team share the learning that this brings.

LLOYD'S REGISTER UNITE

Unite is the LRG programme of improvement to establish a whole new level of maturity in the management system – it is both an improvement project and an important foundation for all improvement. The programme is actively sponsored by the Executive Team, demonstrating their commitment to improvement and integrating performance with the processes. Figure C3.4 shows how Unite provides:

- *True Process Focus:* an 'outside-in' approach to a true process-based management system starting with a clear understanding of the desired executive-defined performance outcomes and ongoing improvements.
- *A Single Standard:* development of an LRG Standard providing a single source of all of the required external and internal compliance outcomes.
- *Creating process line of sight:* clear line-of-sight between the outcomes (the end) and the required cross-functional processes to deliver them (the means).

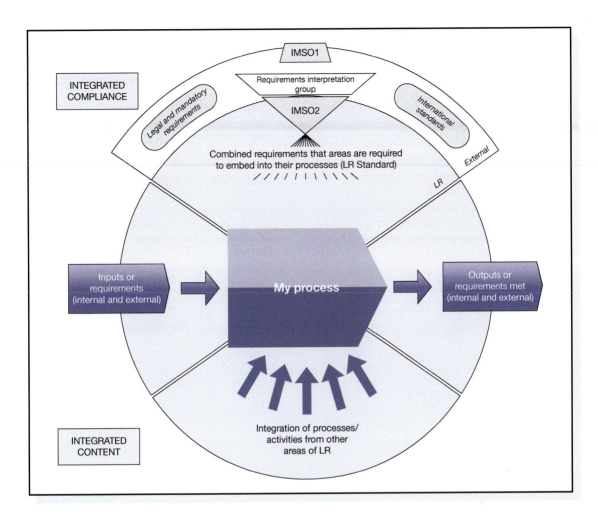

Figure C3.4
Unite

- *Clear processes that enable improvement:* incorporation of a scalable implementation framework that defines ownership and accountability for the management system at each level of the organization, including a clearly defined route for continual improvement through the plan-do-check-act (PDCA) cycle.

THE RESULTS

The non-financial performance section of the monthly operating report aligns to the themes on the Group Strategy map: drive external focus, step change in efficiency and effectiveness and right people, right place, right time. As part of recent strategy refreshes LRG have created a new fourth theme – 'Add Value to Society'. The results from these inform the decisions made by the Executive at the quarterly performance reviews.

PHYSICIAN HEAL THYSELF; APPLY THE APPROACH TO GBA

- **Understand what we are & doing it better:** GBA believe that, by understanding what they are doing and how well they are doing, they will be able to make it better. With this in mind they have spent time developing their part of the management system and appropriate control metrics, through the use of the portfolio office and VOC through sponsors and end users.

- **Getting out of the ivory tower:** they have moved from a centralized to a decentralized deployment and taken people with them.
- **Regular review:** GBA review on a weekly basis performance against agreed milestones.
- **Adapt and be accepted:** GBA continue to adapt and adopt solid principles to the environments they encounter in order to help gain acceptance.
- **Continued evolution:** GBA continue to anticipate, identify and satisfy customers' needs and wants effectively, through the use of solid methodologies, sympathetic approach, strong governance and talented people.

ACKNOWLEDGEMENT

The author is grateful for the help and information provided by Estelle Clark, Group Safety and Business Assurance Director in the preparation of this case study.

DISCUSSION QUESTIONS

1. Why is business assurance so important in an organization such as Lloyd's Register? Describe other business sectors which have needs for similar assurance systems and identify how they would need to be developed to meet the needs of those businesses and sectors.
2. Evaluate the Lloyd's Register closed loop improvement cycle approach in relation to the nature of their business; could this type of approach be adapted for a government department?
3. Discuss the approach to KPIs used in Lloyd's Register and prepare a presentation for a bank on the why, what and how of such a system .
4. What role could benchmarking play in the development of the OMP methods used?

Case study 4   *TQM implementation and policy deployment at STMicroelectronics*

COMPANY BACKGROUND AND TQM

STMicroelectronics is one of the world's largest semiconductor companies with net revenues of nearly US$ 9 billion. Offering one of the industry's broadest product portfolios, ST serves customers across the spectrum of electronics applications with innovative semiconductor solutions by leveraging its vast array of technologies, design expertise and combination of intellectual property portfolio, strategic partnerships and manufacturing strength.

ST focuses its product strategy on sense and power technologies, automotive products and embedded-processing solutions. The sense and power segment encompasses MEMS (Micro-Electro-Mechanical Systems) and sensors, power discrete and advanced analog products. The automotive portfolio covers all key application areas from powertrain and safety to car body and infotainment. The embedded-processing solutions include microcontrollers, digital consumer and imaging products, application processors and digital ASICs.

ST products are found everywhere micro-electronics make a positive and innovative contri-bution to people's lives. The company's world-class products and technologies serve to:

- deliver compelling multimedia experiences to consumers anytime, anywhere – in the home, in the car, and on the go
- increase energy efficiency along the energy chain, from power generation to distribution and consumption
- provide all aspects of data security and protection
- contribute to helping people live longer and better by enabling emerging healthcare and wellness applications.

ST is among the world leaders in a broad range of segments, including semiconductors for industrial applications, inkjet printheads, MEMS, MPEG decoders and smartcard chips, automotive integrated circuits, computer peripherals, and chips for wireless and mobile applications.

STMicroelectronics was created in 1987 by the merger of two long-established semiconductor companies, SGS Microelettronica of Italy and Thomson Semiconducteurs of France, and has been publicly traded since 1994; its shares trade on the New York Stock Exchange (NYSE: STM), on Euronext Paris and on Borsa Italiana. The group has approximately 48,000 employees, 12 main manufacturing sites, advanced research and development centres in ten countries and sales offices all around the world.

Corporate Headquarters, as well as the headquarters for Europe, the Middle East and Africa (EMEA), are in Geneva. The Company's Americas Headquarters are in Coppell (Texas); those for Greater China and South Asia are based in Shanghai; and Japanese and Korean operations are headquartered in Tokyo – it is truly a global operation.

RESEARCH AND DEVELOPMENT

Since its creation, ST has maintained an unwaver-ing commitment to R&D. Almost one quarter of its employees work in R&D and product design and the company spends nearly 30 per cent of its revenue on R&D. Among the industry's most innovative companies, ST owns about 16,000 patents, about 9,000 patent families and over 500 new filings. The company draws on a rich pool of chip fabrication technologies, including advanced FD-SOI (Fully Depleted Silicon-on-Insulator) CMOS (Complementary Metal Oxide Semicon-ductor), mixed-signal, analog and power processes, and is a partner in the International Semiconductor Development Alliance (ISDA) for the development of next-generation CMOS technologies.

MANUFACTURING

To provide its customers with an independent, secure and cost-effective manufacturing machine,

ST operates a worldwide network of front-end (wafer fabrication) and back-end (assembly and test and packaging) plants. ST's principal wafer fabs are located in Agrate Brianza and Catania (Italy), Crolles, Rousset and Tours (France) and in Singapore. The wafer fabs are complemented by world-class assembly-and-test facilities located in China, Malaysia, Malta, Morocco, the Philippines and Singapore.

ALLIANCES

From its inception, ST established a strong culture of partnership and through the years has created a worldwide network of strategic alliances with key customers, suppliers, competitors, and leading universities and research institutes around the world.

SUSTAINABILITY

STMicroelectronics was one of the first global industrial companies to recognize the importance of environmental responsibility and, over recent years, the company's sites have received more than 100 awards for excellence in all areas of sustainability, from quality and product responsibility to corporate governance, social issues, employee health and safety, and environmental protection. The approach to sustainability is set out in ST's Principles for Sustainable Excellence, while performance is reported in detail in the annual Sustainability Reports.

In 1997 STMicroelectronics won the European Quality Award EFQM. This marked the progress made in developing as a world-class organization and also coincided with the tenth anniversary of the formation of the company created by the merger. In order to fully appreciate the achievement of the company since 1987, it is first necessary to describe some of the dynamics of the semiconductor industry since these shaped ST's TQM programme. Microelecronics is one of the most competitive industries in the world with several hundred merchant suppliers, most of them being global players, servicing a huge market worldwide. The economic law of microelectronics is 'when the demand goes up – prices fall; when the demand goes down – prices fall'. Technological advance is very rapid and capital intensity is high. Spending on R&D overall runs at about 14–16 per cent of sales, two to four times higher than most other industries. Every dollar of incremental sales requires a dollar of incremental investment, with the investment usually one year ahead of the sales.

In this environment companies tend to polarize into two groups: the broad line global companies with market shares in the range of 4–7 per cent; narrow niche companies with market shares of less than 1 per cent. A notable exception to this group structure is, of course, Intel, which has a narrow product base but a high market share.

In 1991 ST launched a TQM initiative, based on the European Foundation for Quality Management (EFQM) model, with total commitment from the CEO and all his executive staff. In fact in December 1991 Pasquale Pistorio, CEO stated that: 'TQM is a mandatory way of life in the corporation. SGS-Thomson will become a champion of this culture in the Western world'. These words needed to be backed by action and resource – both financial and people. Very quickly there was a framework put in place, based on an analysis, which determined that the key components of successful implementation of TQM should be:

- Organization
- Common Framework
- Local Initiatives
- Culture Change
- Mechanisms
- Policy Deployment.

Also the programme needed to be driven from the top down, not by dictat, but by example.

There was already in existence a corporate mission statement but it was not closely linked in the minds of the staff with their day-to-day activities. Furthermore it had been written shortly after the merger and did not totally reflect the needs of the company, the shareholders, the employees or the customers. It was, therefore, revised and became the key launching point for all the decisions which affect the future of the corporation.

The mission statement was both short and clear reading:

> To offer strategic independence to our partners world wide, as a profitable and viable broad range semiconductor supplier.

Following the revitalization of the Mission Statement there quickly followed publication of the corporation's:

- Objectives
- Strategic Guidelines

- Guiding Principles
- TQM Principles
- Statement of the Future.

All of these were published in a leaflet titled 'Shared Values' which was circulated to all employees worldwide.

These initial efforts by the corporate management team would have been in vain if the necessary resources had not been provided to support the implementation of TQM. A corporate TQM support group was established, budgets were allocated and the executive management, including the CEO, allocated significant time to TQM implementation. In the initial phase most of the time and effort went into training and communications with regular bulletins, emails and brochures.

The policy deployment process allowed the corporate goals to be cascaded into local goals which were both realistic and challenging. The training programmes, targeted at 50 hours per employee per year, ensured that people had the skills to accept the goals and translate them into local action plans. The management were encouraged to recognize achievements at local, national and international level. Finally strong efforts were made to break down the walls between the various parts of the organization and create an atmosphere in which cross fertilization was not only accepted but actively encouraged, until it became a way of life.

These changes were not easily or readily accepted in all parts of the corporation. Whilst the benefits could be seen on an intellectual plane at a cultural level some groups found it easier to move faster than others. The corporate TQM Vice President described the process as 'pulling down the walls and using the bricks to build bridges'. The difficulty of achieving success cannot be underestimated. ST started with the advantage that many of its European staff had a fundamentally Latin culture and many of the managers had been exposed to American culture, either as a result of working in American companies or interfacing with American customers. Also the semiconductor industry had its own culture which was and still is very strong. None the less cultural barriers still existed and ST had to find ways of working with many different cultures whilst trying to overlay a common corporate culture, ways of working and vision of the future.

POLICY DEPLOYMENT AT ST

Policy deployment (PD) was the primary method used in ST to make TQM 'the way we manage' rather than something added to operational management. In order to make it effective, ST used a simplified approach, combining as many existing initiatives as possible, to leave only one set of key improvement goals deriving from both internal and external identified needs. In this process the management of ST also provided a mechanism for 'real time' visual follow-up of breakthrough priorities to support very rapid progress.

In ST policy deployment is regarded as:

- The 'backbone of TQM'
- The way to translate the corporate vision, objectives and strategies into concrete specific goals, plans and actions at the operative level.
- A means to focus everyone's contributions in support of employee empowerment.
- The mechanism for jointly identifying objectives and the actions required to obtain the expected results.
- A vehicle to ensure that the corporate quality, service and cost goals are given superordinate importance in annual operational planning and performance evaluation.
- The method to integrate the entire organization's daily priority activities with its long-term goals.
- A process to focus attention on managing ST's future, rather than the past.

Policy deployment's place in ST's overall TQM scheme of continuous improvement is illustrated in Table C4.1.

A policy deployment manual, addressed to all managers at any level of STMicroelectronics, was developed as a methodological and operative user guide for those charged with planning and achieving significant improvement goals. Examples, detailed explanations, and descriptions of tools/forms were included in the manual. Policy deployment operates at two levels: continuous focused improvement and strategic breakthrough – referred to as Level 1 and Level 2. The yearly plan is designed by assembling the budget and improvement plan, but also taking into account the investment plan.

All these elements must be consistent and coherent. Current year business result goals are defined in the budget and the underlying

Table C4.1 Policy deployment's place in ST's overall TQM scheme of continuous improvement

Concern	Vehicle	Focus	Responsible	Frequency
Current year Business year	Budget and Operations Reviews, Accounting control	The past/results (Mainly 'What')	Operations management	Monthly/ Quarterly
Prioritise operating results and total capabilities improvement	Policy Deployment Level 1	The future/processes (Mainly 'How')	Steering committees Action owners. All involved personnel	Monthly/ Quarterly
Breakthrough improvements	Policy Deployment Level 2	The near future	Steering committees and all levels of involved personnel	Daily/Weekly
Ad-hoc improvements and problem solving	QIT/PST/NWT	Past, present and future	Steering committee and team members	As necessary
Continuous improvement at departmental level	Next operation as customer (NOAC). Defects per million opportunities (DPMO)	The internal customer The work product	All department managers and personnel	As necessary

operations and capability improvement goals have to be approached using policy deployment. Among all the improvement goals, a very few (one to three per year) are then selected for a more intensive management. These are the breakthrough goals and must be managed using special attention and techniques. Policy deployment goals have to be consistent with long-term policies, and finally, everything must be consistent with and must be supported by the investment plan.

Continuously improving performance and capabilities, and especially achieving 'breakthroughs', i.e. dramatic improvements in short times, was the main task that each manager was asked to face and carry out in his/her activities. Once the importance of achieving dramatic goals was clear, the problem arose of how to identify and prioritize them. To assist ST fixed four long-term policies (broad and generic objectives):

- become number one in service
- be among the top three suppliers in quality
- have world class manufacturing capabilities
- become a leader in TQM in the Western business world.

These long-term policies reflected the need to improve **strategic capabilities**. They were implemented progressively by achieving sequential sets of shorter-term goals focused on **operational capabilities, operational performance** and urgent requirements, as illustrated in Figure C4.1. ST recognized that a successful enterprise ensures consistency between its short-term efforts and long-term goals.

POLICY DEPLOYMENT FLOW IN ST

Figure C4.2 is a high level schematic of the yearly planning flow, relating Budgeting and Policy Deployment in STM.

Figure C4.3 shows the sequential deployment at different levels, of policy deployment goals and action plans, linked to the STM TQM 'Management Amplifier'. This illustrates four key requirements for good policy deployment:

- a negotiation at each level to agree means and goals or targets, illustrated by the 'catch ball' in Figure C4.3

URGENCIES	SHORT TERM	MEDIUM TERM	LONG TERM

• Removal of problems
 	
• Catching of opportunities
Examples:
  - Quality problem
  - Process breakdown
  - Customer complaint
  - Important opportunity
	• Improvement of
	  operational performance
	
	Examples:
	  - JIT
	  - Cycle time
	  - Inventory turns
	  - Defectiveness
	  - Yield
	  - Productivity
		• Improvement of
		  operational capabilities
		
		Examples:
		  - Concurrent engineering
		  - TPM
		  - Logistics
		  - Self-managing teams
		  - Planning/Scheduling
			• Improvement of strategic
			  capabilities
			
			Examples:
			  - Human Resources
			    capabilities
			  - Technological
			    breakthroughs
			  - Multi-processing
			  - Mixed design know-how
			  - Time to market

Figure C4.1
Example of objectives by different horizon

- the creation of action plans to achieve goals or targets
- review of action plan progress, and adjustment as necessary
- standardization of improvement to 'hold the gains'.

Figure C4.4 illustrates some of the policy deployment tools used to help with means analysis, ownership assignments and progress assurance. These tools were explained in detail in the manual for managers.

APPROACHES TO MANAGE AND ACHIEVE THE ST GOALS

The yearly plan comprised all the goals and the performances the company had to reach during the year. Goals related to sales volume, profit and loss, inventories, standard costs, expenses etc. were generally managed by management control through the budget. In order to be more and more competitive, however, more challenging goals had to be identified each year and these goals – the ones that constitute the improvement plan – need 'Special Management' through a specific approach. This approach is Policy Deployment, in which a policy can be fully defined as the combination of goals/targets and means (Figure C4.5). The characteristics of the different approaches to manage the different goals (budget level and Policy Deployment level) are illustrated in Table C4.2.

Policy deployment applies both to 'What' goals, i.e. mainly results oriented, and 'How' goals that are more related to operational, technological,

organizational and behavioural aspects, mainly process oriented (Figure C4.6).

'How' is mainly concerned with improving capabilities and 'What' is mainly concerned with improving results, deriving from improved capabilities.

Drivers for 'What' goal are mainly corporate standards, prior results and vision statements.

Drivers for 'How' goals are mainly vision statement, climate survey, self-assessment, customer feedback and strategic plans.

Each level of the company (Corporate, Group, Division) must perform its own 'Whats' deployment and 'Hows' deployment.

'Whats' deployment means both targets and means deployment, where means deployment must be supported and must be coherent with 'Hows' deployment, that is generally related to a longer term vision.

In STMicroelectronics they believe that, to be a total quality company, strategy, philosophy, values and goals must be transmitted throughout the organization, from level to level in a systematic way, to provide focus, clarity direction and alignment. For them, policy deployment is the process through which goals, and the action plans to achieve them, in support of and consistent with the top level corporate mission, strategic guidelines and objectives, are cascaded to all levels of the organization. Effective policy deployment ensures that ST's goals and actions are aligned 'from top floor to shop floor'.

The goal cascade involves decomposition at each level to get to detailed goals that are readily

Case 4 Policy deployment: STMicroelectronics

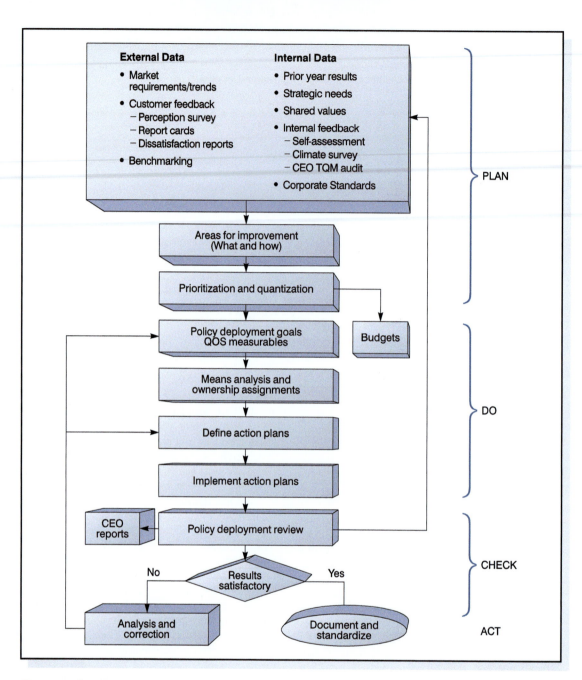

Figure C4.2
Policy deployment management process

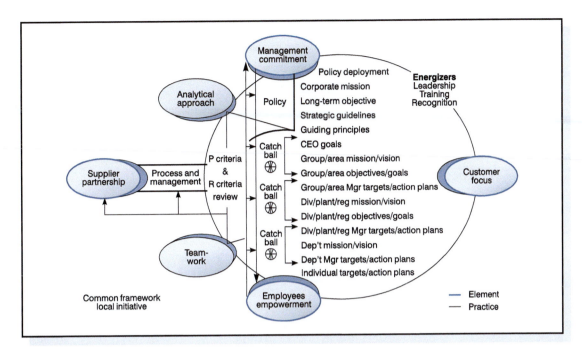

Figure C4.3
The elements and practice of policy deployment in ST

obtainable. The x-matrix is a tool to aid this decomposition and fix ownership for the detailed goals.

It is not by accident that STMicroelectronics is among the world leaders in a broad range of sectors and segments for semiconductor applications. Through its outstanding TQM approach and clear goal deployment strategy, it has achieved decades of business success and growth in a very competitive market.

ACKNOWLEDGEMENT

The author is grateful for the contribution made by Georges Auguste, and Fabio Gualandris, Executive VPs in STMicroelectronics, in the preparation of this case study.

DISCUSSION QUESTIONS

1. Discuss the TQM implementation framework developed in this case and its application to other organizations, including those in the service sector.
2. How does the 'goal translation' approach relate to TQM – what are the linking factors?
3. Show how the approach used by STM could be applied to any change management problem.

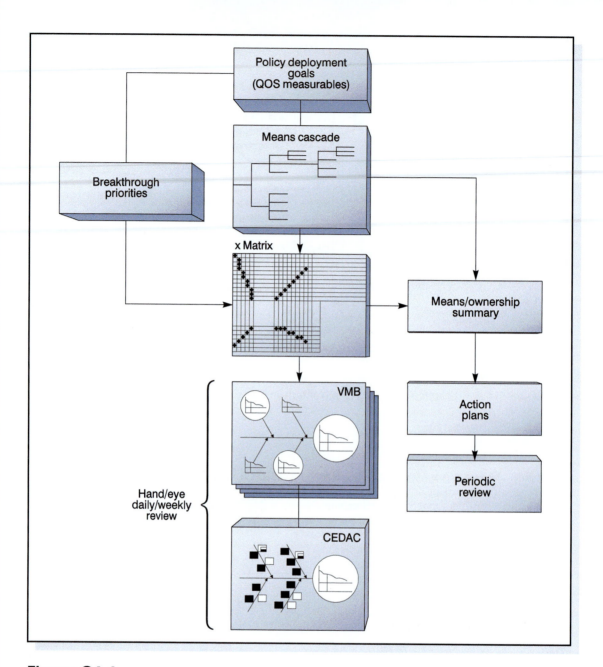

Figure C4.4
Policy deployment tools (VMB – virtual management for breakthrough)

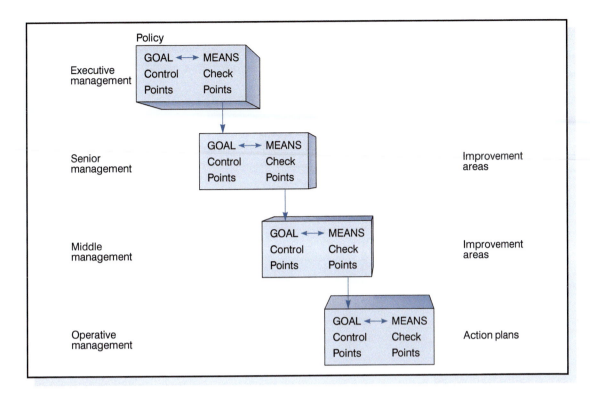

Figure C4.5
Policy deployment terminology illustrated

Table C4.2 Characteristics of different approaches to manage different goals

	Improvement goal	Improvement approach	Drivers
Budget level	Business as usual	Maintenance growth sporadic or undefined	Budgets Competition Customers Stops routine Tactical opportunities
2 levels of policy deployment	Focused improvement (policy deployment) Breakthrough (Policy deployment and visual management)	Kaizen Quantum	Shared values Corporate standards Strategic focus Self-assessment Benchmarking Vital priorities Benchmarking

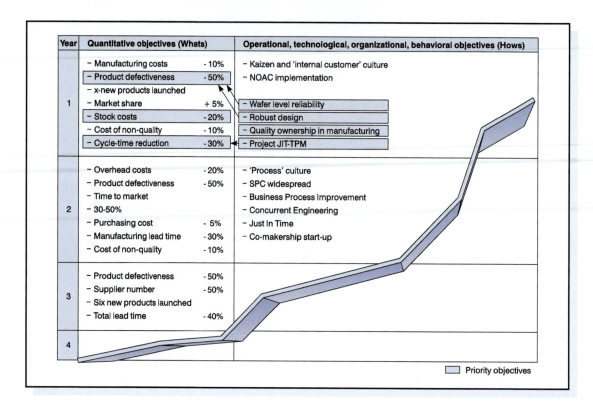

Year	Quantitative objectives (Whats)	Operational, technological, organizational, behavioral objectives (Hows)
1	− Manufacturing costs − 10% − Product defectiveness − 50% − x-new products launched − Market share + 5% − Stock costs − 20% − Cost of non-quality − 10% − Cycle-time reduction − 30%	− Kaizen and 'internal customer' culture − NOAC implementation − Wafer level reliability − Robust design − Quality ownership in manufacturing − Project JIT-TPM
2	− Overhead costs − 20% − Product defectiveness − 50% − Time to market − 30-50% − Purchasing cost − 5% − Manufacturing lead time − 30% − Cost of non-quality − 10%	− 'Process' culture − SPC widespread − Business Process Improvement − Concurrent Engineering − Just In Time − Co-makership start-up
3	− Product defectiveness − 50% − Supplier number − 50% − Six new products launched − Total lead time − 40%	
4		

☐ Priority objectives

Figure C4.6
'What' and 'How' goals (examples)

Case study 5 Business process management within TNT Express

BACKGROUND AND EVOLUTION

TNT Express is a global Express Parcels Company which offers time definite and day definite pick-up and delivery. It delivers an average of 4.4 million parcels, documents and pieces of freight every week to over 200 countries worldwide.

TNT's origins date back to after the Second World War in Australia where Ken Thomas started K W Transport with a single vehicle. In 1958 the name of the company was changed to Thomas National Transport and TNT was born.

In 1978, the growth of TNT really began to take off when it purchased County Express in the UK. TNT UK continued to grow and offer new services until 1996 when TNT was taken over by the Dutch telecommunications company KPN. TNT UK became one of the most important contributing businesses to the overall company performance.

The culture of TNT was always described as entrepreneurial with an emphasis on speed of action. This was further exaggerated by the growth of the company into a major player in the global express parcels market with a particular strength in Europe. The strength of the company was a road and air network which connected the majority of the major economies within Europe. The different national cultures and an entrepreneurial attitude created a successful but reactionary company.

In 1998, TNT UK won the EFQM European Quality Award. This was under the leadership of Alan Jones and was important for the company, in terms of recognition as a quality organization within the Transport and Logistics industry sector, because of the value proposition of TNT. The award was made for role model performance in business excellence. The EFQM model helped TNT UK develop an ethic of continuous improvement.

The EFQM model and the European Quality Award were well respected within the global organization. As Leadership and Processes are key areas within the EFQM a key question became: 'how do you align these areas globally and build on the success of the European Quality Award?'

With the appointment of a new CEO, Marie Christine Lombard (MCL) in 2004, her strategy was to change the company from a traditional and entrepreneurial transport company to a customer-centric service based organization.

The company had rapid organic and acquired growth in the late 1990s. Indeed, MCL joined TNT through the acquisition of the French company, Jet Service, in 1998. The result of this was shown in feedback from several large customers. They stated that TNT marketed itself as a global organization but variation in the way it operated locally affected their experience.

One of the major reasons for this was the number of disparate IT systems that were in existence because of the legacy systems inherited through various acquisitions. This situation generated enormous maintenance costs and interface issues. It was recognized that before TNT Express could implement a set of customer service or operational systems that could be deployed globally, they needed a 'common process'. Therefore, MCL decided that TNT, as a global organization, should map the existing processes and agree a global standard.

THE BUSINESS PROCESS MODEL (BPM)

TNT formally launched a BPM programme to enable the new strategy of differentiation through customer touch points. A level 0 to 5 framework was agreed – see Figure C5.1.

The nine key process owners appointed process expert teams and the teams were assembled to build and then deploy the model. After much process mapping and negotiation with key stakeholders as to what should be the standard, the Business Process Model emerged some months later.

The model in Figure C5.2 shows the link from the TNT strategy, through the core processes at the centre, to the key objective of delighting customers. The model supported leadership as it created a sense of purpose for the organization and a visual link for everyone to see how they are connected to the customer. This was also true for people in the supporting processes around the 'edge' of the model.

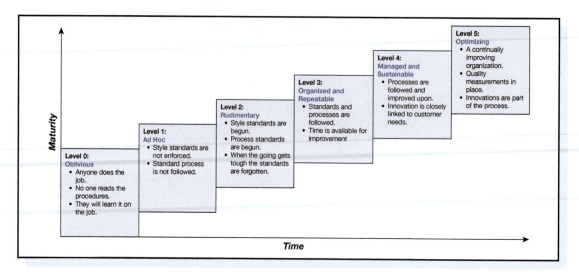

Figure C5.1
BPM 5 level assessment framework

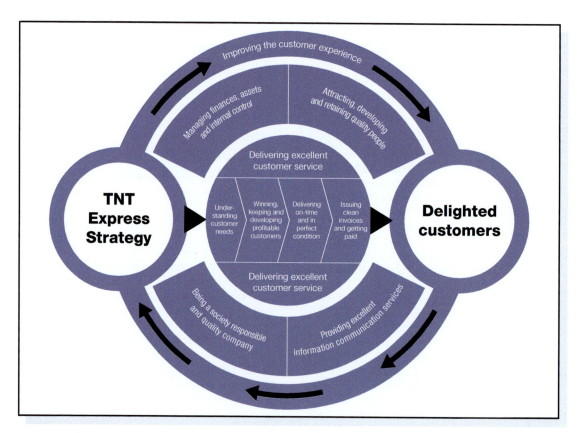

Figure C5.2
TNT Express business processes

Although the model had been created and agreed, it only existed on paper. Implementation generated its own challenges and an implementation processes had to be created. A decision was taken to focus on the core operational process of 'Delivering on Time in Perfect Condition (DoTPC)' as it was recognized that the majority of the costs were incurred within operations.

A team of auditors were assembled to visit each business unit to observe a sample of operational processes and perform gap analysis to find what was not being done according to the agreed standard process. Also where good practice was observed, these ideas were fed back into the standard process and the standard updated.

Figure C5.3 illustrates an example of the results of the six key criteria.

Action plans per TNT OU were then drawn up to ensure they reach a state of readiness for common process and system implementation.

THE BEGINNINGS OF THE PROCESS IMPROVEMENT PROGRAMME

With DoTPC reaching a level of process maturity of level 3, where the standard process was being followed, it was decided that there needed to be a standard method of process improvement. MCL instructed the Process Excellence Team to create a

DOTPC Alignment score – Divisional versus Country processes			
Benelux – March 2009			
Category	Scoring: no alignment = 1, partial alignment = 2, broadly aligned = 3 / 4 = full alignment		Average score
1	Process		2.8
	Process Management	2.7	
	Process Design	2.7	
	Process Presentation	2.4	
	Process Measures	3.2	
1	Process Systems		
1	People		
	Overall score		

DOTPC Alignment score – Divisional versus BU processes

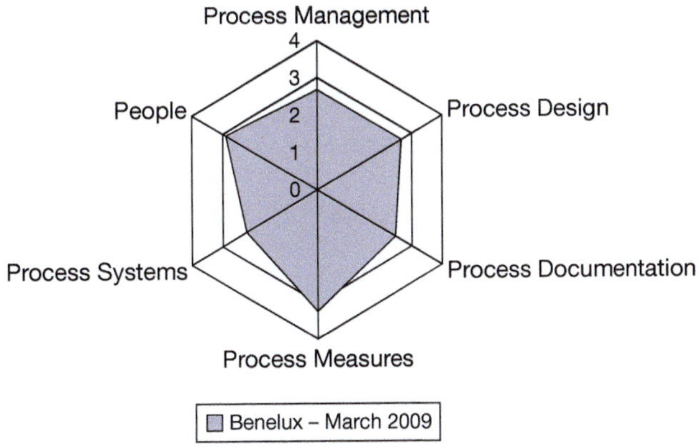

Figure C5.3
Example assessment results against six key criteria

Case 5 Process management: TNT Express

'process improvement programme'. Two external appointments were made to bring in experienced resources to create the programme from scratch.

At that time the financial performance of the company was good and there were good volumes within the network making the necessary contribution to cost recovery. TNT had become used to growth and financial success and, within this environment, gaining any traction for a process improvement programme was extremely difficult. Management at many levels could not see the reason to improve as their thinking was that 'We are good enough already.'

As is often the case in change management, the programme seemed to need a crisis to gain recognition as to the power of structured process improvement. A crisis did in fact arise when one of TNT's largest customers became dissatisfied with the level of service and threatened to take its business elsewhere. This crisis created the sense of urgency for action required.

A team was created to find the root causes of the problem and implement corrective actions. The results of this first large project were far reaching. First, it achieved its initial goal of retaining the large customer; the customer's suppliers rating turned around to the point where TNT could not score any higher in their rating for innovation; the customer's representatives spoke enthusiastically about the method used and the results achieved – indeed the customer was so impressed with the turn around in performance it awarded TNT a substantial amount of new business. Second, the project got the management talking about the results and how to do more of the same. From this, the programme found a champion at a senior level within operations.

It was agreed that a pilot should take place in Germany. Training would be provided for a small team from the selected depot and some people from the German head office. Within the training, it was evident that there was resistance to the Process Excellence team visiting Germany to instruct very senior head office operational experts how to improve the operation.

When talking about the target for improvement, there was a lot of negative discussion about all the reasons why it would not work and what had been tried before. The project involved improving the departure times of the vehicles in the morning and a modest target of 5 minutes was set for the last vehicle to leave. The team were sceptical that this would be achieved.

The Process Excellence team supervised the application of the well-proven DMAIC (Define, Measure, Analyse, Improve, Control) methodology and achieved an improvement of 75 minutes in the last vehicle leaving. The team were astounded how smoothly their operation then ran in the morning. Two years later, the biggest sceptic within the training had become one of the biggest advocates of the methodology.

Hence, the Process Excellence team generated another success for creating momentum and acceptance of the value of process improvement within the senior and middle management levels.

THE GROWTH OF THE PROCESS IMPROVEMENT PROGRAMME

From the start, the intention was to train both full and part time staff in the DMAIC methodology and subsequently training materials and programmes were developed.

In order to be allowed to attend courses, the Process Excellence team insisted that participants came to the training with a project to apply their learning to. This did not always happen, of course, with some people turning up to training at short notice and without any real sense of why they were there. This indicated that there was still a weakness in understanding or real commitment to the programme from some areas of management.

The financial crises of the time, in many economies, were beginning to bite deep. TNT is very much a barometer of economic activity with the volume of trades between countries its life-blood. TNT experienced customers switching to slower but more economical services and the remaining volumes falling dramatically from previous years' high points; it was described as the perfect storm.

Hence, there was an even bigger crisis to be managed, with competition between carriers creating downward cost pressure and deterioration of profit margins. This really created the sense of urgency to get things moving on performance improvement within TNT.

The Process Excellence team were instructed to increase the speed of deployment throughout six countries within Europe. Starting in the Netherlands a programme was initiated that focused on deployment of process improvement techniques at selected important hubs and depots within the TNT network.

A typical intervention would start with a training session of the staff and then a diagnostic assessment would be carried out on the operational processes. This is where having the DoTPC maturity defined to level 4 was of great value to the programme; it gave a managed and sustainable process to observe and improve. From the results of the diagnostic assessments, multiple small scale Kaizen Events would be scoped.

TNT UK took an even more structured route as one of the larger operating entities. Their small team discovered that a lot of the opportunities seen were the same in the first few locations visited. This team changed their strategy to quickly replicate the improvements to increase the speed of benefits realization. The cultural change aspect was done by a team following up afterwards.

Many of the benefits delivered were in terms of improved process flow and productivity improvements but it also delivered substantial cost savings. However, under closer scrutiny, the changes were not being sustained in certain places. The changes were taking people out of their comfort zone as the old way of working was deeply entrenched in some people and they couldn't wait for the process improvement teams to leave!

This situation was analysed and the first learning point was that there was engagement with the front line employees and supervision but very little real engagement with the depot management, because of an aggressive deployment schedule. The depot management were helpful when the initial interventions took place but did not understand the commitment required afterwards to monitor and sustain the changes. To supplement the process improvement training for the employees and front lines supervision, a two-day sponsor training was developed and delivered to middle management to close the gaps in understanding.

THE LINK TO PROCESS EXCELLENCE

The journey had taken TNT Express from understanding its standard processes to improving them. The pieces that were missing were how to bring innovation into the domain and how to take process management to the next level.

One of the good aspects of TNT is its reputation of being very receptive to the needs of large customers and managing these relationships. However, the management and provision of these additional needs created a lot of exceptions to the standard DoTPC processes and increasing variation in the key output metrics.

The question was then – how to innovate the new services and processes without creating disruption to the standard processes. This would have the advantage of economies of scale rather than complete specializations that cater for niche customers which are offered at a premium price. Figure C5.4 shows the elements of process excellence in TNT Express.

THE DEVELOPMENT OF THE PROCESS INNOVATION

The Global Account Management teams had initially been trained in the DMAIC methodology. There were some notable projects delivered around service improvements. Never the less there was a realization that there was a need to develop a structured innovation process.

The chosen methodology was DMEDI (Define, Measure, Explore, Develop, and Implement), the bedrock of which was the application of the four houses of quality used in quality function deployment (see Chapter 6).

The key part of the methodology was developing the customer needs and wants and translating them into measurable criteria. TNT had done extensive work on understanding the customer experience and the resulting needs and wants. The needs and wants are shown in Figure C5.5 and describes the whole life cycle of a customer doing business with TNT.

This fits in perfectly with QFD 1 as these become the 'What's'. This helps the innovation process at the earliest phase as a voice of the customer is well understood. These wants and needs hold true for many of the customers but some may prioritize some wants and needs differently to others. TNT only has to understand the priority placed on them by a particular industry of customer demographic. Innovation can be centred on QFD 2 and QFD 3 when looking for alternatives and moving to a detailed process design.

MOVING TOWARDS PROCESS MANAGEMENT

With the process improvement programme well established the Process Excellence team also

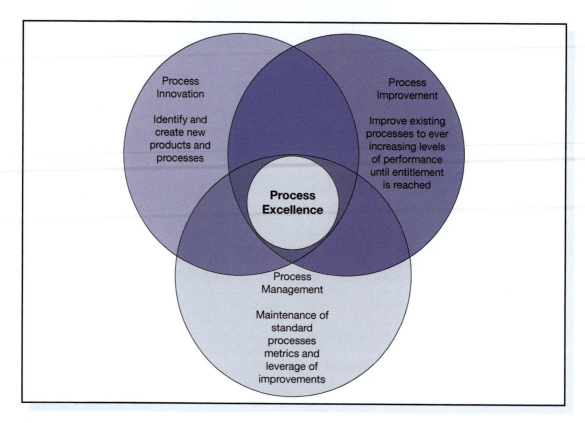

Figure C5.4
Elements of process excellence in TNT Express

wanted to establish a measure of what 'Good' looked like. For this the team developed a scoring scale based on several criteria and using a radar diagram with the scores for each category on each axis (see Figure C5.6).

These scores could be applied at the highest country level or down to a single depot. It was noticeable in many cases that sustainability and process management criteria scored low. Also, when reviewing the improvement projects selected, some were very good but others were solved by just ensuring adherence to the process.

The goal going forward is to develop a focused attitude to managing the processes at the lowest level so that the most critical inputs are controlled.

CONCLUSIONS AND LEARNING POINTS

The EFQM Excellence model provides a good vision of how an organization should work together to deliver successful outcomes for both the business and the customer. In the EFQM model the processes are part of the identified enablers and the effectiveness of a company's processes is important in delivering successful outcomes.

The excellence and effectiveness of a process comes from being able to achieve the highest level where optimization takes place. However, maturity of the processes is critical for optimization to take place. Therefore, the first learning point is to describe the processes and institutionalize them to the point where they are sustainable and repeatable.

The next learning point is to engage with middle management early in the process of setting up a process improvement programme. The importance of the understanding how processes interact across different parts of the business, sometimes in many varied locations within large global companies, cannot be overestimated. Middle management has a key role and breaking

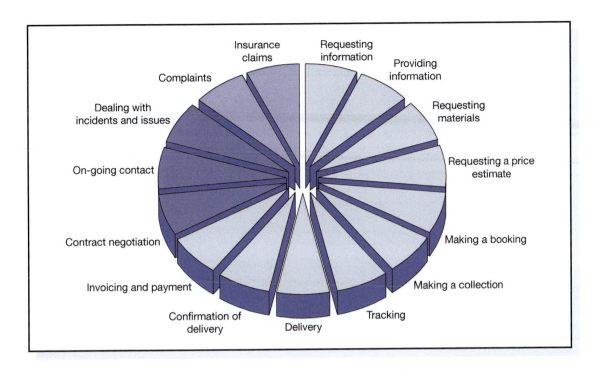

Figure C5.5
Life cycle of a customer doing business with TNT

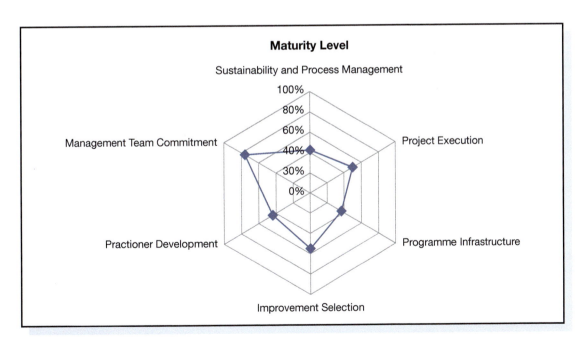

Figure C5.6
Maturity level radar diagram

Case 5 Process management: TNT Express

out of the 'silo' mentality can be difficult. Therefore, a programme should include an element of sponsor training to explain their role in process improvement and what sustainable change means.

Finally, key to process excellence is process and performance management – a lot of problems can be prevented by good process management. If the front line supervision knows that the inputs to the process and the controls are critical, then the outputs of the process will be delivered with a lot less variation in the customer experience.

Good process management and improvement comes from a regular routine of measuring, analysing and then taking corrective action on the process.

ACKNOWLEDGEMENT

The author is grateful for the help and information provided by Ian Kendrick, Process Excellence Development Manager, Global Networks and Operations, TNT Express, in the preparation of this case study.

DISCUSSION QUESTIONS

1. Why is business process management (BPM) so important in a company such as TNT Express? Evaluate their core business process framework and make suggestions for improvement.
2. How important was the background, in terms of awards and other approaches to the development of business process management in TNT Express? Discuss the resulting leadership and commitment aspects in the company and use this to develop a route map for BPM implementation.
3. What role can innovation play in the introduction and development of BPM methods used in companies in other industries or organizations in the public sector?

Case study 6   *Process management and improvement at the heart of Fujitsu UK & Ireland BMS*

COMPANY BACKGROUND

Fujitsu is the leading Japanese information and communication technology (ICT) company offering a full range of technology products, solutions and services. Over 170,000 Fujitsu people support customers in more than 100 countries. The company uses its experience and the power of ICT to shape the future of society with Fujitsu's customers. For more information refer to: www.fujitsu.com.

Fujitsu UK and Ireland is a leading IT systems, services and products company employing over 10,000 people with an annual revenue of £1.6 billion. Additionally, Fujitsu's other operations in the UK bring its total employee numbers to over 14,000 and its total revenues to £1.8 billion. Its business is in enabling customers to realize their objectives by exploiting information technology through its integrated product and service portfolio. This includes consulting, applications, systems integration, managed services and products for customers in the private and public sectors including retail, financial services, telecoms, government, defence and consumer sectors. For more information refer to: www.uk.fujitsu.com.

QUALITY MANAGEMENT IN FUJITSU

Fujitsu's approach to quality has, for many years, been based on an integrated Business Management System (BMS). This fulfils the requirements of all the external standards upon which it bases its management and control approach. These standards include the following certifications, some of which are specific to particular business or capability units:

- TickIT (primarily in UK)
- CMMI-Dev
- E-GIF (UK Government inter-operability framework)

- IT-CoBP (UK Government code of best practice)
- CLAS (UK Government accreditation of individual security practice consultants)
- CHECK (IT systems penetration testing standards in our security practice)
- Association of Project Managers
- Information Systems Examination Board.

The Business Management System is registered to ISO 9001 for design, development, production, installation & servicing and registration is to the revised Standard EN ISO 9001 that covers UK & Ireland and is continually updated. Fujitsu's data centre, networks and internet managed-services businesses have ISO/IEC27001 (Information Security) certification.

All services are based on Fujitsu's best practice standards, informed by the IT Infrastructure Library (ITIL) and ISO/IEC20000 that are held in the BMS to underpin the drive to maximize service quality and to promote re-use. Its policy is to comply with all applicable EU, international and national environmental legislation and to achieve ISO14000 throughout the company according to a planned implementation.

Fujitsu has achieved registration to ISO14001 for all locations it directly controls on the UK mainland and in Northern Ireland. It has a programme of continual assessment in place to undertake regular audits across the organization to ensure it adheres to ISO14001.

It is Fujitsu's policy to support its customers in their reporting under the Sarbanes–Oxley Act, in particular section 404 relating to internal controls.

Fujitsu's London North data centre was the first in Europe to achieve the coveted Tier III status (99.98 per cent site availability) from The Uptime Institute. This rating is the commercial optimum specification – the provision of high-availability data centre as a key component for non-stop computing systems.

483

THE BUSINESS MANAGEMENT SYSTEM BLUEPRINT

At the core of the BMS are mandatory master policies and key business processes that are owned by members of the senior leadership team. These, along with related sub-policies, local processes and procedures, are structured in logical, functional and operational views, making relevant processes easy to find and therefore apply (Figure C6.1).

Master policies are at the top of the hierarchy, being the fundamental principles and standards adopted by the top management team for their governance of the company. A senior executive, reporting to the Chief Executive Officer, owns each master policy and is tasked with the responsibilities which are fully described in a 'Standard for Policy and Process Management' which covers:

- policy management
- key principles of the Fujitsu UK and Ireland process model
- process model – key business processes
- process management governance

- roles, accountabilities and responsibilities in process management
- process change control
- process documentation
- process definition template
- process maturity and measurement
- process improvement repository
- quality assurance.

The purpose of this document is two-fold:

1. to define the methodology for process management in Fujitsu UK & Ireland, including the standards for documenting process information in a manner that is easily understood by all, particularly those people who operate the company's processes;
2. to define the role and responsibilities of policy ownership; which sets boundaries, principles and standards that must be followed in the definition of the company's processes.

Compliance with master policies is mandatory and process owners are responsible for ensuring

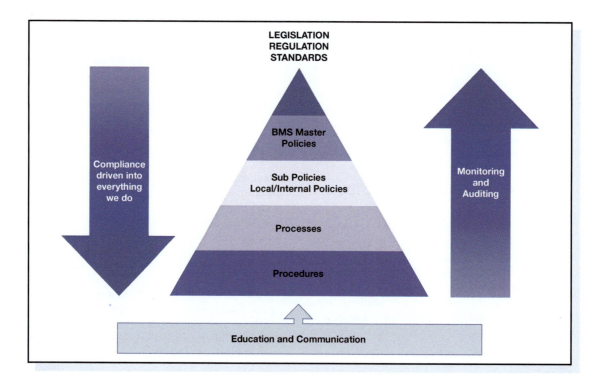

Figure C6.1
Embedding compliance – policies, processes and procedures

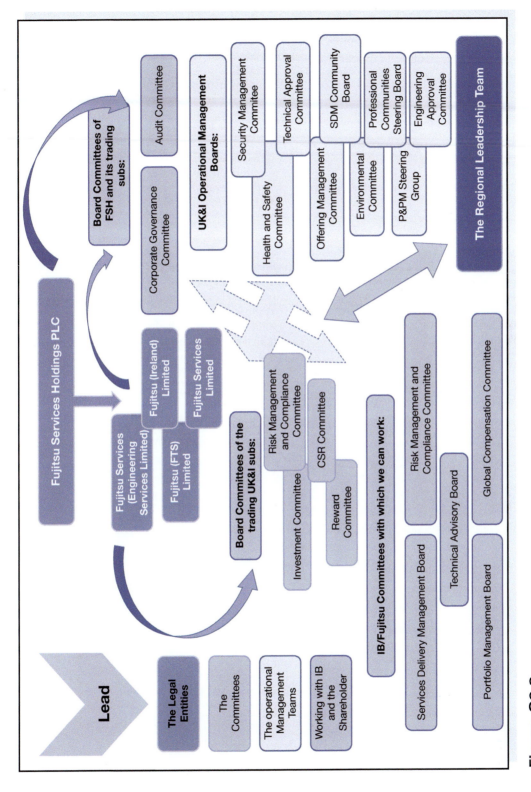

Figure C6.2
UK&I committees & management boards

that compliance with master policies is a consequence of adherence to key business processes, thereby ensuring the top-down approach to compliance shown in Figure C6.1. Figure C6.2 shows the context within which governance, risk management and compliance works.

Figures C6.3–C6.5 show the BMS blueprint, process governance and management.

THE FUJITSU UK & IRELAND PROCESS MODEL

The key principles of the Fujitsu process model are as follows:

- Aligned to the Operating Model, but independent of organization structure
- Single process owner for each Key Business (standard company) Process
- Key Business Processes span the whole business, supporting regional governance by implementing master policies
- Key Business Processes are managed using the Key Process Management (KPM) Process

- Aligned to all standards, models and codes of practice applicable to the scope of the Business Management System (e.g. ISO 9001, ISO 20000, ISO 27001, ISO14001, CMMI, IiP, UKGov. IT COBP)
- Minimal number of Key Business Processes, which are formally managed, measured, reviewed and improved
- Units may use Local Processes if there is no published Key Business Process covering the requirement; such processes must conform to the requirements and have been approved by the relevant Key Business Process owner(s)
- To be accessible via a single BMS Portal.

Process assets and examples provide a ready means of access to techniques, tools, templates, checklists, lessons learned, reference material and links to other sources of relevant material, with options of viewing dependent on role being performed or stage of the activity being addressed. Techniques that are essential to the proper following of process and, in some cases, to the adherence to specific external standards, models or codes of practice are entitled 'Standard for...'

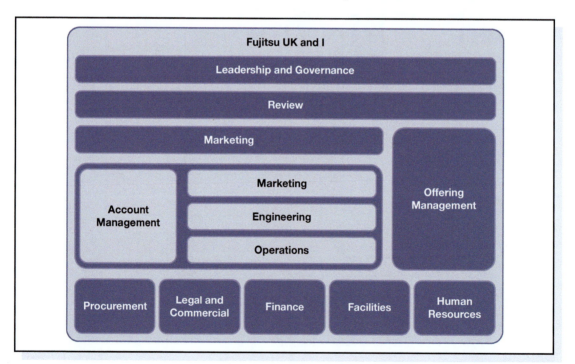

Figure C6.3
BMS blueprint

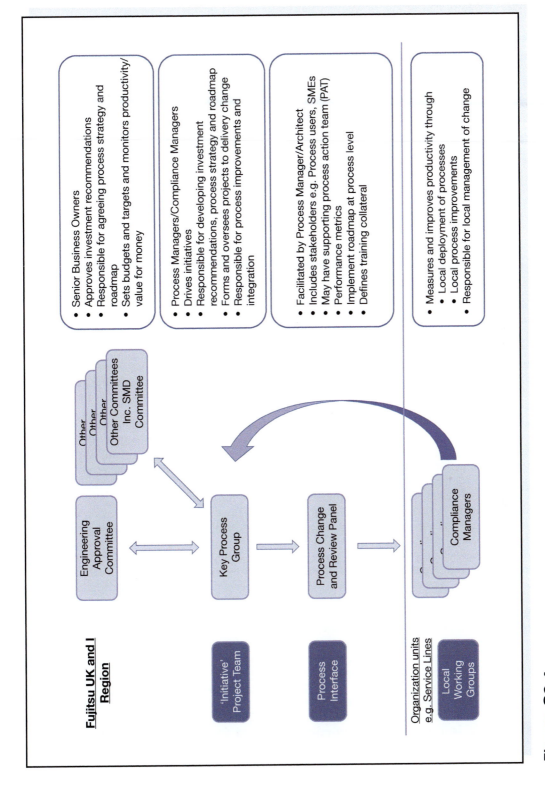

Fujitsu UK and I Region

- Senior Business Owners
- Approves investment recommendations
- Responsible for agreeing process strategy and roadmap
- Sets budgets and targets and monitors productivity/value for money

- Process Managers/Compliance Managers
- Drives initiatives
- Responsible for developing investment recommendations, process strategy and roadmap
- Forms and oversees projects to delivery change
- Responsible for process improvements and integration

- Facilitated by Process Manager/Architect
- Includes stakeholders e.g. Process users, SMEs
- May have supporting process action team (PAT)
- Performance metrics
- Implement roadmap at process level
- Defines training collateral

- Measures and improves productivity through
  - Local deployment of processes
  - Local process improvements
- Responsible for local management of change

Other
Other
Other
Other Committees
Inc. SMD
Committee

Engineering
Approval
Committee

Key Process
Group

Process Change
and Review Panel

Compliance
Managers

'Initiative'
Project Team

Process
Interface

Organization units
e.g. Service Lines

Local
Working
Groups

Figure C6.4
Process governance

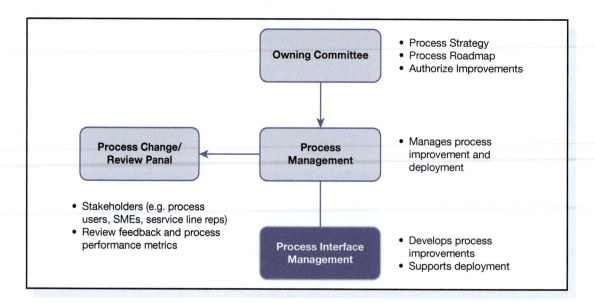

Figure C6.5
Process management

Figure C6.6 shows the Fujitsu UK & Ireland **Customer Solution Lifecycle Model**, a 'role based' representation of how the customer facing processes integrate to provide the means by which contracts for customer business are assessed, proposed, delivered and managed. This allows users to locate appropriate processes dependent on the customer facing activities they are undertaking. The benefit of having this view is that it does not require a detailed level of specific process knowledge in order to locate the correct collateral.

These customer facing processes, together with the internal management processes, constitute the Business Management System. Several of the processes shown comprise a number of constituent processes where the level of detailed content requires it.

Figure C6.7 shows the BMS Processes. Working with accountable owners to amend/ improve processes across the whole Region and to embed Lessons Learnt increases engagement with operational teams, professional communities and RLT (Regional Leadership Team).

In an ever more complex and rapidly changing world, to consistently achieve desired results, business processes in Fujitsu must be actively owned, managed, measured, communicated clearly and supported. Continually challenging

and improving key business processes gives the company competitive advantage. It is also via such continual process improvement that Fujitsu fulfils the requirements of external standards to demonstrate continual improvement.

Every regional key business process has a single, senior executive owner who may appoint a process champion to manage the process on their behalf. The process owner receives guidance on strategic objectives from the regional leadership team or one or more of the region's management committees. Owners of processes are required to ensure appropriate integration of their processes and co-ordinate developments and changes in order to ensure maximum effectiveness for customer benefit and to maintain registration with the external standards. This is achieved by a regular forum attended by their process champions, the Key Process Group (KPG). Process owners/ champions are expected to work with a number of experienced practitioners in the definition and subsequent improvement of processes – through change/review panels and process action teams.

Roles, accountabilities and responsibilities in process management are fully listed and described for: regional process owner, process architect, process manager, process interface manager, service line compliance manager,

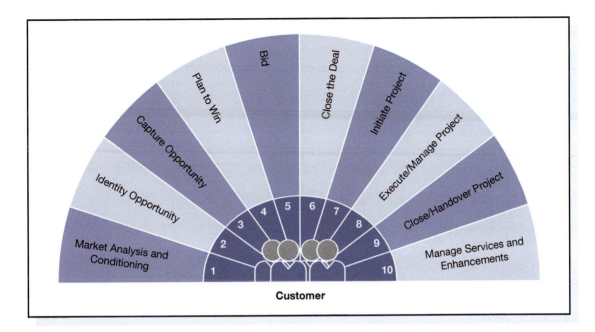

Figure C6.6
The customer solution lifecycle model

Business Excellence director, process operator and process facilitator.

Specifications are given for process change control to provide clarification of responsibilities regarding changes to local and regional key business processes, including the role of key/local process group(s). Change control is exercised to ensure processes comply with ISO9001 requirements and to maintain logical linkages with other processes.

Requests for a new process to be added to the standard company set will be directed to Governance & Compliance as administrators of the BMS portal, who then verify the requirement with the appropriate management committee, ensure a process owner (and champion if required) is appointed and briefed and, through the KPG, help ensure that any new process is properly aligned with existing processes when introduced.

Process change can be initiated:

- as a result of formal process review by the process owner;
- by anyone in Fujitsu UK & Ireland submitting a change proposal to the process owner or Governance & Compliance.

- through Lessons Learnt feedback which is analysed for related process improvement actions.

Records of changes requested and completed are maintained to support the company's ISO9001 registration.

Governance & Compliance will check that the changed process:

- conforms with ISO9001 requirements
- conforms with any other relevant standard or code of conduct to which UK&I Region adheres
- conforms with the Fujitsu UK & Ireland Standard for Policy & Process Management
- has appropriate links to and from other processes by reference to their process modelling tool
- identifies changes made from the previous version
- is approved by KPG/management committee (where applicable).

If the revised process is acceptable it is published on the BMS portal and a copy filed in the relevant section of the process improvement repository on SharePoint. If for any reason the

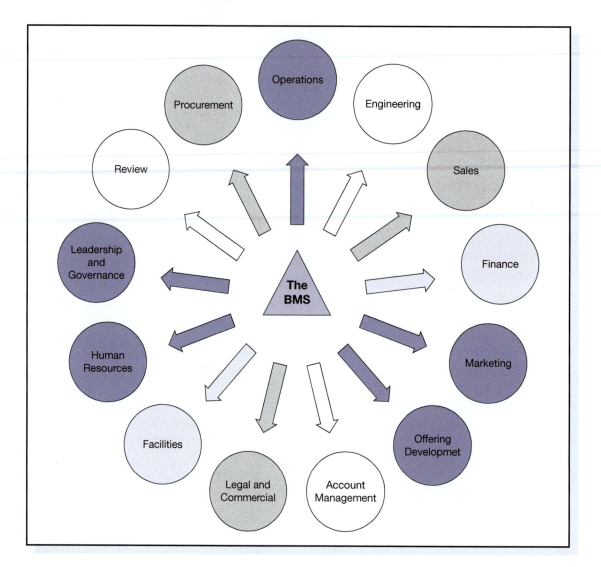

Figure C6.7
BMS processes

process fails the conformance checks, the identified problem, together with a recommended alteration, is passed back to the owner for resubmission.

The standard for process documentation and collateral consists of:

- brief descriptive text addressing overview and scope, aims and objectives
- list of inputs and outputs, preferably showing source and destination respectively
- text/tables defining steps and accountabilities (optional if flowchart included)

- a process flowchart is a mandatory element of the process documentation
- links/cross-references to process assets/ collateral (also known as templates, tools, checklists and procedures)
- reference to records that are required to be maintained
- control points
- definition of minimum measures of the process and any reports generated during process execution

- reference to applicable policies and standards and related processes
- process improvement repository – the process's KPM project on SharePoint
- brief change history
- definitions included in Vikipedia with the process owner set up to receive alerts should another employee attempt to change the definition.

A 'Process Maturity Tool' is used to assess conformance with good practice, identify priorities for improvement action and summarize evidence of improvement successes. This may be used at any time but is especially relevant during process reviews. The aim is to achieve and maintain a score in excess of 75 per cent on all axes within two years of a process being introduced and to at least maintain this level. Individual owners may impose more stretching targets, provided the cost of doing so is justified. More specific targets may be set from time to time by the management committees or key process group.

Maturity of process management is highly dependent on measurement being used to inform the process review activity. Accordingly, a crucial part of a process owner's responsibility in Fujitsu is to specify and subsequently review a set of measures that will form the basis of improvement decisions and evidence of successful implementation of improvement initiatives. Following completion of a 'Mid-Term Plan,' owners of key business processes are expected to review the basis of the measurement and analysis that they will conduct of their processes in the coming company year.

Each Fujitsu key business process has a SharePoint-based project library, forming a process improvement repository. Within this common library structure are folders for:

- email log
- working files, including archive of published versions of process and associated documents
- process strategic development plan and release schedule
- process measures – plans, actuals and results of analysis or links to alternative records
- process feedback, awaiting review
- process review records, including records of maturity assessments
- process improvement plans and progress

- process tailoring matrix
- archived review records and other materials.

The process owner or appointed champion is responsible for maintenance of this repository with assistance and support from Governance & Compliance. It is a requirement that at least three year's worth of historical data be retained for the purposes of providing evidence of continual improvement.

Quality Assurance in Fujitsu UK & Ireland comprises a number of key aspects:

- Objective evaluation of processes and work products is achieved by (a) quality audits and (b) quality reviews respectively. The points at which quality reviews must be conducted are specified in process descriptions. Projects and organizational units must arrange quality audits, to be conducted by auditors who are independent of the project, and which are scheduled into project plans.
- Additional quality audits can be arranged by business operations in order to monitor compliance with corporate policy and process as well as relevant external standards.
- The frequency of quality audits is determined by business need; it is expected that auditing should be more frequent at the time of introduction of significant process change. Projects of duration exceeding six months have at least one quality audit conducted.
- The controls to provide the assurances required are built into the relevant processes and the product description standards for the work products.
- The outputs from quality audits are key inputs to process improvement activities; accordingly formal reporting and monitoring of corrective actions is required – at project, unit and corporate levels, depending on the nature of the improvements. Quality audit reports are recorded on the Fujitsu UK & Ireland Assessors Database.
- Quality audits are carried out only by individuals who have been specifically trained on the relevant standards and approved by the Head of Quality. Guidance on the conduct of quality audits is available in a document: 'Standard for Conducting Quality Audits'.
- Managers are expected to plan and monitor compliance with the quality controls that are

built-in to processes with much rigour as they ensure that the activity/project is progressing on target and within budget. This is achieved through informal reviews with team members and via regular checkpoint reviews. Guidance on the conduct of quality reviews is available in a document 'Standard for Conducting Quality Reviews.'

THE FUJITSU PROCESS MANAGEMENT CYCLE

The Fujitsu Process Management Cycle is a representation of the Key Process Management process and has four phases – Design-Deploy-Review-Improve (See Figure C6.8). Once a process is designed and deployed, the review and improvement phases form a continuous cycle until there is a need to fundamentally redesign the process.

DESIGN	DEPLOY
Define the process. Define the business need, the purpose, the scope and the owner. *Define* the customers & suppliers. Who are the customers and what do they need from the process? What inputs are needed and who will supply them? *Determine* the constraints & resources Are there any constraints (e.g. legislation) and what resources will be needed (people, equipment etc.)? *Define* the roles and responsibilities What roles are required to operate & manage the process and what responsibilities will they carry? *Design* and document the process Deployed flowchart and supporting documentation. *Define* performance measures and targets What measures will be used to check the effectiveness of the process and what are the target values for those measures? *Design* and document the sub-processes.	*Determine* what needs to be done To implement the process and who will do it? *Determine* the blockers And how to overcome them. *Design* and develop the training. Who will need to be trained, how and when? *Design* and implement a pilot. Include a pilot of the training. *Determine* the degree of success. Review the pilot and make any necessary changes to the process. *Design* the data collection mechanisms. *Do it!* Publish the process and associated material Brief/train all who need to know (users, contributors, recipients, quality assurance reps) Establish measurement collection Ensure feedback mechanisms in place Determine review point
IMPROVE	**REVIEW**
Isolate the area to work on. *Identify* and analyse appropriate data. Additional specific data may need to be collected on this area of the process. *Identify* the root cause(s) of the problem. Involve representatives of all groups directly involved in the problem area. *Identify* a potential solution Which addresses the root cause(s). Conduct an internal or external benchmark to identify best practice. *Implement a pilot* - and evaluate the results. Modify the solution if necessary. *Institutionalise* the solution. Revise the process documentation, train all those involved and devise new process measures to monitor the effectiveness *Improve* the process improvement process. Share learning and experience with other process improvement groups.	*Review* the need. Does the process still meet the business need? *Review* conformance. Is the process being operated as planned in all areas? *Review* roles & responsibilities. Are the roles still appropriate, are any changes needed? *Review* and analyse the performance data. Is the process operating to specification, is it meeting targets? *Review* and analyse feedback. Collect and review feedback from stakeholders. *Review* opportunities for improvement. Have any problems arisen related to this process, what elements of the process can be improved? - Definition - Asset material (templates, checklists, guides) - Tailoring Guidelines/local processes - Education/Training - Measurement Might benchmarking yield further opportunities? *Rank* the opportunities. Prioritise, taking into account the importance, urgency and cost of improvements.

Figure C6.8
Fujitsu process management cycle

THE ROLE OF THE BUSINESS MANAGEMENT SYSTEM

The BMS fulfils two roles: firstly, to make readily available and accessible all essential information about the business's purpose, objectives, structure, policies, processes, standards, acquired knowledge and key performance indicators as direction to employees (as such, it constitutes the company's Quality Manual in compliance with the requirements for a Quality Management System (BS EN ISO 9001), as well as the requirements for a Service Management System (ISO/IEC 20000-1), an Information Security Management System (ISO27001), an Environmental Management System (ISO14001) and an Occupational Health and Safety Management System (OHSAS 18000)). The second role is to support standardization and best practice in the delivery of all operations and functions in order to deliver the business objectives. The former role

could be seen as 'Fit for Audit', the latter 'Fit for Purpose'. Given this, the BMS is a fully Integrated Management System (IMS).

BENEFITS OF THE BUSINESS MANAGEMENT SYSTEM

The benefits of operating in the way described in this case study are coming about through reduction in process related incidents, alerts, service credits, audit non-conformities etc. It has taken Fujitsu a long time to capture the baseline data but they now have in place an improvement cycle which not only fixes issues but tackles root cause (in most cases processes being not fit for purpose or not fit for use) and embeds that change in the business (Figure C6.9).

Most of the issues arising can be traced back to process but not all are corrected through processes improvement alone. It is recognized that

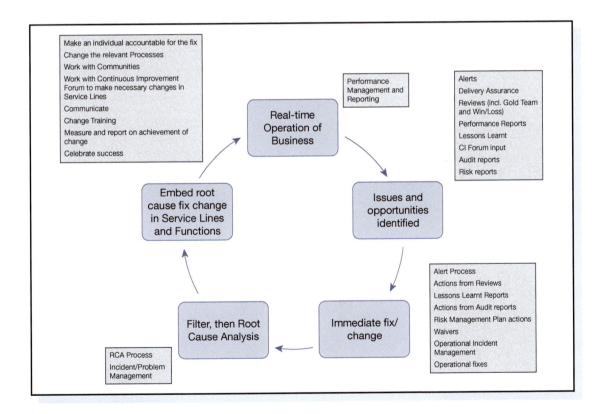

Figure C6.9
Standardization and value-add

Case 6 Process management: Fujitsu

493

some of problems are cultural and some unique to a specific incident. So the policy deployment – process approach to the BMS is not taking credit for fixing everything but it is recognized that there is evidence of reductions in incidents, etc.

ACKNOWLEDGEMENT

The author is grateful for the help and information provided by Simon Dennis, Head of The Business Management System (BMS), Commercial, Legal & Assurance, Fujitsu UK & Ireland, in the preparation of this case study.

DISCUSSION QUESTIONS

1. Evaluate the business management system (BMS) approach adopted by Fujitsu UK and Ireland, in relation to the size and complexity of the business; how may this need to be adapted to provide a suitable approach for an organization in the public sector?
2. Discuss the links between the process frameworks developed in Fujitsu and the BMS deployment.
3. What role could benchmarking play in the development of the BMS in Fujitsu?

Case study 7   Simplifying business processes to secure competitive advantage for Car Care Plan

BACKGROUND AND HISTORY

Car Care Plan (CCP) is one of the world's leading providers of vehicle warranty, GAP, MOT and other after-sales motoring programmes. They work with major motor manufacturers, franchised and independent dealers and have several major affinity partners. They are the leading provider of vehicle warranties and GAP insurance in the UK working with over 1 million customers annually.

Established in 1976, they've grown by developing successful long-term relationships with their clients and through a dedication to customer care. CCP has a significant and growing global presence. From their offices in the UK, Moscow and Shanghai (through a subsidiary SFR) they service the European and other territories. With their sister company Motors Insurance Company Limited (MICL) they are developing global reach whilst ensuring that exacting standards are being maintained.

CCP is FCA regulated and is owned by AmTrust Europe. They are the preferred supplier to more than 25 motor manufacturers. They benefit from working closely with their sister company, Motors Insurance Company Limited (specialist underwriters).

CCP offers a range of products for both FCA authorised and non-FCA authorised dealers. Their online registrations and claims system is one of the most streamlined, reliable and hassle-free administration services available in the automotive industry. They offer full dealer training to ensure high standards are always maintained, and have a straightforward approach to handling claims: 'If it's a valid claim, we will pay it.'

THE BUSINESS CHALLENGE

To help achieve this market leading position, and keen to maintain their competitive advantage, CCP

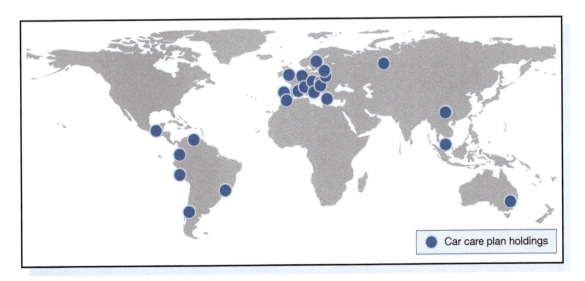

Figure C7.1
Car Care Plan – global reach

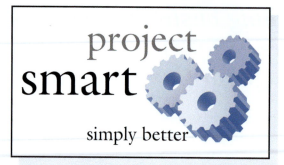

Figure C7.2
Project Smart

decided to replace and future proof their legacy IT systems used for policy and claims administration. They recognised that this was an excellent opportunity to drive simplification into their business processes, in order to maximize the benefits of the IT replacement and to achieve significant improvements in customer service and efficiency. They branded the project to give it an identity and logo that could be used for project documentation and to help in consistent communications. The project was titled *Project Smart* and the 'strapline' was 'Simply Better'.

OVERVIEW OF THEIR APPROACH TO SIMPLIFYING THEIR BUSINESS

They adopted a phased approach to *Project Smart* in order to achieve 'business simplification':

- First they identified the key changes – simplifications – required to simplify a wide range of business processes and secured senior commitment to change.
- The second phase designed these changes and tested their feasibility, with the next layer of management owning the changes.
- The third phase was to implement a set of improvements that were not reliant on the future IT system and assess the impact of the change.
- The fourth phase was to implement the new IT system and processes.

AN EXAMPLE OF A SIMPLIFICATION

CCP analysed the number of pricing variables available in their existing system, they then examined using simple Pareto Analyses which pricing variables were actively used in their policy pricing decisions. This allowed them to identify candidates for removal from their pricing algorithms – a simplification.

MORE ABOUT HOW CCP APPROACHED THE PROJECT PHASES

Phase 1: Using continuous improvement techniques (and broadly following a Plan-Do-Check-Act cycle) a team of business analysts and subject matter experts examined problems and improvement opportunities, to identify potential simplifications throughout the business (in Operations, Sales and Marketing, Finance, Underwriting, Actuarial and Compliance).

The key activities in this phase were to assess the issues the business was experiencing, gather and summarise supporting data and to develop recommendations for improvement. CCP then conducted an externally facilitated series of workshops to refine those recommendations with the directors and senior managers to obtain buy-in to potential improvements. In these workshops they asked managers to self-nominate by voting using red cards to own and lead a simplification and blue cards to be part of the focus group delivering the simplification.

Phase 2: CCP designed an approach to test the feasibility of the improvement recommendations using a 'core team' of business analysts and a group of business subject matter experts. Over 70 potential simplifications/improvements ('mini-projects') were identified. Work plans were created for the implementation of around 30 system independent simplifications. In parallel detailed requirements were developed for the systems dependent simplification of the business and the replacement IT system.

Phase 3: CCP worked on their system independent simplifications until a series of quick wins had been achieved (e.g. rationalization of several major paper flows).

Phase 4: CCP used an impact analysis approach to identify the impacts of the new system and processes, internally and for Car Care Plan's customers. Working with small focus groups of interested managers and employees, the planned and in progress changes were reviewed and a framework was used to identify the impacts of change (see Table C7.1).

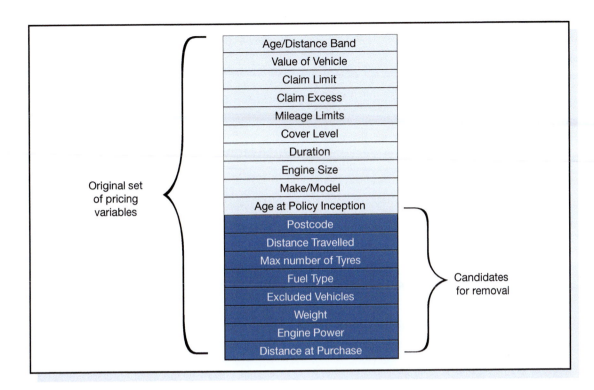

Figure C7.3
Example of a simplification

Table C7.1 Example of impact areas with descriptions

Impact in	Description
Knowledge	Identify what 'knowledge' is required by whom (where knowledge could be information, guidance etc.)
Skills	Identify what additional skills are required
Performance Management	Identify the desired performance and evidence of that performance
Workflow or Process or Procedure	Identify processes that will need to change
Policy or Compliance	Identify any policy or compliance issues/changes
Roles & Responsibilities	Identify any changes to individual roles/responsibilities
Organizational Structure	Identify any changes to organizational structure, reporting relationships etc
Staffing Level	Identify the driver(s) for reduced (or increased) effort

This involved the organization identifying and quantifying the changes it needed to adapt to, while ensuring the highest service levels for customers would be maintained. Transition plans were created for each key change identified to achieve this. To ensure a smooth transition and readiness to make the changes, a team was formed to programme manage the impacts and deliver a project communications and involvement strategy.

Case 7 Simplifying business processes: CCP

RESULTS FROM PROJECT SMART

Over 70 specific simplifications/improvements were identified and implemented, these included significantly reducing product complexity and pricing making products easier for their customers to understand and simpler for CCP to administrate; eliminating unnecessary paper flows and introducing a new, 'intelligent claims settlement' and self-billing approach which has resulted in quicker payment and reduced rework from invoice-claim reconciliation. The replacement IT system was 'cut-over' with very few issues. This was at least in part due to the simplification thinking and impact analysis that had been conducted.

SUCCESS FACTORS IN THIS PROJECT

CCP created a repeatable process for identifying and defining business simplifications, that was scalable to allow numerous simplification 'mini-projects' to run in parallel. They achieved excellent levels of buy-in and engagement with the business through a continuous process of two-way communications and involvement of the wider business in the impact analysis process for example.

To hold the gains they had made in simplification, CCP developed and established a process for product governance and innovative tools for helping guide the business – for example they developed an interactive 'smart-guide' to communicate the details of simplifications across the business.

ACKNOWLEDGMENT

We would like to thank Paul Newton, Chief Operating Officer of Car Care Plan for his input in helping develop this case study.

DISCUSSION QUESTIONS

1. Identify three ways in which CCP ensured that the people change aspects of this project were successfully managed.
2. List the ways in which you think that giving a project a 'visual identity' can help to accelerate the change process.
3. Review the CCP framework used in their impact analysis approach. Do you think that there are any items which have been missed off the framework? If you had to prioritize the items on the framework for a change that you were managing, how would you go about it?

Case study 8   Building quality and operational excellence across ABB

COMPANY BACKGROUND AND HISTORY

ABB is a global leader in power and automation technologies. Based in Zurich, Switzerland, the company employs 150,000 people and operates in approximately 100 countries. The firm's shares are traded on the stock exchanges of Zurich, Stockholm and New York.

ABB is the world's largest builder of electricity grids and is active in many sectors, its core businesses being in power and automation technologies. The company has one corporate division and five production divisions.

POWER PRODUCTS

Power Products are the key components for the transmission and distribution of electricity. The division incorporates ABB's manufacturing network for transformers, switchgear, circuit breakers, cables, and associated high voltage and medium voltage equipment such as digital protective relays. It also offers maintenance services. The division is subdivided into three business units – High Voltage Products, Medium Voltage Products and Transformers.

POWER SYSTEMS

Power Systems offers turnkey systems and service for power transmission and distribution grids, and for power plants. Electrical substations and substation automation systems are key areas. Additional highlights include flexible AC transmission systems (FACTS), high-voltage direct current (HVDC) systems and network management systems. In power generation, Power Systems offers the instrumentation, control and electrification of power plants. The division is subdivided into four business units – Grid Systems, Substations, Network Management and Power Generation.

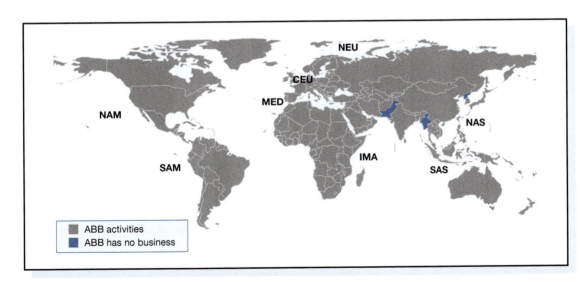

Figure C8.1
World coverage by ABB

DISCRETE AUTOMATION AND MOTION

Discrete Automation and Motion provides products and services for industrial production. It includes electric motors, generators, drives, programmable logic controllers (PLCs), analytical, power electronics and industrial robots. ABB has installed over 200,000 robots. In 2006, ABB's global robotics Manufacturing headquarters moved to Shanghai, China. Also, wind generator and solar power inverter products belong to this division.

LOW VOLTAGE PRODUCTS

The Low Voltage Products division manufactures low-voltage circuit breakers, switches, control products, wiring accessories, enclosures and cable systems to protect people, installations and electronic equipment from electrical overload. The division also makes KNX systems that integrate and automate a building's electrical installations, ventilation systems, and security and data communication networks. Low Voltage Products also incorporates a Low Voltage Systems unit manufacturing low voltage switchgear and motor control centres. Customers include a wide range of industry and utility operations, plus commercial and residential buildings.

PROCESS AUTOMATION

The main focus of this ABB business is to provide customers with systems for control, plant optimization and industry-specific automation applications. The industries served include oil and gas, power, chemicals and pharmaceuticals, pulp and paper, metals and minerals, marine and turbocharging.

CORPORATE AND OTHER

The Corporate and Other department of ABB deals with the overall management and functioning of the company as well as asset management and investment. It supports MNCs.

The company in its current form was created in 1988, but its history spans over 120 years. ABB's success has been driven particularly by a strong focus on research and development. The company maintains seven corporate research centres around the world and has continued to invest in R&D through all market conditions. The result has been a long track record of innovation. Many of the technologies that underlie our modern society, from high-voltage DC power transmission to a revolutionary approach to ship propulsion, were developed or commercialized by ABB. Today, ABB stands as the largest supplier of industrial motors and drives, the largest provider of generators to the wind industry, and the largest supplier of power grids worldwide.

THE BUSINESS CHALLENGE

To enhance ABB's ability to delight customers with the products and services they provide, a global Operational Excellence (OPEX) program was launched. In summary, the aims of this were to provide:

- A common methodology for transforming improvement ideas into business reality
- Training on tools and techniques to build improvement competencies
- Coaching of improvement projects to turn real business issues into real business results.

The fundamental idea was that OPEX would equip ABB employees with the skills and ability to generate process improvements to ensure On-Time, On-Cost and On-Quality delivery. It was felt that this approach would also create a culture of 'Continuous Improvement' within ABB through widespread use of its established '4Q' improvement system; hence the ABB OPEX:

- provides a *common approach to problem solving* based on the **ABB 4Q Methodology**
- has a *common structure to facilitate participation* by all employees regardless of location and function
- *provides training materials, instruction and coaching* to assure a quick start and transition to local ownership.
- is the programme promoted by the OPEX and Quality networks which *creates value in operations*.

ABB OPEX provides the training and tools cascade for employees to quickly resolve problems and implement improvements in their workplace (Figure C8.2).

The 4Q Program supports the ABB Quality Policy and aims to grow continual improvement competencies and culture on a global scale. These competencies help to eliminate waste, reduce variation and enable people to work in a safe and

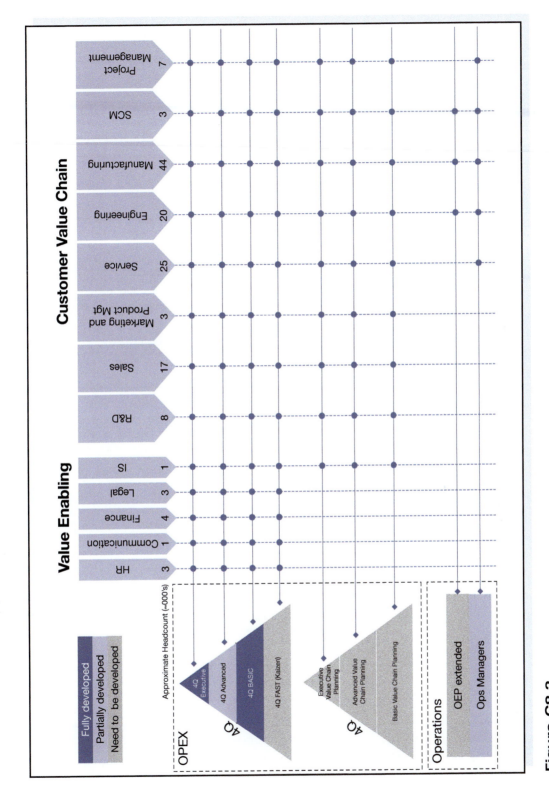

Figure C8.2
ABB OPEX competence development overview

efficient manner. They support the primary business objectives of:

- Delivering to the customer products and services that meet their expectations of quality, cost and delivery date
- Sustainable profitable growth
- Long-term marketplace competitiveness.

Through training, coaching and practical 4Q projects the program engages employees in the drive for process improvement. ABB business leaders select urgent improvement areas in their organizations; they then select individuals from their organizations to be trained in the 4Q methods and become 4Q Project Leaders; those Project Leaders then lead 4Q project teams (Figure C8.3).

Figure C8.4 is an overlay on C8.3 to show the benefits of this approach at the various stages and levels in the programme.

ABB 4Q METHODOLOGY

4Q is the standard ABB improvement methodology (Figure C8.5) and is applicable to all types of improvement projects. It is a systematic, four step method that ensures processes are mapped and understood, performance data is gathered, real root causes are determined and improvements are built into the local processes to sustain long-lasting results.

ASSESSMENT AND SELECTION OF 4Q PROGRAM TRAINERS AND COACHES

The training for 4Q is deployed by internal change agents that deliver both 'Basic' and 'Advanced' level workshops globally across ABB. These trainers require the necessary skills and capability to deliver effective training in the improvement methodology and then coach improvement projects that deliver benefits to the business. They also require the enthusiasm and energy to help drive the program, establishing credibility and momentum. Designed, built and deployed into the programme is a complete candidate assessment process that:

- Assesses and validates the suitability of potential trainers and coaches against a specific criteria of key competences

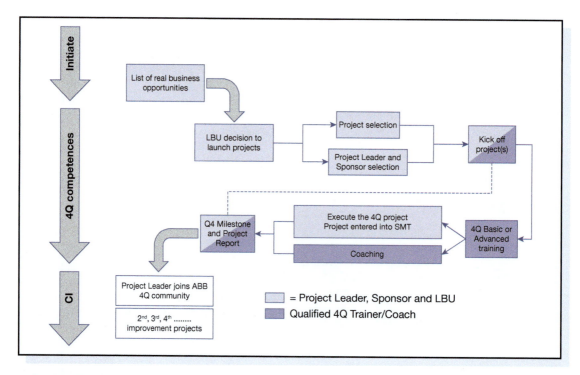

Figure C8.3
High level process flow of programme

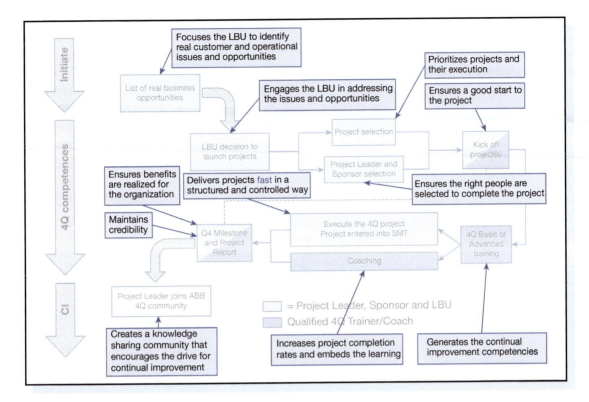

Figure C8.4
Benefits of the approach at the various stages and levels of the programme

- Identifies the individual's learning needs to be addressed and provides guidance on their development requirements, as a program trainer and coach.

DEVELOPING INTERNAL TRAINERS AND COACHES

For the initial rollout, approximately 20 trainers were identified to run courses across all four regions of the business. An external consultancy designed and conducted a comprehensive 'Train the Trainer' program that rapidly equipped the trainers with the capability to run training courses and coach improvement projects.

The training included:

- Technical knowledge of quality and performance improvement tools and techniques
- Coaching skills for improvement projects to ensure they delivered real business results

- Workshop delivery and group coordination skills in different cultures
- Presentation and inter-personal skills training.

Figure C8.6 gives an overview of the ABB 4Q certification and trainer/coaches pathway

QUALITY AND CONTINUOUS IMPROVEMENT TRAINING MATERIAL

The 'Advanced' level workshops provide training for experienced improvement practitioners who have the capability to lead medium/large-scale improvement projects on cross functional value chains. A level of expertise was, therefore, required to structure and develop this training material. The consultancy led the development of this 'Advanced' training material, combining best practice lean and six sigma principles. The training modules included: value stream mapping, measurement system analysis, statistical methods,

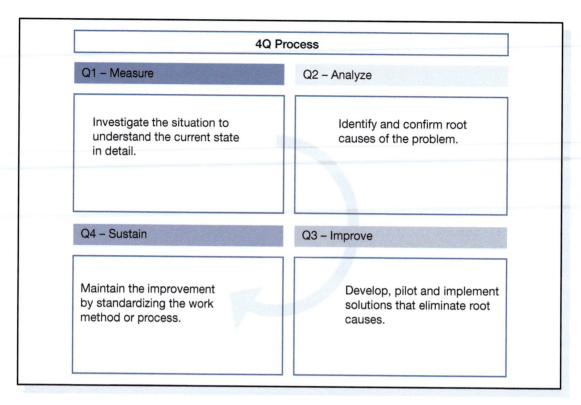

Figure C8.5
Overview of the ABB 4Q methodology

project management, cost benefit analysis and change management.

KEY ELEMENTS OF THE OPERATIONAL EXCELLENCE PROGRAMME

The critical success factors of programme are shown in Figures C8.7 and C8.8, which includes the 'A.R.O.W.' structured approach to technical coaching.

A – ASSESS

At the start of each coaching session, determine where the manager is in the improvement process using the coaching review checklist to assess progress – you are looking for supporting evidence.

Then ask the question: 'What would you like from this session?'

R – REALITY

What is happening at the moment?
How do you know that this is accurate?
When does this happen?
How often does this happen? Be specific/precise where possible
What effect does this have?
How have you verified, or would you verify this?
Who else is involved?
What is their perception of the situation?
What have you tried so far?

O – OPTIONS

What are the relevant constraints and can they be removed?
What else could you do – options?
Think of approach/actions you have seen in similar situations
Who might be able to help?
Would you like suggestions from me?

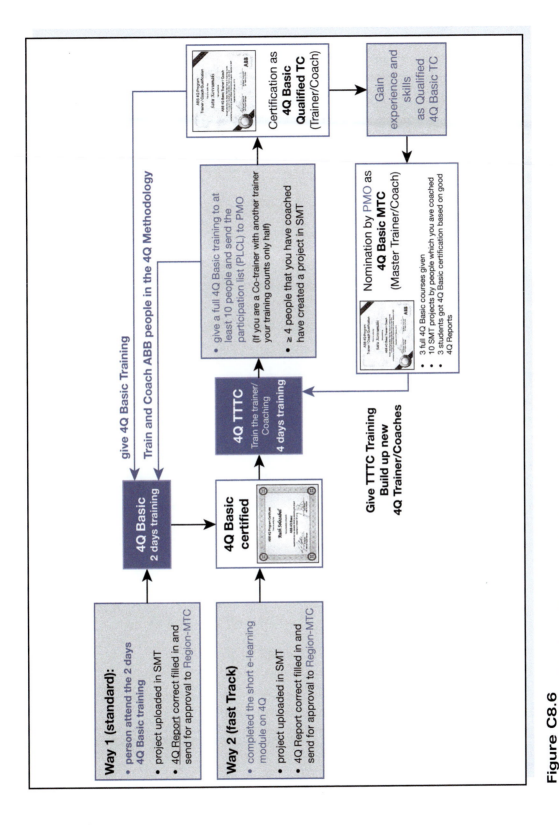

Figure C8.6
Overview of the ABB 4Q certification and trainer/coaches pathway

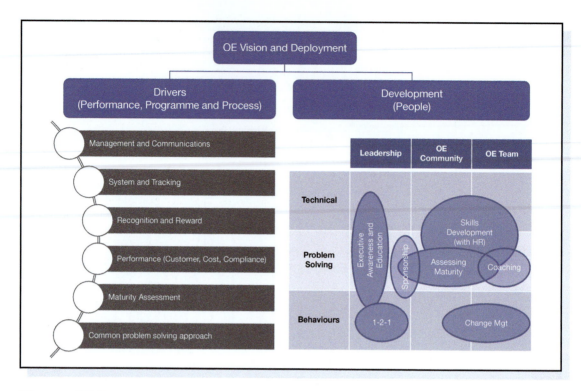

Figure C8.7
Key elements of the Operational Excellence Programme

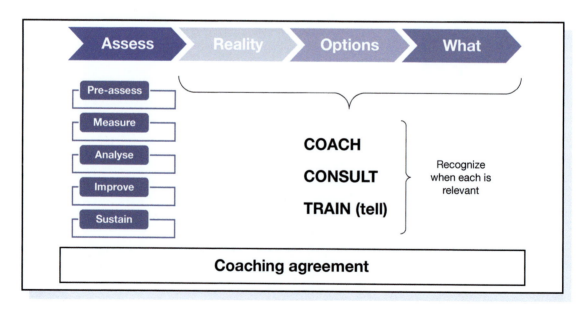

Figure C8.8
A.R.O.W. approach to technical coaching

What are the benefits and down sides of each option?

What factors will you use to evaluate the options?

W – WHAT

What are your next steps?

When?

What might get in the way?

How will you overcome?

What support do you need?

How and when will you get that support?

Do you need another coaching session?

Manager to record actions and agreements

REWARD AND RECOGNITION – THE CEO EXCELLENCE AWARDS

ABB employees make an invaluable contribution to the company's success everyday so a CEO Excellence Award was launched to recognize outstanding achievements accomplished through the talent, passion, and drive of employees and teams around the world. The award acknowledges people and teams that have made significant contributions to operational excellence in ABB and recognized improvements to the business, in terms of customer satisfaction, on-time delivery and cost opportunities.

At the outset it was recognized that award schemes already existed across ABB so the CEO Excellence Award was designed to complement rather than replace existing schemes, providing a harmonized approach to acknowledging contributions to operational excellence across ABB.

The award framework is shown in Figure C8.9.

BENEFITS DERIVED FROM THE PROGRAMME

In very large complex organizations such as ABB, success in improvement programmes like the

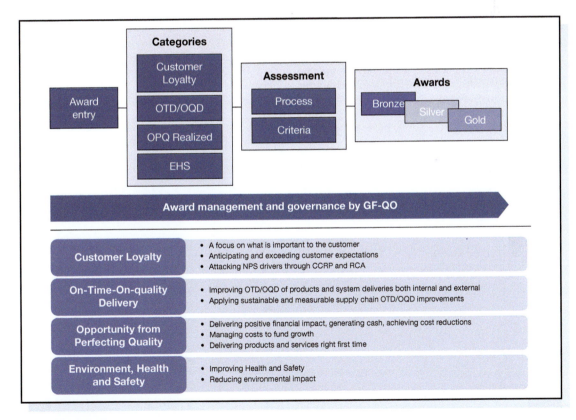

Figure C8.9

The ABB CEO excellence awards framework

OPEX are more likely to be qualitative than quantitative, as the latter are specifically related to the particular industry sectors, geographical and even political situations. The whole OPEX approach was designed and developed to meet the future scale requirements of the ABB and is now fully documented and deployed across the organization.

The standardized trainer/coach assessment process has become embedded within the programme infrastructure as a sustainable approach to qualifying internal trainers and coaches in the divisions, regions and local business units, ensuring a standardized global approach whilst meeting local needs. This has accelerated development of the initial wave of trainers required to meet the language, cultural and geographical demands of the ABB global organization. Qualified trainers and coaches provided the early momentum required to establish the program credibility and are distributed across the entire organizational footprint, supporting all divisions and regions in ABB.

Nearly 6,000 were trained by this method fairly quickly, including 350 trainer/coaches thereof 190 are active trainer/coaches with total savings from 4Q projects ramping up very quickly to over $500m. Objectives such as having at least one qualified ABB 4Q Basic Trainer/Coach under every ABB roof helped accelerate progress.

The 'Advanced' workshop material, including Lean and Six Sigma techniques, provided a sound basis on which to develop expert quality improvement practitioners within ABB and formed a necessary platform to deliver advanced projects with improvements in on-time, on-cost and on-quality delivery.

ACKNOWLEDGEMENT

The author is grateful for the help and information provided by Bill Black, Senior Vice President, ABB and Dr Robert Oakland in the preparation of this case study.

DISCUSSION QUESTIONS

1. Consider the complexity of the ABB business and provide a rationale for the development of the approach described in this case study; identify and describe an alternative approach that could have been considered by the senior quality and operational excellence group.
2. Explain the role coaching played in the success of the ABB improvement programme and suggest additional methods and tools that might have led to even greater success.
3. Evaluate the CEO Excellence Award used by ABB and explain how this could be used in another organization of your choice to develop capability and sustainable improvement.

Case study 9   The EADS (Airbus Group) Lean Six Sigma approach to performance improvement

INTRODUCTION TO EADS

EADS is Europe's largest Aerospace, Defence and Security Group and includes Airbus, Astirum Space, Cassidian (defence systems) and Eurocopter. Superior designs and effective marketing have driven up global market share and given an order book that is the largest in the world. To enhance their ability to deliver sophisticated products and services on time and on quality, a Quality and Operational Excellence Programme (QOEP) was launched, championed by the then Chief Quality Officer (CQO).

The programme had three main elements: a customer review process, improvements in the end-to-end supply chain and in product realization, and the development of improvement skills and knowledge in the key personnel who lead improvements in operational performance, based on a new propriety methodology.

Like many large corporations EADS has grown significantly through acquisition and merger but what is unique about EADS is that this came about almost as a product of the great European experiment, not currency unification but a programme to transform an extremely large, pan-European, pseudo-public sector collection of aeronautics, defence and space businesses into a dynamic industry leader that could compete in the global markets.

In 1997, the then Managing Director of British Aerospace Defence, commented that 'Europe, with defence spending of $125 billion a year, is supporting three times the number of contractors on less than half the budget of the U.S.'[1]

It is arguable that the current EADS is the result of a massive Lean Six Sigma exercise to create a new and very competitive player in the field!

'Day One' for EADS was 10 July 2000 with the merger of DASA and Aérospatiale-Matra to become the second largest aerospace company in the world, after Boeing, and Europe's second largest military arms manufacturer, after BAE. Over the intervening years, ownership has evolved to become predominantly French and German, with the expected dose of political intervention in such an important business at country levels. Further acquisitions, notably from BAE Systems, has led to the company as it is today – a global leader in aerospace, defence and related services, employing over 130,000 people at over 170 sites worldwide (www.eads.com).

With revenues of ca €50b there are four main divisions[2] –

- Airbus (€33.1b, 69,000 employees) – leading commercial and military transport aircraft manufacturer
- Astrium (€5b, 17,000 employees) – European leader in space programmes, from large-scale space systems to satellites services
- Eurocopter (€5.6b, 21,000 employees) – the top civil sector helicopter manufacturer offering the world the largest range of civil and military helicopters
- Cassidian (€5.8b, 21,000 employees) – partners in Eurofighter (with the Italian, Spanish, Germans and British – not the French) is a world-wide leader in the state-of-the-art solutions for armed forces and civil defence.

Although the common theme amongst the four is the deployment of innovative highly technical solutions, the culture of each is very different. Indeed with c.170 sites, the proportion that are non-European is increasingly being located near their main markets (BRICS). Clearly, the extending range of different cultures provides ever-increasing challenges to management.

It is in fact a testament to the management of the organization that it has been able to merge so many well-known organizations, each with its own products and cultures, and then to continue to grow and deploy an enormous range of products and services that have enriched human existence over recent years. At the German Head Quarters in Ottobrunn near Munich there is a presentation of the EADS heritage stretching back to the dawn

509

of flight, military armaments and space exploration. The company really does have a hundred year legacy!

DEVELOPMENT OF THE EADS LEAN SIX SIGMA PROGRAMME

In 2002 Oakland Consulting were involved in the development of the Airbus Improvement Management Systems (AIMS), a precursor to the EADS Quality & Operational Excellence Programme (Q&OPEX) a few years later. As part of this programme a portfolio of improvement tools were developed.

The then Head of Quality, Bill Black, recognized the need to equip Airbus people with the capabilities to improve performance (see Aims below). This led to the development of the Airbus Black Belt (ABB) and Green Belt (AGB) programme to deliver these competencies. The methodology adopted was not the widely recognized DMAIC but a uniquely modified version of Oakland's *DRIVER* (see below).

Following his move to Corporate EADS as Chief Quality Officer (CQO) Bill Black developed the EADS Black Belt (EBB) programme with Oakland's help and introduced it to the other divisions. This also included a 'Silver Belt' for project sponsors and 'Executive Black Belt' aimed at making senior managers aware of the value of this programme to their business units. To pump-prime the engagement of the programme at the divisions the cost of the training was shared equally with the divisions and corporate EADS. This continued until 2009 at which point it was felt that benefits of the programme far outweighed the overall costs by a factor of 15:1 and so EADS ceased the corporate stimulus funding.

Convergence between the ABB and EBB programme was also completed in 2009 and the ABB was dropped, followed in 2010 by the programme being relaunched as the EADS Lean Six Sigma (EADS L6S) programme. This included the introduction of new material to reflect the evolving maturity of the improvement culture at EADS, a wider portfolio of tools and more emphasis on change management to enhance the sustainment of the benefits delivered.

AIMS OF THE L6S PROGRAMME

The key objectives of the programme have not changed since its launch and are as follows:

- to equip EADS people with the competencies, skills and knowledge (tools, techniques and methods) to improve On Time, On Quality, On Cost performance where ever they may work and hence enhance EADS's reputation its customers and other stakeholders
- to establish a standard language and problem solving methodology (*DRIVER*) across different business functions and divisions – *prevents jumping from the problem to a convenient solution*
- the resulting EADS L6S performance improvers competencies shall be flexible enough to be transferrable to other complementary improvement and functional activities, e.g. development, HR, IM, procurement, risk management, project and programme management.

Over the years each division has launched its own improvement initiatives and the above aims have seamlessly dovetailed into these, to assist in delivering the benefits. This has included the global POWER 8+ campaign and the global harmonization exercise recently completed called FUTURE EADS.

APPROACH

It is recognized that all at EADS are involved with processes, transforming inputs from suppliers into outputs – products and services – for customers, some internal to the business. In such a large and complex organization it is not surprising that it is a significant challenge to provide people with an understanding of these relationships and of the opportunities that lie in their hands to deliver improvements, including better utilization of the resources under their control.

Several divisions have strong Lean Academies and similarly robust Quality functions, so it has been important for these not to have become two distinct entities – they need to work in harmony and avoid competing with each other. Oakland's experience with other clients that adopted a Lean and/or Six Sigma approach convinced EADS that what was needed was a holistic programme that utilized both the traditional Lean and Six Sigma tools and techniques, blended in such a manner as to be applicable to all functions and all levels of project complexity. Hence the Lean Six Sigma approach was adopted.

The programme approach is best represented by Figure C9.1.

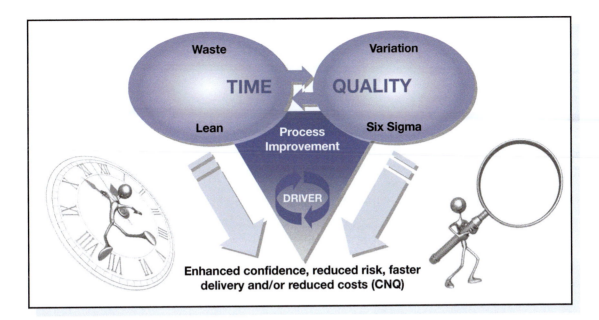

Figure C9.1
EADS Lean Six Sigma programme

Those involved with processes are taught and shown how to recognize and minimize waste (lean thinking) and to recognize, measure and reduce variation (quality or six sigma thinking) in their processes to deliver enhanced customer reputation and at lower costs. Addressing these issues is then shown to increase resource utilization, reduce risks and impact positively on the introduced corporate metric – Cost of Non Quality.

Underpinning all this is the *DRIVER* methodology – uniquely developed for EADS (see Figure C9.2) – note the four company languages used to represent the principle quartet.

This is similar to the more widely recognized DMAIC but with more emphasis on V, E and R for optimizing sustainability and sharing the lessons learnt. *DRIVER* has been recognized by users as an extremely valuable addition to their portfolio of techniques. The main benefit is that 'it tells me where to start and what to do next in a disciplined way'.

METHODOLOGY

Training is based on an effective pedagogical approach of *tell, show and experience* using bespoke

training materials, case studies and practical application of what has been taught and supported by a portfolio of >100 tools in the EADS L6S Toolkit and Toolbox.

The programme has been developed to be modular and builds to suit the various skills levels to be attained. It is delivered to multi-divisional groups with qualification predominantly via examination. For Black Belts and Advanced Green Belts an improvement project is completed, supported with coaching, to deliver tangible and validated benefits – improvements in On-Quality; On-Time; On-Cost delivery performance. All but Yellow Belt training is delivered by experienced and certified Black or Master Black Belts from Oakland Consulting. EADS have chosen not to take in house the development and delivery of the training believing that outsourcing the expertise ensures more consistent outcomes.

There are six current levels of qualification and certification is similar to programmes at other large global organizations. This is summarized in Figure C9.3.

All levels lead to **qualification** either on successful completion of the course or by successfully passing an exam. Further **certification** is

D	**Define**	**Define the problem, scope and goals** of the improvement project in terms of **customer and/or business requirements**
R	**Review**	Map the current state process and measure baseline process performance, including CNQ
I	**Investigate**	**Analyse** the gap between the current and desired performance, prioritize problems and identify root **causes** of problems
V	**Verify**	Generate **improvement solutions** to **permanently fix root cause** and **prevent reoccurrence**
E	**Execute**	**Implement the solutions** and **'sustain the gain'. Standardize improvements** via systems such as EN 9100 and implement **control stategies** such as Statistical Process Control
R	**Reinforce**	Capitalize the improvement by **'sharing lessons learnt'** and establish process re-assessment for **continuous improvement**

Figure C9.2
EADS DRIVER methodology

accomplished by demonstrating capability as an improver (at Green and Black Belt levels) by the successful completion and approval of an improvement project that has to meet one or more of the strict thresholds. These have also to be presented at a Panel Assessment.

The structure and content is comparable to other programmes but initially reflected the level of improvement maturity at Airbus and within EADS. Over the years the programme has evolved as the improvement culture developed but has also changed to suit the needs of the divisions. This evolution will continue, including through benchmarking exercises to compare and contrast the programme with others from both within and outside the aerospace sector. This in itself is an encouraging sign of confidence the company has

in the benefits of the Lean Six Sigma approach and the *DRIVER* methodology.

IMPACT

At the outset of the programme it was understood that this was to be, at least initially, a corporately funded initiative. If there was to be a wide acceptance and use of Lean and Six Sigma, the introduction had to be carefully planned and controlled to demonstrate value to the diverse divisional and functional activities.

Success was to be measured in terms of number of people trained, the value of the benefits delivered from Black Belt certification projects, as well as the breadth of application of the programme to drive improvements.

Training Course	Objective	Roles and Responsibilities	Training Days
Yellow Belt	To provide an introduction to the DRIVER methodology, tools and techniques likely to be locally deployed	• To support the performance improvement team • To provide local skills, experience and knowledge	0.5 to 2
Green Belt	To provide all employees with basic skills to improve OTOQD performance	• Lead local improvement actions using DRIVER methodology • Participate with Black Belt improvement projects	4 or 9
Black Belt	To equip selected employees with detailed skills and knowledge to improve OTOQD performance	• Lead improvement projects using DRIVER • Coach and disseminate DRIVER methodology across EADS business units	4 + 12
Silver Belt	To give participants the knowledge to identify and support L6S projects	• Sponsor EADS L6S projects • Conduct regular Black Belt project reviews • Understand the DRIVER methodology	2.5
Executive Black Belt	To provide senior managers with an understanding of the EADS L6S programme	• Support and deploy EADS L6S improvement	0.5
Master Black Belt	To provide the pool of expert practitioners inside EADS	• Provide expertise on advanced techniques • Train and coach EADS L6S candidates • Provide strategic improvement guidance for executive/senior management	Under review

Figure C9.3
EADS L6S skill levels

Over an approximate ten-year period, nearly 500 Black Belts, 4,500 Green Belts, 900 Silver Belts and over 1,000 Executive Black Belts have been trained and qualified. These, together with an untold number of locally trained Yellow Belts, have contributed to nearly €200m tangible annual benefits. The programme has been deployed on a truly global basis with courses and projects delivered throughout Europe, India, China, Brazil, Middle East, Australasia, Japan and of course the United States. Even after almost ten years the appetite for developing these skills is growing.

As will happen in large complex organizations, there have been changes in senior personnel, with three different sponsors of the programme and four changes in the L6S project management role. An indicator of the strength of the programme and the increasing recognition of the benefits delivered is that it continues to evolve, grow and add value to EADS.

Since the start of the programme the common value metric used has been the annual benefits of the Black Belt certification projects. Green and Silver Belts have mostly had an involvement with these BB projects so it was felt to be a reasonable measure of success. Unlike in some other programmes, EADS resisted counting the 'total' benefit of a project so as not to 'inflate' the project value.

Figure C9.4 shows a classic ramp up of performance – in this case the total cumulative annual value of the Black Belt certification projects since the start of the programme. These are just the tangible financial benefits (usually very conservative) but clearly there are other benefits of equal value but not so easy to quantify. These include enhanced customer satisfaction (resolved issues, improved feedback scores, increased orders), improvement of the On-Time delivery of the process (getting to the internal or external customer quicker – this usually leads to better resource utilization and hence reduced costs) and reduced risks (in terms of safety, security, quality, programme delays etc.).

As the programme has matured the average value of each Black Belt certified project has risen

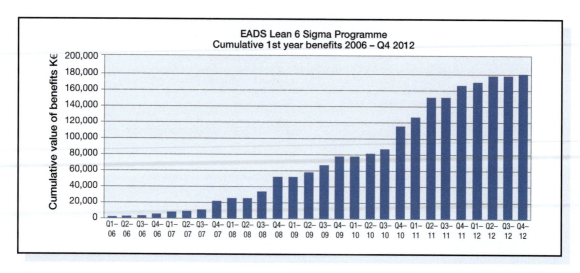

Figure C9.4
Cumulative benefits from the EADS L6S programme

from an initial target of €150k to over €500k. This is as a result of better project selection, project leader selection, team and sponsor selection as well as the formal introduction of change management and statistical tools. This is again further evidence of a maturing programme. One of the key decisions taken early on was to engage the finance community in the validation of each project. This served two purposes: first, it provided confidence that the claimed benefits were real and, second, it engaged the finance community and provided them with tangible evidence of the value of the training programme. This established the perception of EADS Lean Six Sigma less as a training cost and more as an investment. Returns on Investment of this programme, i.e. the benefits versus the total cost of the training (including for all those that do not lead and deliver a project, the sponsors, etc.) is >15 times the first year costs. This does not include following year benefits, the benefits of other projects undertaken, increased moral and better communications, etc.

Finally, the use of EADS L6S and *DRIVER* has broadened to most functions in addition to the traditional manufacturing and quality arenas (see Figure C9.5).

It is clear that the principles of lean and six sigma thinking are increasingly being appreciated in the areas of engineering, supply chain, MRO, finance and elsewhere. This has been shown in Black Belt Panel Assessments which now include

supply chain, logistics and finance projects that have deployed *DRIVER* with the EADS L6S programme.

THE FUTURE FOR LEAN SIX SIGMA IN EADS

EADS continues to undergo significant changes with the centre of gravity perhaps moving towards Toulouse in South West France, the home of its largest division Airbus. Further divisional changes are bound to happen in a business this size and in this industry; this is sure to involve further mergers and rationalizations. However, the recognition of the value of Lean and Six Sigma tools and techniques is now so widespread and that the company is seeing the increased reuse of the community of improvers in all divisions.

Harmonization of quality tools and techniques across all divisions has taken place and not surprisingly many of those in the EADS Toolkit and Toolbox are being recognized as corporately valuable. Furthermore, exercises to map the skills and competencies of the EADS L6S programme with the EADS Competency Management group has demonstrated the complementary value of the programme to project and programme management, risk management, engineering, development and other functions further supporting the 'hidden values' of the programme to both the individuals and divisions within EADS.

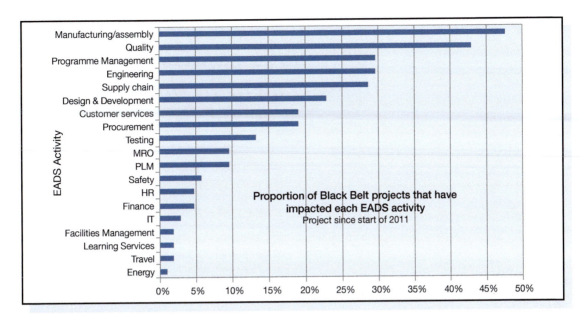

Figure C9.5
Proportion of Black Belt projects that have impacts on each EADS activity

All divisions are adopting an improvement culture but in differing guises with titles ranging from Operational Excellence, Business Excellence, Lean Quality Academy, etc. This will lead to the creation of a central portal for improvement to which all employees associated with any improvement activity will be directed to obtain guidance about all matters related to lean, six sigma, DRIVER, quality and other techniques that can leverage improved On-Time, On-Quality and On-Cost performance.

ACKNOWLEDGEMENT

The author is grateful for the help and information provided by Niall Cleary in the preparation of this case study.

NOTES

* On 1 January 2014 EADS was renamed Airbus Group to capitalise on the strong brand of Airbus. The four divisions were rationalised into three and renames as Airbus, Airbus Helicopters, and Airbus Defence and Space. The latter being a merger of Astrium, Cassidian and Airbus Military in order to optimise common technologies and skills and to respond to the global decline in military markets. The EADS Lean 6 Sigma programme continues to evolve following a large benchmarking exercise supported by Oakland Consulting. It is anticipated that this will be renamed with the Airbus brand.

1. http://community.seattletimes.nwsource.com/archive
2. Data from EADS 13.06.13

DISCUSSION QUESTIONS

1. Explain why Lean and Six Sigma approaches have been so widely adopted in product based industries; in what way has EADS adapted the Lean Six Sigma approach to suit its own requirements?
2. Choose a service sector business and explain in detail how this type of approach could be successfully deployed in that environment; show the main areas of difference between the two programmes.
3. Discuss the issues of tracking the benefits from an approach and programme such as the one deployed by EADS. What are the pitfalls to be avoided in a programme such as this and why have so many organizations failed to deliver the expected benefits?

Case study 10　Establishing a capability for continuous quality improvement in the NHS

BACKGROUND AND HISTORY

In March 2009 County Durham and Darlington Community Health Services (CDDCHS) formulated a strategy to improve patient outcomes, safety and service efficiency by developing and implementing a large-scale quality improvement programme across the organization. CDDCHS was the primary care provider for a large, semi-rural area around Durham and Darlington, serving a diverse community of around 600,000 people across the region. The CD&DPCT employed approximately 3,000 staff and had an annual budget of £1.1 billion to spend on services. They provided over 50 services, such as community nursing, speech and language therapy, podiatry, school nursing and the provision of out-of-hours services. CDDCHS had recently formed from six previous Primary Care Services covering the County Durham and Darlington operational region. The programme and its supporting infrastructure would also play a key role in helping CDDCHS with their journey towards a culture of continuous quality improvement.

APPROACH

CDDCHS developed a simple and pragmatic approach to the improvement programme, based on the 'quality journey' that organizations typically progress on (Figure C10.1). The approach was highly customised to meet their service needs (Figure C10.2) which involved the following.

Executive Engagement – they developed a clear vision for quality, created executive buy-in through two-way interviews and workshops and developed guidelines to clarify the role of leadership in driving the programme. The Executive team also received training to ensure that they were well versed in the concepts and methodologies associated with the improvement programme.

Quality Improvement Training – representatives from each of their directorates were trained as 'improvement leaders', providing them with in-depth knowledge about the *DRIVER* improvement methodology – a proven approach to continuous improvement (Figure C10.3). Part of this training involved mentoring the participants as they tackled a key project in their directorate so that they could learn from experience and start to quickly reap the benefits of the training. This created a community of people who could share experiences and best practices and actively coach others.

They also provided for training for second and third level managers across all directorates to provide them with quality tools and techniques and a robust approach to implementing continuous quality improvement (CQI) projects. Managers were also provided with CQI toolkits to enable them to introduce the methods to their teams and encourage a continuous improvement mindset.

Breakthrough Wins in Service – CDDCHS developed the criteria and process for selecting priority areas to reengineer and designed standardized materials and guidelines to ensure a consistent approach to process improvement across the organization in line with the local Primary Care Trust's existing policies and frameworks. They also identified and delivered breakthrough wins in service that improved patient pathways and created exemplars for the approach.

Quality Infrastructure – the approach was underpinned by the deployment of a culture assessment to evaluate the extent to which the culture of the organization supported a continuous improvement orientation, development and deployment of an integrated communications programme to engage the wider PCT community, the establishment of a measurement system to monitor progress and the development of a governance process to ensure that the programme continued to maintain momentum.

PROGRAMME DELIVERABLES

The Executive team in CDDCHS identified a common view of the 'vision' for quality

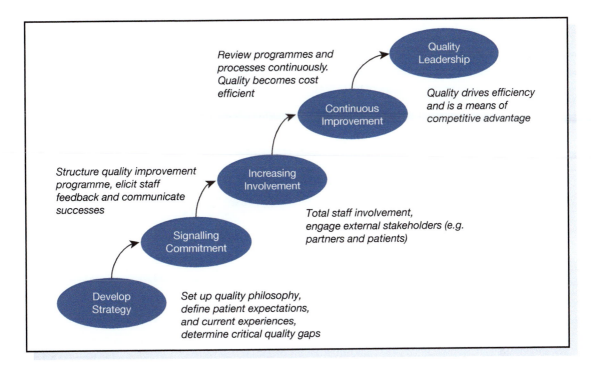

Figure C10.1
Organizations progress on a 'quality journey'

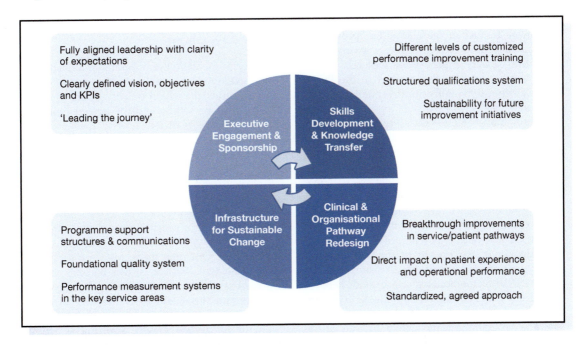

Figure C10.2
Addressing four areas helps create a sustained culture of continuous improvement

Case 10 Capability for continuous quality improvement in the NHS   517

DRIVER is a proven approach for improving the efficiency ***and*** effectiveness of healthcare services:

- **D**EFINE – the scope and goals of the patient process to be redesigned
- **R**EVIEW – understand the current processes and measure performance
- **I**NVESTIGATE – analyse the gap between the current and desired performance
- **V**ERIFY – generate the improvement solutions
- **E**XECUTE – implement the improved processes in the pathway
- **R**EINFORCE – capitalize the improvement by learning the lessons

'I've seen many different improvement methodologies come and go in my 33 years working in the NHS. For me, *DRIVER* is an approach that makes good practical sense and stops you jumping from the problem to the solution without considering the improvement options. My team are using *DRIVER* to make our work more efficient, more effective and ultimately to get more children into our improved services.'
Head of Children's Services, PCT

Figure C10.3
DRIVER structured approach for improvement

improvement and this provided the basis for developing a PCT wide approach to quality improvement. They also determined pre-requisites for the achievement of that vision and aligned quality improvement initiatives to the PCT's strategic agenda.

All of the executives and senior management team were trained in the *DRIVER* improvement methodology and were actively involved in supporting quality improvement projects.

Over 130 people were trained in the PCT's highly customized quality improvement approach and were actively involved in improvement projects and over 75 *DRIVER* continuous improvement projects were in progress across all directorates delivering significant benefits to patients, staff and the organization.

Four major breakthrough projects were delivered with significant improvements to waiting times, non-attendance rates, wound care management costs, resource utilization and access for patients. They were widely perceived to have not only delivered real insight to the business but they have provided a meaningful example of the way the *DRIVER* methodology can provide benefits to a primary care environment in the NHS (Figure C10.4).

The programme team were conscious of the need to build a solid foundation for the PCT by creating a critical mass of employees that were engaged, inspired and enabled to transform the business through process improvement. A number of initiatives were put in place to help the PCT to start to establish this foundation and continue to be a main area of focus for the programme as it developed into maturity.

EVALUATION OF THE PROGRAMME

In order to measure the programme a dashboard was developed to enable the team to monitor progress and take action quickly if any aspect of the programme fell behind the agreed timelines. The dashboard featured activity measures and outcome measures for each element of the programme and was reviewed regularly by the project team.

Executive Engagement – activity measures comprised availability of the executive training and vision workshops within agreed timescales, proportion of the executive team that attended and quality of the training as perceived by the participants. Outcome measures were concerned with the extent that Executives became actively engaged as sponsors on key quality improvement projects and the extent to which they provided active support to the programme.

Quality Improvement Training – activity measures included timely development of courses, numbers of participants on courses, distribution of participants across the directorates and quality of

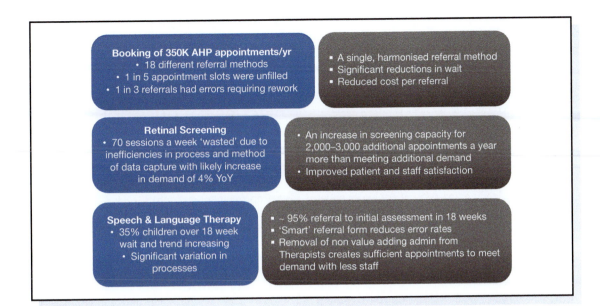

Figure C10.4
DRIVER in Primary Care

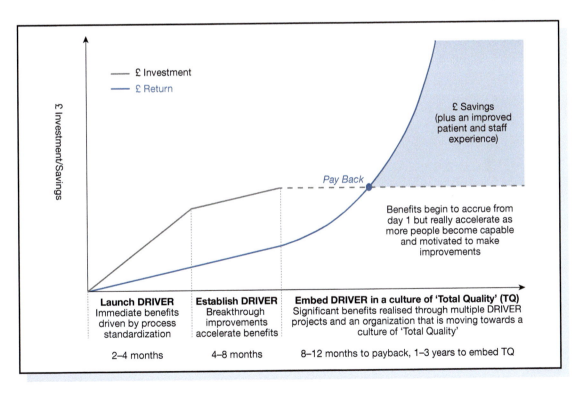

Figure C10.5
Typical profile for RoI

Case 10 Capability for continuous quality improvement in the NHS

courses as perceived by participants. Outcome measures were focused on the proportion of trained employees involved in and completing improvement projects and proportions by directorate (Figure C10.5).

Breakthrough Wins in Service – activity measures included the number of breakthroughs, the value of the improvement as perceived by the leadership team and satisfaction with the support provided by Oakland Consulting on the breakthrough. Outcome measures were concerned with the benefits realized by the breakthroughs in terms of efficiency gains and patient experience.

Quality Infrastructure – measures on the culture assessment related to timeliness and response rates and measures on the communications programme related to coverage and quality of the communications.

ACKNOWLEDGMENT

The author would like to thank the team from CDDCHS for their agreement to publish this case study.

DISCUSSION QUESTIONS

1. Discuss different possible approaches to establishing a CQI capability in an organization. In your opinion, which parts of the CDDCHS approach had the greatest impact on sustainable improvement and why?
2. Review the measures CDDCHS used to evaluate their programme. Why is it important to have both activity and outcome measures for such a programme?
3. Discuss how the approach to CQI might need to vary between a service based organization such as CDDCHS and a manufacturing organization.

INDEX

Page references to Figures or Tables will be in *italics*.

Ishikawa (Japanese quality guru) 270, *280*

Japan: Deming Prize 22, 29, 156, 176; just-in-time (JIT) management 79, 80; policy deployment process 140; quality function deployment developed in 94

JDM (joint design and manufacturing) 71

JIT *see* just-in-time (JIT) management

Johnson, G. 69

joint design and manufacturing (JDM) 71

Jones, Daniel 305, 309, 310, 312

judging personality type 372

Juran, Joseph M. 19, 20, *21–2*

JUSE (Union of Japanese Scientists and Engineers) 22, 349

just-in-time (JIT) management 62, 79–82, 86; aims 80; Kanban system 81; material movements/accumulations, tracking 80; operation 80–1; outcomes 79; in partnerships and supply chain 81–2; process flexibility 80

Kaizen (rapid improvement processes): 'Kaizen Blitz' events 351–2; Kaizen Teams 348–52, 355; operation of teams 350–1; role of 319; structure of teams 349–50; training of teams 350

Kanban system, just-in-time (JIT) management 81

Kanri, Hoshin 140

Kaplan, R.S. 57, 124, 141–2

key performance indicators (KPIs): balanced scorecard 141, 142; defining 56–7; implementation of TQM 412–13; Lloyds' Register Group (LRG) 456, 458; performance measurement framework 142, 154

key performance outcomes (KPOs): balanced scorecard 141, 142; performance measurement framework 139, 142

knowledge: explicit, as information 399–400; and information 398–9; learning-knowledge management cycle 400, *401*; modes of conversion *399*; sharing, practicalities 401–2, 404

KPIs *see* key performance indicators (KPIs)

KPOs *see* key performance outcomes (KPOs)

labour intensity, service sector 102

Larkins, T.J. and S. 387

lateral thinking 239

leadership: action-centred 362–6, 384–5; business process re-engineering or redesign (BPR) 239–40, 242; clear and effective strategies, developing 41; and commitment 31–50; continuum of leadership behaviour *366*; critical success factors and processes, identifying 41, *42*; and customers 46; effective, requirements for 40–4, 48; empowerment of workers 41–4; excellence in 44–6, 48; management structure, reviewing 41; people skills 45–6; performance 45; planning 45; processes 45; process management 221–3, 225; quality management systems, responsibility in relation to 251–3; self-assessment enablers 157–8; situational 365–6; style 108–9; teamwork 359, 370; TQM approach 31–3, 47; vision (corporate beliefs and purpose) 40–1

Lean approach 222, 305–26; building blocks 317–19, 326; business process re-engineering or re-design (BPR) 226–8, 241; definition of 'Lean Production' 226; DMAIC and Lean toolkit *311*; 5S system 317–18; implementation as an investment 307; interventions 308–12, 326; Kaizen, role of 319; Lean Manufacturing concept 305; loss of people not necessarily the case 307; myths and facts *306*; nine-step approach (Womack and Jones) *311*; persistence of 307; and Six Sigma improvement model 266–7, 308, 325, 510; transferability 306–7; value stream mapping 312–16, 326

Lean Six Sigma 266–7, 308, 325, 510; in EADS 514–15

Lean Thinking (Womack and Jones) 309

legal services, evolution *106*

liability, costs 127

Lloyds' Register Group (LRG) 433, 456–64; background 456; closed

loop improvement cycle 457; engagement with business 461; everyday principles 461; GBA, applying approach to 463–4; improvement programme 456–8, 461; local skills and improvement ability, developing 461–2; management information, reporting and review 458; results 463; selection and delivery, managing 459, 461; strategy, linking to improvement 458, *459*, *460*; Unite programme 462–3

Machine That Changed the World, The (Womack and Jones) 309

MBNQA (Malcolm Baldrige National Quality Award) 23

management: assigning quality manager 345–6; assigning quality manager adviser 346; and design 108–11, 113; education and training 396–7; operations management 43–4; of quality 11–14, 18; quality management systems, responsibility in relation to 251–3; review of structure 41; style 108–9; *see also* middle management; senior management

management by objectives (MBO) 144

management systems: audit and review 170–5, 177; checking 171, *172*; error or defect investigations and follow-up 171–2; internal and external system audits and reviews 172–5, 177

marketing 8

marketing-led innovation 89

Maslow, Abraham 362

Master Black Belts 298, 304

Material Requirements Planning (MRP) 81

matrix data analysis 291

matrix diagrams 290–1

maturity level radar diagram *481*

Mazda 228

MBO (management by objectives) 144

MBTI (Myers-Briggs Type Indicator) 371–2, 373

McGregor, Douglas 362

MCPs (most critical processes) 59

middle management: change curves and stages 415; education and training 396–7

excellence 44–5; model 424–5; 'Oakland TQM Model' (four Ps and three Cs) 27–8, 30, 45, 48; user-driven 40

Toyota-style production 226, 227, 305

TQM *see* total quality management (TQM)

training: communication 390–3, 403; cycle of improvement 391–3; errors, defects or problems 394; human resource management 340–2, 354; implementing and monitoring 393; leadership, effective 43; needs 392; prevention costs 126; programmes and materials, preparing 392; quality training cycle *391*; record keeping 395; results, assessing 393; reviews 393, 394–5; starting points 395–8, 403; systematic approach to 393–5; turning into learning 398–400, 403–4; *see also* education

transferability, Lean approach 306–7

transformation process 12, *13*, 15

transparency 73, 122

T-shaped matrix diagram 291, *292*

Union of Japanese Scientists and Engineers (JUSE) 22, 349

United Kingdom (UK), 'Quality Campaign' and 'Managing into the 90s' programs (DTI) 22

United States (US): American 'gurus' of quality management 19, *21–2*, 29; American Productivity and Quality Center 179, 203–16, 224; Baldrige Performance Excellence Program 23–4, 29, 156; 'Criteria for Performance Excellence,' National Institute of Standards and Technology 25, 27; Federal Information Processing Standards (FIPS) 206

value-added management (VAM) 124

value stream mapping (VSM) 309, 312–16, 326; four step procedure 316; process families 313, *314*, 315

value stream scopes (end-to-end processes) 315–16

vendor rating, appraisal costs 126

verification, appraisal costs 126

vertical functions 230

virtual management for breakthrough (VMB) *472*

virtual reality (VR) technologies 101

vision (corporate beliefs and purpose) 53; developing clear and effective strategies/plans for achieving 41; effective leadership, requirements for 40–1; operational 239; process management 199–201, *202*, 224; shared, developing 54–5; *see also* mission

VSM *see* value stream mapping (VSM)

Wal-Mart 233

warranty claims 127

waterproofing, construction industry 100

Welsh, Jack 268

Whittington, R. 69

Womack, Jim 226, 227, 305, 309, 310, 312

World Class Manufacturing (WCM) 351

Xerox 228, 229

Zemke, Ron 108